Ecology and Restoration of Red Spruce Ecosystems in the Central and Southern Appalachians

Donald J. Brown · Melissa A. Thomas-Van Gundy ·
Corinne A. Diggins · Alexander Silvis ·
Kathryn M. Shallows
Editors

Ecology and Restoration of Red Spruce Ecosystems in the Central and Southern Appalachians

Springer

Editors
Donald J. Brown
USDA Forest Service
Pacific Northwest Research Station
Amboy, WA, USA

Melissa A. Thomas-Van Gundy
USDA Forest Service
Northern Research Station
Parsons, WV, USA

Corinne A. Diggins
US Fish & Wildlife Service
Science Applications
Albuquerque, NM, USA

Alexander Silvis
West Virginia Department of Natural Resources
Elkins, WV, USA

Kathryn M. Shallows
The Nature Conservancy
Appalachians Program
Fairmont, WV, USA

ISBN 978-3-032-19615-6 ISBN 978-3-032-19616-3 (eBook)
https://doi.org/10.1007/978-3-032-19616-3

Cover art by Carly Thaw

This Springer imprint is published by the registered company Springer Nature Switzerland AG
The registered company address is: Gewerbestrasse 11, 6330 Cham, Switzerland

We dedicate this book to the community that has worked hard to conserve and restore red spruce ecosystems in the central and southern Appalachian Mountains. A combination of federal, state, and tribal agencies, numerous universities, and non-governmental organizations have collaborated for decades to make conserving these rare and unique ecosystems possible. Many volunteers, students, and professional planters have devoted their time and resources to planting red spruce across its historic range to restore these forests. The collective work of these entities and individuals has enabled the persistence of red spruce and spruce-fir forests across the region. We look forward to continued collaboration on conservation of one of the most endangered forested ecosystems in the USA.

Foreword

Science and science-based action must be the basis of defining and understanding the natural resource targets we hope to conserve. This is the bedrock of a good conservation approach. From this solid foundation land managers, biologists, ecologists, scientists, conservation program leads, stewardship teams, and implementation teams can develop actions to protect these resources. In the 1700s and into the 1800s the Appalachian red spruce forests were considered by white settlers as forbidding, frightening, dark and formidable. Red spruce forests were so dense with laurel brakes and giant trees, both standing and fallen, that only the most intrepid, adventurous explorers dared to penetrate this "ill-omened region" (Mt. Porte Crayon as conveyed in Tumult on the Mountains (Clarkson, 1964)).

Today, these high elevation relicts ranging from Maine to Georgia are recognized as ecological wonders. These amazing habitats serve a myriad of functions for nature and people. The high mountain peaks red spruce forests inhabit create a cool, moist environment for a richness of species as the conjunction of northern relics and southern plants and animals into unique communities, including endemics found nowhere else, making the central and southern Appalachians a biodiversity hot spot in the US. Blocks of high elevation red spruce act as refugia for current and future migrations of species. The amazing ability of red spruce forests to lock up and store carbon makes these areas one of the most regionally important carbon sequestration locations and contributes to national mitigation needs. They are home to a long list of species of conservation concern, including several species federally listed under the Endangered Species Act.

These areas serve equally as magical places. Stately red spruce boughs draped in mist or clouds and deep organic layers of moss and ferns muffle sounds to create a sense of quietude and serenity not often found elsewhere. What happened to our human view of them and why should we care? What happened to them from a scientific perspective and where should we focus science in the future? It is imperative that we learn from the past to care for these unique resources in the future.

Shortly after the Civil War, the central Appalachian spruce forests were explored for their wood and other products and broad-scale lumbering and extraction began. The high elevation spruce became not something to be feared, but a resource to be exploited and from which vast profits could be made. The aftermath of the major lumbering period (1870–1920) was a greatly diminished spruce-fir forest with a remnant tree cover of only about 10% of its pre-lumbering extent in the central and about 50% in the southern Appalachians. The deep organic soils were eroded or burnt off by post-logging fires, to bedrock in places, and the majority of old-growth spruce forests were gone in the central and reduced in the southern Appalachians.

Once the timber was removed, these areas in the central Appalachians were rarely considered as having value and became ignored waste lands. Larger remnants of spruce survived elsewhere in the northern and southern Appalachians where less severe logging had taken place compared to the central Appalachians. Particularly in the southern Appalachians and the Adirondacks large areas of spruce-fir ecosystems were set aside in the 1920s and 1930s as parks or reserves. These iconic natural areas, including the Great Smoky Mountains National Park, became beloved as natural areas sought after for recreation, respite, and inspiration.

In the early 1980s, there were published reports of unexplained mortality in red spruce, particularly in the Adirondacks and northern Appalachians. Concerns arose about the concurrent decline and apparent mortality in the southern Appalachians as well. These were linked to the effects of acidic deposition, and a new relationship between humans and spruce-fir forests began. Because of world-wide awareness of air pollution and concern about its effects, the National Acid Precipitation Assessment Program (NAPAP) was created and funded by the US Congress to research the effects of acidic deposition (acid rain) on human health, agriculture, materials, and on natural resources, including spruce-fir forests. The Spruce-Fir Forest Research Cooperative was one of the programs created within NAPAP, and investment flowed into those forests which had become so important to citizens for their intrinsic and spiritual values.

During the research of the Spruce-Fir Research Cooperative much was learned about the basic ecology of montane spruce-fir forests. However, there remained a relative paucity of research conducted in either the central or southern Appalachians during the acid rain research period. It was this need, recognized by scientists, public land managers, and conservation organization leaders that led to the development of first the Central Appalachian Spruce Restoration Initiative (CASRI) and then the Southern Appalachian Spruce Restoration Initiative (SASRI). Scientists and naturalists have produced maps and data that have allowed conservationists, land management agencies, and conservation practitioners to capture attention and align actions to protect this locally, regionally, and globally important asset. The attention that led to the motivated rallying cry creating CASRI and SASRI, has driven forward a concerted effort to garner the attention and actions of agencies and partners to prioritize bold and innovative action to restore an entire ecosystem for the benefit of nature and people.

But there is much more beyond the science that makes an action like red spruce restoration and the groups and partners who focus on it successful. These forests are simply *natural magic*. Not a term that is based in science. Rather it is born out of our human connection to special places in nature. It is this human joy in nature that is as important to convey to individuals, partner organizations, funding organizations and donors as the ecological values these lands hold. As we continue to build and pass on the strong efforts of CASRI and SASRI to the next generation of leaders, as conservationists and managers of these red spruce forests we must build on the body of science that justifies, identifies, and outlines effective action. We also should not forget to engage ourselves and others to appreciate and use the *natural magic* these amazing forests offer to inspire action and feed our souls.

Dr. Mary Beth Adams is an Emeritus Research Soil Scientist with the USDA Forest Service. She started her Forest Service career as the Assistant Program Manager for the Spruce-Fir Research Cooperative, part of the National Acid Precipitation Assessment Program.

Kate Goodrich-Arling is a retired Staff Officer for Public and Legislative Affairs, Monongahela National Forest, USDA Forest Service. She was involved in the development and activities of the Central Appalachian Spruce Restoration Initiative (CASRI).

Katherine Medlock is a Conservation Director for the Appalachians Program of The Nature Conservancy and was involved in the origins of the Southern Appalachians Spruce Restoration Initiative (SASRI).

Thomas Minney is the State Director of the West Virginia chapter of The Nature Conservancy. He has been involved in the activities and development of the Central Appalachian Spruce Restoration Initiative in several roles over his career.

Seymour, IN, USA — Mary Beth Adams
Presque Isle, MI, USA — Kate Goodrich-Arling
Knoxville, TN, USA — Katherine Medlock
Morgantown, WV, USA — Thomas Minney

Reference

Clarkson R (1964) Tumult on the mountains: lumbering in West Virginia 1770–1920. McClain Printing Company, Parsons, West Virginia

Preface

The first efforts to restore red spruce (*Picea rubens*) ecosystems in the central and southern Appalachians began nearly 100 years ago, following broad scale declines that resulted from industrial logging activities and associated wildfires. While many localized efforts have occurred over the last century, the scale of red spruce restoration has dramatically increased over the last two decades. In the mid-2000s, the Central Appalachian Spruce Restoration Initiative (CASRI) was formed to bring restoration partners together, followed by establishment of the Southern Appalachian Spruce Restoration Initiative (SASRI) around a decade later. The visions and missions of CASRI and SASRI serve as guiding lights for red spruce restoration across the central and southern Appalachians, promoting collaborative relationships among federal and state management agencies, non-governmental organizations, restoration businesses, and scientific researchers, and encouraging restoration practitioners to plan at the scale of landscapes.

As the interest in red spruce restoration grows, new organizations and individuals are continually joining CASRI and SASRI and engaging in red spruce restoration efforts. The idea for this book arose out of the need for a single resource to introduce new restoration practitioners to contemporary red spruce ecosystems and restoration goals and practices. As the book concept crystalized, we expanded the scope of content to include the history and potential future of red spruce and increased the depth of material covered to provide a valuable resource for seasoned practitioners and research scientists. The book begins with a focus on the historical and contemporary geographic distribution of red spruce, including the influence of anthropogenic activities on its decline. Chapter 2 synthesizes our understanding of the biology of red spruce. The following four chapters cover ecological communities and ecosystem dynamics associated with red spruce and spruce-fir forests, with focal topics ranging from soils to wildlife. Chapter 7 reviews the influence of climatic conditions on the physiology and distribution of red spruce and discusses potential responses of the species to rapid contemporary climate changes. The final two chapters focus on restoration of red spruce in the central and southern Appalachians, including considerations and approaches, historical efforts, and major accomplishments to date.

Over 40 scientists and managers working in red spruce ecosystems contributed to writing the nine chapters in this book, with over 30 additional individuals contributing their knowledge and expertise as content reviewers and editors. The result is not only a robust synthesis of the ecology of red spruce ecosystems in the central and southern Appalachians, but also a story of their dramatic historical decline and current efforts to restore these unique and majestic forests.

Amboy, WA, USA — Donald J. Brown
Parsons, WV, USA — Melissa A. Thomas-Van Gundy
Albuquerque, NM, USA — Corinne A. Diggins
Elkins, WV, USA — Alexander Silvis
Fairmont, WV, USA — Kathryn M. Shallows

Acknowledgments

Funding for open access was provided by the National Forest Foundation and The Nature Conservancy in West Virginia. Carly Thaw created the beautiful cover art for the book. Nearly all maps were designed and created by Meryl Friedrich. We are grateful to the chapter authors, many of whom worked on this book off the clock, for their devotion to creating a valuable and engaging product. We appreciate the dozens of individuals who contributed photographs for the book. We thank Doug Manning and Jason Fisher for providing reviews of the full book, and all of the individuals who provided content suggestions, edits, and in some cases original text for individual chapters, including Rodney Bartgis, Elizabeth Byers, Susan Cameron, Charlie Cogbill, Meredith Cornett, John Dalen, Colette DeGarady, Mack Frantz, Kelly Holdbrooks, Shane Jones, Christine Kelly, Anthony Khiel, Andrea Larrivee, Gregory Nowacki, Adrienne Nottingham, Thomas Pauley, Jason Rodrigue, Dave Saville, Michael Schafale, John Schmidt, Thomas Schuler, Sandra Simon, Craig Stihler, Jack Tribble, James Vanderhorst, James Van Gundy, Robert Whetsell, Peter White, and Anja Whittington.

Contents

Contributors

Chris D. Barton University of Kentucky, Department of Forestry and Natural Resources, Lexington, KY, USA

Anna M. Branduzzi Green Forests Work, Lexington, KY, USA

Donald J. Brown USDA Forest Service, Pacific Northwest Research Station, Amboy, WA, USA

John R. Butnor USDA Forest Service, Northern Research Station, Burlington, VT, USA

Alton C. Byers University of Colorado, Institute of Arctic and Alpine Research, Boulder, CO, USA

Elizabeth A. Byers West Virginia Division of Natural Resources (retired), Elkins, WV, USA

Thibaut Capblancq University of Grenoble, Alpine Ecology Laboratory, Grenoble, France

David R. Carter Michigan State University, Department of Forestry, East Lansing, MI, USA

Sophan Chhin West Virginia University, School of Natural Resources and the Environment, Morgantown, WV, USA

Hannah L. Clipp USDA Forest Service, Northern Research Station, Delaware, OH, USA;
Appalachian Mountain Club, Jackson, NH, USA

Charles V. Cogbill Harvard University, Harvard Forest, Petersham, MA, USA

Beverly Collins Western Carolina University, Department of Biology, Cullowhee, NC, USA

Corinne A. Diggins US Fish & Wildlife Service, Science Applications, Albuquerque, NM, USA

C. Andrew Dolloff USDA Forest Service, Southern Research Station (retired), Blacksburg, VA, USA

Will Evans The Nature Conservancy, West Virginia Chapter, Elkins, WV, USA

Matthew C. Fitzpatrick University of Maryland Center for Environmental Science, Appalachian Lab, Frostburg, MD, USA

W. Mark Ford US Geological Survey, Virginia Cooperative Fish and Wildlife Research Unit, Blacksburg, VA, USA

Jakob T. Goldner West Virginia Division of Natural Resources, Elkins, WV, USA

Stephen R. Keller University of Vermont, Department of Plant Biology, Burlington, VT, USA

Susanne Lachmuth University of Maryland Center for Environmental Science, Appalachian Lab, Frostburg, MD, USA

Deborah Landau The Nature Conservancy, Maryland/DC Chapter, Rockville, MD, USA

Chad M. Landress USDA Forest Service, Monongahela National Forest, Bartow, WV, USA

James E. Leonard USDA Natural Resources Conservation Service, Morgantown, WV, USA

Matthew McKinney West Liberty University, College of Sciences, West Liberty, WV, USA

Katherine Medlock The Nature Conservancy, Appalachians Program, Knoxville, TN, USA

Danika Mosher USDA Forest Service, Southern Research Station, Eastern Forest Environmental Threat Assessment Center, Research Triangle Park, Durham, NC, USA

Kevin M. Potter USDA Forest Service, Southern Research Station, Eastern Forest Environmental Threat Assessment Center, Research Triangle Park, Durham, NC, USA

Jay E. Raymond Virginia Western Community College, Integrated Environmental Studies, Roanoke, VA, USA

Benjamin M. Rhodes Ruffed Grouse Society & American Woodcock Society, Barbourville, KY, USA

Craig Roghair USDA Forest Service, Southern Research Station, Blacksburg, VA, USA

David Saville Appalachian Forest Restoration LLC, Morgantown, WV, USA

Michael P. Schafale North Carolina Department of Natural and Cultural Resources, North Carolina Natural Heritage Program, Raleigh, NC, USA

Kathryn M. Shallows The Nature Conservancy, Appalachians Program, Fairmont, WV, USA

Alexander Silvis West Virginia Division of Natural Resources, Elkins, WV, USA

S. Jason Teets USDA Forest Service, Monongahela National Forest, Morgantown, WV, USA

Melissa A. Thomas-Van Gundy USDA Forest Service, Northern Research Station, Parsons, WV, USA

James A. Thompson West Virginia University, School of Natural Resources and the Environment, Morgantown, WV, USA

Saskia L. van de Gevel Virginia Tech, College of Natural Resources and Environment, Blacksburg, VA, USA

Peter S. White University of North Carolina at Chapel Hill, Department of Biology, Chapel Hill, NC, USA

William A. Whitter USDA Forest Service, National Forests in South Carolina, Columbia, SC, USA

Abbreviations

ac	Acre
ca.	Circa
CASRI	Central Appalachian Spruce Restoration Initiative
cm	Centimeter
CO_2	Carbon Dioxide
DBH	Diameter at Breast Height
ft	Feet
ha	Hectare
in	Inch
km	Kilometer
m	Meter
mi	Mile
mm	Millimeter
Mt.	Mount
NF	National Forest
NRCS	Natural Resources Conservation Service
ppb	Parts per Billion
SASRI	Southern Appalachian Spruce Restoration Initiative
U.S.	United States
USDA	United States Department of Agriculture
YBP	Years Before Present

Chapter 1
History and Biogeography

Corinne A. Diggins, Charles V. Cogbill, and Melissa A. Thomas-Van Gundy

The flora of the eastern United States is a fortuitous grab bag of species, the survivors of 16 to 18 glacial-interglacial cycles.
-Davis 1981

1.1 Introduction

Red spruce (*Picea rubens*) is a coniferous tree native to the eastern U.S. (Fig. 1.1), ranging from eastern Canada and New England south along the Appalachian Mountains, where it occurs in disjunct populations in Pennsylvania, West Virginia, Virginia, North Carolina, and Tennessee. Red spruce-dominated forests (i.e., $\geq$ 50% of the canopy trees are red spruce) occur in near-boreal regions in the northern U.S. and Canada, temperate forests in the central part of their range, and montane *sky islands* in the southern part of their range. Temperature and moisture drive the distribution of red spruce, with latitudinal dynamics driving the distribution in the northern part of the range and elevational and climatic complexes driving the distribution in the southern parts of the range (White and Cogbill 1992). For example, red spruce occurs at sea level north of the St. Lawrence River and at over 1,700 m (5,500 ft) in elevation in the southern Appalachians.

The distribution of red spruce has oscillated over time with climatic changes and disturbance factors—the latter having profound effects over the last 150 years due to Euro-American activities, specifically logging and human-ignited wildfires (Korstian 1937; Clarkson 1964; Pielke 1981). How these disturbances have affected red

C. A. Diggins (✉)
US Fish & Wildlife Service, Science Applications, Albuquerque, NM, USA
e-mail: holyflyingsquirrel@gmail.com

C. V. Cogbill
Harvard University, Harvard Forest, Petersham, MA, USA

M. A. Thomas-Van Gundy
USDA Forest Service, Northern Research Station, Parsons, WV, USA

D. J. Brown et al. (eds.), *Ecology and Restoration of Red Spruce Ecosystems in the Central and Southern Appalachians*, https://doi.org/10.1007/978-3-032-19616-3_1

Fig. 1.1 Red spruce (*Picea rubens*) on Cabin Mountain, West Virginia (photo by Marquett Crockett)

spruce genetics are described more fully in Sect. 7.2.3 of Chap. 7. Active management and restoration of red spruce forests requires an understanding of their history and factors influencing their biogeography. For example, this information may help to determine how stressors such as climate change may influence the future distribution of red spruce forests (see Chap. 7). In this chapter, we discuss past and current geographic distributions of red spruce and review how environmental factors influence occurrence of the species throughout its range.

1.1.1 *Distribution*

1.1.2 *Paleohistory*

During the last glacial maximum 18,000 YBP, ice sheets covered the continent north of Pennsylvania and New Jersey including parts of the continental shelf exposed by the lowered sea level. Beyond a narrow treeless zone in front of the ice, the land in eastern North America was covered with an open boreal woodland of spruce (*Picea* spp.) species (Delcourt and Delcourt 1981; Davis 1983). Closed-canopy conifer forests occurred further south to at least the Carolinas, with pine (*Pinus* spp.) becoming dominant. In the southeastern U.S., mixed hardwood forests were found consisting of all the common tree species surviving today, with the addition of another temperate spruce (*Picea critchfieldii*), which was common and widespread during the Pleistocene, but did not survive through the Holocene (Jackson and Weng 1999). With climatic warming and glacial retreat marking the end of the Pleistocene (Fig. 1.2), the trees closest to the retreating glacial edge subsequently migrated slowly northward. Pollen core and macrofossil data indicate that primarily white spruce (*Picea glauca*), followed by black spruce (*Picea mariana*), invaded the recently deglaciated land. This forest passed the former glacial front at 14,000 YBP in present-day Pennsylvania and expanded to northern latitudes, arriving north of the St. Lawrence River around 9,000 YBP.

In about 9,000 years, the spruce forest had migrated < 750 km (466 mi) from the maximum glacial extent to the present boreal region. Nearly all areas along this northward expansion zone were dominated by a boreal spruce ecosystem for approximately 2,000 years, including co-dominant species such as jack pine (*Pinus banksiana*) and aspens (*Populus* spp.). As this boreal spruce forest retreated north, pollen counts of any spruce species dropped precipitously starting at 9,000 YBP within the southern U.S. Across the eastern U.S., all areas then had a zone of temperate species (white pine [*Pinus strobus*], eastern hemlock [*Tsuga canadensis*], and northern hardwoods), with a marked hiatus of any spruce pollen from 8,000 to 2,000 YBP (Davis 1983). Spruce pollen then rose dramatically in New York, northern New England, and the coastal regions at northern latitudes, indicating the rapid expansion of red spruce in the last 2,000 years (Spear 1989; Jackson and Whitehead 1991). This southward expansion of spruce corresponds to (and was undoubtedly driven by) a cooler and wetter climate, known as the neoglacial period (see Fig. 1.2 and Fig. 1 in Abrams and Nowacki 2015).

Red spruce is closely related to black spruce, but diverged presumably through geographic isolation when driven south by colder climates and glacial advances in the mid-Pleistocene (ca. 400,000 YBP; Jaramillo-Correa and Bosquet 2003). The

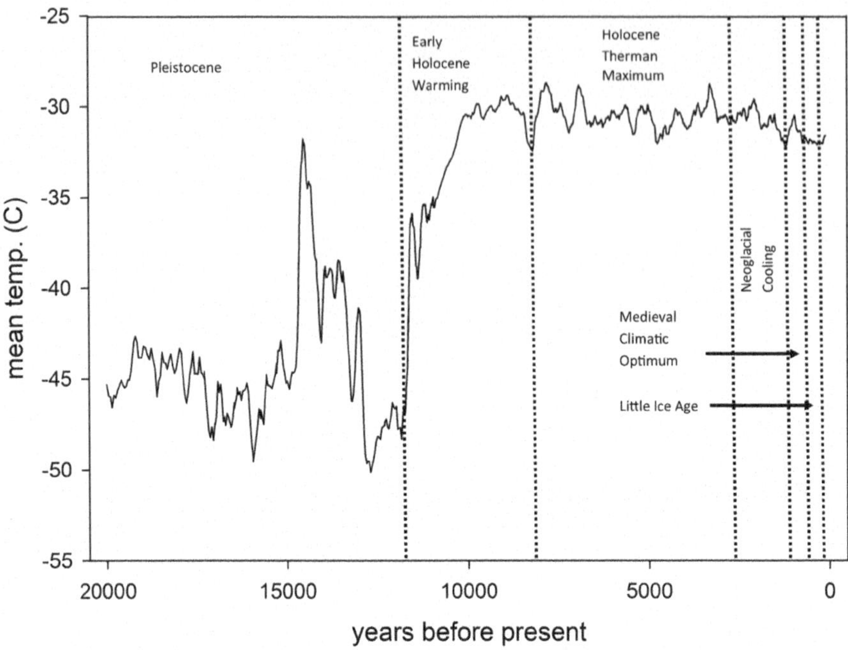

Fig. 1.2 Mean air temperature (°C) in the northern hemisphere based on central Greenland ice core data (GISP2 Ice Core; Alley 2004) spanning 20,000 YBP

species must have gone through subsequent geographic separation during several interglacial cycles. After isolation at the end of the last glacial maximum, red spruce overlapped with black spruce and hybridized during the early Holocene (Bashalkhanov et al. 2023), and now these species readily hybridize, especially in the northeastern extent of its range (Manley 1972; Perron and Bousquet 2003; Jaramillo-Correa and Bousquet 2005). The genetic structure of red spruce shows a geographic pattern confirming the general history of Holocene migrations after the last glacial maximum. There are three ancestry groups of red spruce haplotypes (Capblancq et al. 2020; see Chap. 7). The first genotype covers separated, high-elevation forests from the central and southern Appalachians. The second genotype combines the entire northern contiguous distribution including the Catskill Mountains. The third genotype from geographical intermediate Pennsylvania (Pocono Mountains, Bear Meadows) is divergent from the southern sites but contains elements from the intermediate northern sites. The genetic data, together with biogeographic, paleohistoric, and the non-montane setting, support the idea that the Pennsylvania populations are distinct from either the northern or southern mountains and not remnant populations left behind from a northern migration.

Red spruce pollen was only weakly represented in southern Pennsylvania and Massachusetts in the early Holocene prior to 11,000 YBP (Watts 1979; Lindbladh et al. 2003). Confirmed historical occurrence of red spruce north of Pennsylvania began during the early stages of the Holocene Thermal Maximum, wherein continuous red spruce pollen occurred in several extreme coastal sites in Maine and with macrofossils at mid-elevations in the northern Appalachian Mountains. During the mid and late stages of the Holocene Thermal Maximum (9,000–2,000 YBP), spruce pollen had low scattered abundance in the Adirondack and White Mountains, but sediment samples had significant spruce macrofossils. Although pollen can be a poor indicator of tree abundance (Abrams and Nowacki 2019), macrofossils in both the White Mountains and the Adirondacks indicate that spruce had a continuous, but not dominant, presence at an elevational range of 850–1,150 m (2,800–3,800 ft) through the mid-Holocene (Spear 1989; Jackson and Whitehead 1991).

Although the late-Pleistocene refugia for red spruce are unclear, they were south of the maximum extent of glacial ice, including presently submerged surfaces along the Atlantic shoreline (approximately 120 m [400 ft] below current sea level). Judging from the prominence and current outliers east of the Appalachian Mountains, red spruce survived the glaciations cryptically somewhere between New Jersey and the Carolinas. Red spruce refugia were most likely cool, foggy coastal areas, possibly including the exposed and unglaciated continental shelf that is now below sea level. Red spruce most likely migrated north as part of the post-glacial boreal spruce wave. The spruce-dominated woodlands near the glacial maximum at 13,000–10,500 YBP contained American mastodons (*Mammut americanum*) and predominantly white spruce macrofossils just south of retreating glaciers and in Pleistocene coastal regions (Feranec et al. 2021). Although minor and mixed with the two other spruce species, pollen evidence indicates that red spruce was present in Massachusetts and Maine at 11,000–10,000 YBP (Lindbladh et al. 2003). Limited evidence also indicates red spruce pollen at 10,000–9,000 YBP in south-central Pennsylvania in the rearguard of the post-glacial spruce migration (Watts 1979). Then, for at least the second time, red spruce was restricted to refugia from 8,000–3,000 YBP. This is well after the boreal white and black spruces had migrated north or remained as isolated remnants in wetlands. Spruce pollen deposited in the mid-Holocene in montane and coastal habitats of south-central Pennsylvania was not identified to species, but was most likely red spruce because of the site conditions and because white and black spruces were restricted to isolated wetlands at that time (Schauffler and Jacobson 2002).

One refugium for red spruce in the hiatus was certainly along the Atlantic coast in the Canadian Maritimes and coastal Maine, where spruce pollen was moderate and continuous in the mid-Holocene (Schauffler and Jacobson 2002). It is likely that red spruce survived inland in interior valleys or moist, cool conditions similar to the present-day lakeshore, streamside, swamp, or bog margin habitats of relict populations, such as Mann's bog and War Spur in Virginia. Red spruce also survived in mixed forests such as isolated patches from Maine to New York, where it then had a rapid and synchronized population explosion in the last 2,000 years. Importantly, there was too little time for spruce to have migrated from known refugia along the coast, so spruce must have been in place before the late-Holocene expansion. Regardless of the refugia, the recent rise of spruce pollen throughout the area south of the current boreal forest zone was not due to black spruce expansion during the Neoglacial Cooling Period of the late-Holocene. The spruce pollen resurgence over the last 2,000 years was undoubtedly red spruce expanding to occupy its current continuous range in the north during this cool period.

According to Watts (1979), spruce pollen is missing from mid-Holocene sediment cores at lower elevation sites in Virginia and Pennsylvania, apparently indicating no evidence of red spruce movement across the gap between the southern and northern mountains known as the Pennsylvania Saddle (Nowacki et al. 2010). The source for the handful of current spruce remnant sites within this saddle is unclear, because both red and black spruce are morphologically, genetically, and palynologically problematic at peripheral populations (e.g., at Bear Meadows, Pennsylvania; Morgenstern and Farrar 1964; Gordon 1976; Bobola et al. 1996; Capblancq et al. 2020). What is certain is that there was no northern migration of red spruce along the continuous uplands. The spruce range in West Virginia occurs beyond the current range of black spruce, thus interpretation of spruce pollen is unambiguous. Unfortunately, the few pollen diagrams available in West Virginia are at low elevations and even those where red spruce presently grows (e.g., Cranesville Swamp [Cox 1968], Cranberry Glades [Watts 1979]) display a spruce pollen hiatus in the mid-Holocene. In poorly drained sites or valleys with cold air drainages, red spruce in West Virginia can be found down to 760 m (2,500 ft) elevation (Cogbill and White 1991), with individual red spruce witness trees found as low as 509 m (1,670 ft; Thomas-Van Gundy et al. 2012). Regardless, red spruce obviously survived in the moist, cool environments on the higher mountains through the Holocene and is now abundant on the four ridges that are over 1,190 m (3,900 ft) elevation in West Virginia.

Unfortunately, there is little pollen or macrofossil evidence to show the migration history of red spruce at the southern end of the range. Therefore, any surmise about the Holocene history of the southern Appalachians is based primarily on the current biogeography. In the early Holocene, the highest elevations of the Appalachian Mountains had a periglacial environment above forests at lower elevations, leaving frost-riven substrates which were presumably treeless (Delcourt and Delcourt 1985). As the climate warmed, the surrounding spruce-pine forest moved north and the only remnant stands, presumably red spruce together with the endemic Fraser fir (*Abies fraseri*), remained in the southern Appalachians.

Given the lack of paleoecological information for the southern Appalachians, the geographic setting of the current forest has been used to interpret the past. Whittaker (1956) hypothesized that the spruce forest in the southern Appalachians expanded to higher elevations during the Holocene Thermal Maximum Period. Climatic fluctuations during this mid-Holocene warming period may have caused the loss of spruce from peaks less than 1,740 m (5,710 ft) in elevation in the Great Smoky Mountains (Whittaker 1956), which hosts the most southern populations of red spruce. Subsequently, a cooler, moister climate of the Neoglacial Cooling Period may have led to downslope expansion of spruce to 1,370 m (4,500 ft) within the southern Appalachians. While this 1,370 m (4,500 ft) adjustment may be locally true, spruce also occurs at this latitude at lower elevations (e.g., Long Hope Creek Bog at 1,320 m [4,330 ft], Alarka Laurel Bog at 1,265 m [4,150 ft]; 35.34° N; both in North Carolina). Regardless, the elevational zone is relatively abrupt latitudinally and by elevation as evidenced by the southwestern limit at Double Spring Gap (1,678 m [5,505 ft]) on the

Tennessee-North Carolina ridge in the Great Smoky Mountains and the lower limit of the small spruce cap on Whitetop Mountain (1,684 m [5,525 ft]) in Virginia. Spruce has apparently persisted on all mountain *sky islands* in the southern archipelago through the Holocene and the spruce community held its ground on rocky summits with shallow soils in the cloudy and cool orographic conditions.

It is speculated that there were at least three Pleistocene refugia for red spruce: near the northeastern coast, to the east of or within the southern mountains, and somewhere near the ice margin in between the southern mountains and eastern coast. From these sites, red spruce expanded, maintaining low abundance especially at the northern limits where today it overlaps with black spruce. All of the northern sites greatly expanded in the last 2,000 years as the cool and cloudy mountain conditions recreated the ancestral moist coastal environment. The genetic heritage indicates a divergence in the early Holocene (roughly 9,000 YBP; Capblancq et al. 2020), which would indicate that all three ancestry groups existed at that time along the East Coast. This is consistent with the pollen data and post-glacial refugia as the three spruce regions became distinct around 8,000 YBP (Gaudreau 1988; see Sidebar 1.1 for a perspective on pollen data). The present-day distribution and the convergence of the montane occurrences with the cool, moist, lowlands clearly reflect evolutionary and biogeographic history of red spruce as it segregated from its boreal, continental parent that has a wide ecological range.

Sidebar 1.1 Individual species migration, lag times, and ephemeral forest communities

Through analysis of pollen preserved in sediment cores, we can make educated guesses about the response of vegetation to the last ice age and changes in species abundances through the Holocene. While we are accustomed to thinking of these forests as groups of species adapted to similar climate and disturbance regimes, it is important to remember that tree species respond in an individualist manner when expanding into new territory and can colonize at differing rates. Given the huge distances involved, and the slow colonization rates of some species (for instance, those with heavy seeds), there are likely periods of time when a species' response lags behind the availability of suitable habitat. The individualistic nature of this response also means that many modern forest communities are very young, in geologic terms, with component species that may have only arrived a few thousand years ago. Also, past species combinations apparent in the pollen and fossil record may also have been relatively ephemeral in geologic terms.

1.1.3 Pre-Euro-American Extent

Before Euro-American settlement (see Sidebar 1.2), red spruce occurred in eastern North America and ranged from the interior of Ontario, Canada to the coast of eastern Canada and New England, then followed the Appalachian Mountains southward, forming disjunct populations in Pennsylvania south through Tennessee and North Carolina (Fig. 1.3), similar to the present-day geographic distribution of red spruce (Bailey and Ware 1990; White and Cogbill 1992). The red spruce forests located in Virginia in the transition between the Ridge and Valley and the Allegheny Plateau physiographic provinces are small and isolated relict stands (e.g., War Spur, Mann's Bog) compared to red spruce forests in the Allegheny Mountains and southern Blue Ridge Mountains. Relict populations also occur in Maryland and Pennsylvania, as well as at lower elevations in North Carolina (e.g., Alarka Bog).

Sidebar 1.2 Indigenous people in red spruce forests

While our discussion of factors affecting the current distribution of red spruce focused on events known to greatly impact the forest—glaciation and broad-scale timber harvesting by European settlers—the history of these forests includes Indigenous people. Pre-European history in the eastern U.S. is divided into five cultural periods by anthropologists and archaeologists: Pre-clovis, Paleoindian, Archaic, Woodland, and Contact. No pre-Clovis sites are known in West Virginia. However, the pre-Clovis Meadowcroft Rockshelter site is approximately 200 km (124 mi) north of spruce-dominated forests in West Virginia. Only one Paleoindian (12,000 to 10,000–8,000 YBP) projectile point has been found on the Monongahela National Forest (host to the largest extent of contiguous red spruce forests in the central Appalachians) on a tributary of the Greenbrier River. Paleoindian occupation in the larger central Appalachian region likely occurred after the less complex terrain of the Coastal Plain and Piedmont regions were occupied (Lane and Anderson 2001). The Archaic Period (10,000–3,000 YBP) is characterized by shifts in climate, vegetation, and technology with an increase in occupation of upland sites by the Middle Archaic (Custer 1990). In the uplands of the central Appalachians, however, much more is known about early humans during the Woodland Period (3,000–1,000 YBP). This multi-part period is defined in part by developments in ceramics, mortuary rituals, native plant cultivars, and the construction of earthworks. In the Late Woodland period, Indigenous settlements in West Virginia increased in number and size.

Two well-documented sites within or adjacent to the Monongahela NF help frame the early human use and occupation of the uplands of the central Appalachians. While the Mouth of Seneca site, a Late Woodland village (1,600–1,000 YBP), is not located within the historical range of red spruce-dominated forests, it shows that people did make their home here. At this

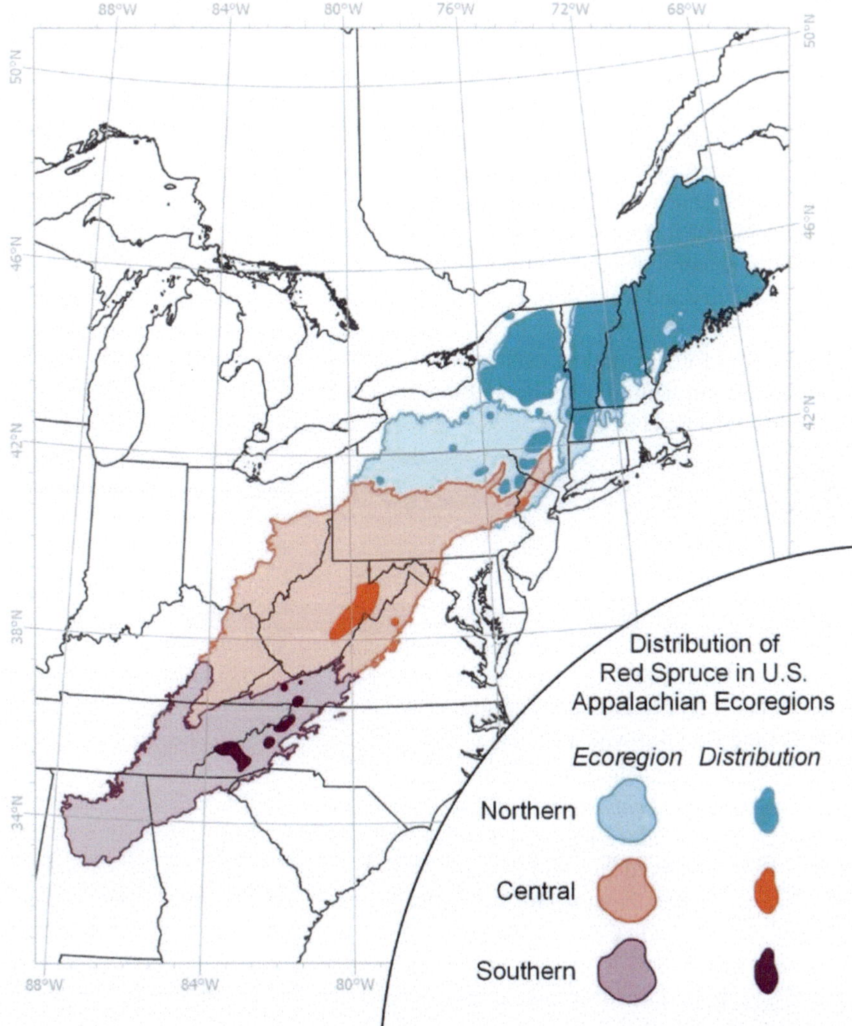

Fig. 1.3 Present-day distribution of red spruce (*Picea rubens*) in the eastern U.S. and Appalachian ecoregions as defined in this book. The depiction of the distribution in the southern Appalachians is exaggerated for visibility at this scale

upland site, the remains of deer and food plants including walnut, hickory, blueberry, huckleberry, cherry, and squash are found. This village site may have been occupied annually for short periods before more substantial shelters were built.

The Hyre Mound site near Huttonsville, West Virginia, lies in the Tygart Valley approximately 10 km (6 mi) from the Cheat Mountain range where red spruce persists. Recent analysis of this Middle to Late Woodland site confirmed extensive human occupation in the valley and suggests that Indigenous people here participated in the broader practices of the Eastern Woodlands culture (Rosencrance and Hirshman 2021). The reevaluation of lithic materials at this site further explores procurement strategies that linked people north and south along major drainages, including an important economic connection between the Greenbrier and Tygart River valleys. An over-land connection between these two drainages would take people across the higher elevations through red spruce-dominated forests. Upland archaeology is hampered by generally shallower sites than those at lower elevations or ones with gentler topography and upland sites often lack stratigraphic context making determinations of cultural chronology difficult. Indigenous people appear to have had less of a presence in West Virginia beginning in the 1700s, possibly because of the introduction of European diseases and the presence of Europeans themselves (Lesser 1993). However, early European settler accounts continued to document the presence of indigenous groups in the uplands of West Virginia engaged in hunting and resisting European settlements (Lesser 1993). European settlement of the area that would become the Monongahela National Forest began about this time (early to mid-1700s) and the low population densities of indigenous people led to the European idea that the uplands west of the Allegheny Front were not permanently settled (Henderson 1992). We now know this is not true and many are striving to recover this important history.

Although the pre-Euro-American geographic distribution of spruce-fir forests were distributed similar to today, red spruce-dominated forests occurred over larger extents in the pre-European landscape. While an estimate of extent for the entire range of red spruce is unknown, regional estimates for the central and southern Appalachians are available. Methods such as witness trees (Thomas-Van Gundy and Strager 2012; Thomas-Van Gundy et al. 2012), the presence of soils with spodic properties (Nauman et al. 2015, see Chap. 3), and historical data from early published reports (Hopkins 1891; Korstian 1937) can be used to determine historical locations of red spruce. Habitat suitability models may also reveal areas that were most likely favorable to red spruce, but where it may no longer occur (Nowacki and Wendt 2010; Beane et al. 2013; Koo et al. 2014; Andrews et al. 2022), potentially giving insight to pre-logging distributions throughout most of the red spruce's range.

Prior to exploitative logging at the turn of the twentieth century, there was an estimated 200,000 ha (494,200 ac) of red spruce-dominant forests in the central Appalachians, although if red spruce-northern hardwood forests are included estimates rise to about 1.2 million ha (2.96 million ac; Hopkins 1899; Fig. 1.4). The historical extent of red spruce in the southern Appalachians was about 405,000 ha

(1 million ac; Korstian 1937), with the largest extent in the Great Smoky Mountains at the most southern tip of red spruce's range. However, other estimates for the combined area of spruce occurrence in the central and southern Appalachians are only about 600,000 ha (1.5 million ac; Minckler 1940). Variation in the estimated area of red spruce occurrence may depend on whether both red spruce-dominated stands and red spruce-northern hardwood stands were included in those estimates (Pyle 1984).

In the central Appalachians, red spruce was seldom found below 700 m (2,300 ft) and most abundant above 900 m (2,900 ft), where soil and moisture conditions are favorable (Hopkins 1899). Red spruce forests before logging were restricted to elevations above 900 m (2,900 ft), with pure stands of red spruce likely above 1,200 m (3,900 ft) in Virginia (Pielke 1981). However, some have noted that Pielke's (1981) description of red spruce extent was based on lumber products which could have included hemlock, but labeled as spruce (Bailey and Ware 1990), potentially explaining the documentation of spruce lumber produced in states where the species was not considered to occur, such as Delaware. An analysis of witness trees from deeds and grants mainly dating in the 1800s documents individual red spruce trees at lower elevations, approximately 500 m (1,650 ft; Thomas-Van Gundy et al. 2012). In the southern Appalachians, red spruce occurred at elevations $\geq$ 1,375 m (4,500 ft; Korstian 1937), with Fraser fir becoming dominant above 1,800 m (5,900 ft; Pyle and Schafale 1988). Although there are many peaks with elevations greater than 1,375 m (4,500 ft), red spruce only occurs on peaks greater than 1,680 m (5,500 ft) in elevation (Cogbill and White 1991), highlighting the idea that past glacial and interglacial events influenced the pre-Euro-American distribution of red spruce in the region. Fraser fir typically dominates summits above 1,820 m (5,950 ft; Cogbill and White 1991), although the most northern extent of Fraser fir on Mt. Rogers in Virginia occurs below 1,700 m (5,580 ft).

1.1.4 Current Extent

Broad-scale clearcut logging occurred between 1870 and 1940 throughout the Appalachians, with sites in the northern Appalachians logged earlier followed by sites in the central and southern Appalachians (Fig. 1.5; Pinchot 1898; Korstian 1937; Pyle and Schafale 1988; Stephenson 1993). Logging in the northern extent of the range still occurs today, but commercial timber harvest in spruce forests in the central and southern Appalachians is limited (Nowacki et al. 2010). Logging since the nineteenth century removed most of the large merchantable spruce throughout the red spruce range (Korstian 1937; Westveld 1953; Hayes et al. 2007). Industrial clearcut logging resulted in an uphill elevational contraction of red spruce in mountainous regions (Pielke 1981; Busing et al. 1993; Hayes et al. 2007). Red spruce's current geographic distribution is similar to pre-Euro-settlement times, although its extent and density within those distributions have been greatly diminished due to anthropogenic activities.

Fig. 1.4 Old-growth red spruce (*Picea rubens*) and eastern hemlock (*Tsuga canadensis*) forest in 1936 on the Nels Wilson Tract, Monongahela NF, Randolph County, West Virginia. Photo by USDA Forest Service

The complete removal of overstory trees and accumulation of logging slash made both logged and adjacent unlogged red spruce forests susceptible to large, high-severity wildfires (Figs. 1.6 and 1.7). Fires were historically rare in the original montane spruce forests (Cogbill 2000), but in extreme multi-year drought events, spruce can fuel large wildfires (e.g., 1947 in southwest Maine and Mt. Desert Island).

Fig. 1.5 Residual red spruce (*Picea rubens*) trees left after logging on the Roaring Plains, Canaan Mountain, Tucker County, West Virginia, Monongahela NF, July 1938 (Minckler 1939). Note the absence of branches on the windward side. Photo by USDA Forest Service

However, this was thought to be even rarer in the southern Appalachians due to near-continual cloud deposition and frequent precipitation (see Sect. 1.4.2). Some sites, such as Dolly Sods in West Virginia in 1930 and Graveyard Fields in North Carolina in 1925, were converted from spruce-dominated forests to grasslands, shrublands, and boulder fields by logging-associated stand-replacing fires. Approximately 100 km^2 (39 mi^2) of spruce forests were consumed in the 1930 Dobbins Slashings Fire at Dolly Sods, a high-severity fire that burnt through logging debris and consumed thick duff layers down to bedrock. Almost a century later, neither Dolly Sods nor Graveyard Fields have recovered to their pre-fire vegetation condition, remaining in early successional states. At sites where fires were not severe enough to prevent forest reestablishment, logging and subsequent fires removed seed sources (i.e., mature trees, seedbanks) and changed soil and microclimate conditions, allowing faster growing northern hardwoods to easily outcompete and dominate sites formerly occupied by old-growth spruce forests (Pinchot 1898; Korstian 1937; Pielke 1981). The anthropogenically-induced species turnover from red spruce to northern hardwoods especially occurred at lower elevations in mountainous regions (Busing et al. 1993; Hayes et al. 2007) and increased mixed forests in flatlands and broad mountaintops (Boucher et al. 2006).

Fig. 1.6 Sherwood Forest burn on the Pisgah NF, April 1938 (approximate elevation of 1,830 m [6,000 ft]; Minckler 1939). The view is from near the top of Black Mountain facing northeast across the burn site. Remnants of spruce-fir can be seen at upper left. The foreground is a grassy bald and the darker areas down slope to the right regenerated with blackberry (*Rubus* spp.) and pin cherry (*Prunus pensylvanica*; photo by USDA Forest Service)

Regional estimates for loss of spruce-fir in the central and southern Appalachian Mountains are 50–95% and 35–60%, respectively (Stephenson 1993; Noss et al. 1995; Nowacki and Wendt 2010). Although losses of red spruce have not been specifically calculated for the northern Appalachians, climatic niche modeling indicates suitable habitat where red spruce may have been extirpated due to Euro-American logging and other land-use activities (Andrews et al. 2022). In West Virginia, red spruce occurred over about 300,000 ha (740,000 ac) in 1865 and was subsequently reduced to about 90,000 ha (220,000 ac) by 1899 and about 24,000 ha (60,000 ac) by the 1990s (Hopkins 1899; Stephenson 1993). In the Great Smoky Mountains, the lowest latitudinal population of red spruce experienced an estimated 45–50% loss of red spruce extent across the *sky island* landscape due to logging (Pyle 1984; Hayes et al. 2007). Estimates of spruce loss in refugia sites, such as Limberlost and War Spur in Virginia, due to logging are unknown. However, it is likely refugia also suffered decreases in extent or complete extirpation due to logging (Pielke 1981), especially since they are in minimally suitable habitats at elevations where hardwood forests are typically dominant. The drastic loss of red spruce-dominant forests is the

Fig. 1.7 Red spruce (*Picea rubens*) regeneration on Black Mountain, Monongahela NF, August 1938 (approximate elevation 1,370 m [4,500 ft]; Minckler 1939). This area was clearcut but not burned. Photo by USDA Forest Service

main reason why these forests in central and southern Appalachians are considered two of the most endangered forested ecosystems in the U.S. (Noss et al. 1995).

When red spruce was removed from mixed woods there was a profound lack of initial spruce regeneration given the competition with fast-growing, early successional hardwoods such as birch (*Betula* spp.) and black cherry (*Prunus serotina*), particularly in the central Appalachians (Nowacki et al. 2010). This "Westveld paradox" (Westveld 1930, 1953) resulted in restriction of the red spruce range and a sharpening of the coniferous/deciduous ecotone after logging. This sharpened ecotone is more obvious in the southern Appalachians, which may also have to do with the prevalence of mixed forests being more common in the flatter areas in the northern latitudes versus mountainous southern latitudes due to strong elevational gradients. In mountainous locations, logging activities shifted the lower elevational distribution upwards throughout its range. For example, in the Great Smoky Mountains prior to logging, red spruce was generally found above 1,513 m (4,964 ft) with minimal elevational limits down to 1,285 m (4,216 ft), but logging caused a 200 m (650 ft) upslope contraction in the upper and lower minimal elevational distribution of red spruce (Hayes et al. 2007). The logging-driven loss of red spruce at lower elevations and the potential for red spruce to occur at lower elevations is apparent in relict populations.

For example, Alarka Bog spruce forest in North Carolina sits at 1,220 m (4,000 ft; Collins et al. 2010), approximately 150 m (500 ft) lower than minimal elevations of red spruce at other sites.

Anthropogenic land uses may have also shifted the dominance of red spruce in the overstory, resulting in mixed forests, especially in the central Appalachians and northern parts of the red spruce range (Korstian 1937; Boucher et al. 2006). Within mountainous regions, the loss of overstory red spruce dominance may be driven by elevation, but not latitude depending on localized factors such as pre-logging forest composition and post-logging conditions (i.e., loss of seedbank, colonization distance of hardwoods). However, where red spruce forests suffered uphill contractions due to logging, regeneration currently indicates a recolonization of areas occupied before Euro-American activities, despite a warming climate (Busing et al. 1993; Nowacki et al. 2010; Rollins et al. 2010; Foster and D'Amato 2015; Wason and Dovciak 2017). Past land use may represent a stronger driver of red spruce elevational shifts than climate change at some sites (Iverson et al. 2008; Foster and D'Amato 2015; Danneyrolles et al. 2019), although some areas have experienced uphill shifts in northern hardwood-spruce ecotones since the 1960s (Beckage et al. 2008).

Current populations of red spruce seem stable throughout the geographic distribution of the species, including in the most southern latitudes (Nowacki et al. 2010). Nevertheless, climatic factors may influence the downhill shift of red spruce, wherein more pronounced shifts downslope to recolonize historical areas may be limited or hindered by warmer temperatures at lower elevations (Wason and Dovciak 2017). Therefore, past anthropogenic activity and climate change may interact and alter the future extent and distribution of red spruce (see Chap. 7). Within the central and southern Appalachians, most red spruce forests occur on state or federal land, are habitat for species of conservation concern, and are largely not available for commercial timber harvest (see Chap. 6). Presently, silvicultural activities on public lands in the central and southern Appalachians are generally restricted to treatments that promote red spruce restoration as guided by management plans (e.g., USDA Forest Service 2006; Rentch et al. 2016).

1.2 Topographical Influences

One of the main drivers influencing the geographic distribution of red spruce is topography (White and Cogbill 1992) and conditions associated with topography like soil depth and moisture (Murphy 1917). Variation in topographical factors interacting with climatic, edaphic, and biotic factors influences where species occur across the landscape (Braun 1950). Topographic features such as elevation, aspect, and landform shape (e.g., sheltered coves, exposed ridgelines) influence microclimate conditions (e.g., temperature, moisture, solar exposure), and may influence factors driving disturbance at local and regional scales (see Chap. 5), and potential climate change refugia (Stark and Fridley 2022).

Elevation interacting with latitude is a major driver of the red spruce distribution with occurrences at lower elevations corresponding to higher latitudes. The lower elevational limit of spruce-fir forests increases in elevation with decreasing latitude (i.e., southward) along the Appalachian Mountain chain (Cogbill and White 1991; White and Cogbill 1992). Red spruce currently ranges from sea level to 1,065 m (3,494 ft) in elevation in Canada and the northern Appalachians, 700–1,100 m (2,300–3,600 ft) in the Adirondack Mountains of New York, 915–1,480 m (3,000–4,850 ft) in the Allegheny Mountains of the central Appalachians, and 1,370–2,037 m (4,500–6,680 ft) in the southern Appalachians. However, relict sites in the Blue Ridge Mountains in North Carolina occur at 1,220 m (4,000 ft; Collins et al. 2010) and in the Ridge and Valley in Virginia at 975 m (3,200 ft; Adams and Stephenson 1991). In the central Appalachians, red spruce has been reported at elevations down to 822 m (2,697 ft). These relict populations and lower elevational distributions indicate potential lower elevational distributions prior to Euro-American disturbance (see Sect. 1.2.3). Indicator species analysis of red spruce witness tree locations in the mountains of east central West Virginia showed significant associations with higher elevations, moderate moisture (based on a moisture index), low topographic roughness (for example, stream channels, wetlands, or broad ridges) toeslope positions, and northeastern aspects (Thomas-Van Gundy and Strager 2012). These associations differed by ecological subsection. For example, red spruce was associated with lower elevations and valley landforms in the western Allegheny Mountain subsection (Thomas-Van Gundy and Strager 2012). It should be noted that elevation is closely correlated with climate, negatively with mean annual temperature and growing degree days, and positively with number of frost days (Nowacki and Wendt 2010). Although categorized as a topographic feature, elevation largely controls vegetation expression through climate.

Aspect influences insolation, moisture, and temperature, wherein a similar elevation on a south-facing slope is drier and hotter (more xeric) than the same elevation on a north-facing slope (more mesic). The cooler, moister microclimates on north-facing slopes likely explain why aspect is a predictor of the lower elevational limits of red spruce. For example, red spruce can occur 120–200 m (400–650 ft) lower in elevation on north-facing compared to south-facing slopes (Davis 1930; Brown 1941; Rheinhardt and Ware 1984; Hayes et al. 2007). Current spruce-fir forests on Mt. Rogers occur above 1,500 m (4,920 ft) on northern aspects, 1,600 m (5,250 ft) on aspects facing northwest, west, or south, and 1,700 m (5,580 ft) on southwestern aspects (Rheinhardt and Ware 1984), highlighting the variation in elevation limits of dominant spruce-fir stands on a single mountain.

1.3 Climatic Environment

The geographic distribution of red spruce corresponds to the Appalachian Mountain chain, an area of ecological diversity and climatic complexity located between the Mississippi River basin, the Great Lakes region, and the Atlantic Ocean. The complexity of the topographic and climatic interactions found here are often simplified to differences in elevation and the orographic effect when prevailing west-to-east air masses intersect with the roughly north–south trending mountain chain. A recent broad-scale analysis across the Appalachian Mountains at airports on an east-to-west transect determined that westerly winds were observed 54.5% of the time on average (Kutta and Hubbart 2019). This data was derived north of our focal areas; however, the Appalachian Mountains could be described as having a bidirectional rain shadow. Flash flooding and winter storms, including icing events, are associated with air currents from the east (Kutta and Hubbart 2019). However, this work is based on surface winds which often do not reflect wind direction at higher latitudes.

1.3.1 Temperature

Within the range of red spruce estimated by Little (1971), the mean annual temperature ranges from 2.0 to 13.5 °C (35.6–56.3 °F; USDA Forest Service Climate Change Atlas, version 4). The gross range of temperature estimates can be narrowed by using climate stations within red spruce sites. The average annual temperature envelope of red spruce, utilizing an extension of Little's range and calibrating for stations at elevation in the mountains, ranges from 0.7 to 9.7 °C (33.3–49.5 °F). Additionally, the annual temperature regime is misleading as it does not account for winter or summer temperatures, which differ in the amplitude of variation around those seasonal means. Based on logistic regression, mean annual temperature almost exclusively explains red spruce occurrence in the northern Appalachians, whereas snowfall is the principal driver in the central and southern Appalachians (Nowacki et al. 2010).

The temperature parameter most closely correlated with spruce occurrence is the peak summer (i.e., July) temperature with a range of 12.9–20.3 °C (55.2–68.5 °F; mean = 18.2 °C [64.8 °F]) across the red spruce range (Cogbill and White 1991; Wason et al. 2017). Many summer temperature parameters covary with the peak, but July temperature was also the highest rated of 38 environmental variables used to predict the importance value of red spruce (USDA Forest Service Climate Change Tree Atlas, version 3). The red spruce climatic envelope is closely aligned with temperature observed at the northern hardwood-red spruce ecotone, which has a peak summer temperature of 16.6–17.5 °C (61.9–63.5 °F) and a corresponding growing season (>5 °C [41 °F]) of 160–238 days (Cogbill and White 1991). Because the southern mountains have a less continental temperature regime (thus lower amplitude) than the northern lowlands, the summer peak (July) temperature is more important in defining red spruce habitat than the annual average.

The mean minimum winter temperatures (January) are substantially warmer in the south (-6 °C [21 °F] at Newfound Gap, elevation 1,536 m [5,039 ft]) than in the north (-17 °C [1 °F] at Mt. Mansfield, elevation 1,205 m [3,953 ft]). Although the absolute minimum temperatures in winter are lower in the north (-40 °C [-40 °F]), a similar minimum temperature (-35 °C [-31 °F]) was recorded at Grandfather Mountain, North Carolina in 1985. However, extremes in the south are likely less frequent and of shorter duration than the north. Since absolute minimum temperatures are typically observed in cold air drainages (e.g., coves, closed valleys), the same phenomenon of frost pockets allows occurrence of red spruce in lower elevation sites in the southern Appalachians, especially on northern aspects.

Despite the shorter summer photoperiod, the southern montane red spruce forests at lower latitudes receive more photosynthetically active radiation in the growing season (around 1,750 $\mu mol{\bullet}m^{-2}{\bullet}s^{-1}$; Reinhardt and Smith 2008) than northern mountains (around 1,300 $\mu mol{\bullet}m^{-2}{\bullet}s^{-1}$; Richardson et al. 2004). In general, red spruce forests have dense evergreen foliage (7–12 m^2/m^2 leaf area indices for unthinned forests in Maine; DeRose and Seymour 2012), which efficiently captures this energy leaving dark and open understories, typically dominated by mosses. Southern red spruce forests, however, have enhanced growing conditions at the same July average temperature with a longer growing season, greater precipitation, and increased energy input than northern forests. Additionally, the less extreme minimum winter temperatures and decreased snowfall reduce the stress on southern trees (Cogbill and White 1991).

Climate is inherently complex in mountainous regions, thus predicting impacts of climate change is equally difficult (see Chap. 7). Paradoxically, some trends might contrast what is expected from rising global temperatures. For instance, in West Virginia, Kutta and Hubbart (2018) found that recent reductions in diurnal temperature range and vapor pressure deficit may lead to increased relative humidity, cloud cover, and soil moisture which, in turn, support cool-adapted, mesophytic species such as red spruce. Similar resistance to climate warming was found on Mt. Washington in New Hampshire due to insulating factors such as thermal inversions and high incidence of cloudiness (Seidel et al. 2009).

1.3.2 Cloud Immersion

Even casual observers of red spruce forests of the central and southern Appalachian Mountains recognize that the topographic position of these forests, higher elevation and moist air masses, results in cloud cover. Red spruce is prominent in cloudy environments along coastal regions or in the mountains. A widely distributed network of regional monitoring stations has shown a cloud base typically forms at 800 and 1,200 m (2,600 and 3,950 ft) in the Adirondacks and northern Appalachian Mountains and between 1,000 and 1,400 m (3,300 and 4,600 ft) in the southern Appalachians (Mohnen 1992). Cloud deposition from orographic clouds (resulting from moist air forced upwards by mountains) provides additional moisture and cooler daytime temperatures to these forests in the summer, along with other ecophysiological benefits (see Chap. 2). At Yarmouth on the coast of Nova Scotia, around 15% of the summer hours are foggy (Environment Canada: Canada national climate data and information archive, https://climate.weather.gc.ca/historical_data/search_historic_data_e.html) and coastal Maine is immersed in fog 9–20% of the time during the summer months (Jagels 1986). Similarly, with orographic clouds, the northeast mountains have increasing immersion from 17–19% of the day at 748–1,025 m (2,450–3,360 ft) in elevation (Mohnen 1992; Miller et al. 1993), rising through 37–50% of the day at 1,000–1,500 m (3,280–4,920 ft; Lovett and Reiners 1986, Mohnen and Vong 1993, Roberti 2011) to a peak of 57% of all hours at 1,917 m (6,289 ft) on Mt. Washington (Seidel et al. 2007).

In the southern Appalachians, cloud immersion occurs during 55–75% of growing season days (Reinhardt and Smith 2008; Berry and Smith 2012). Like the mountainous areas in the northern Appalachians, the number of cloud-immersed days in the southern Appalachians varies with elevation, with cloud immersion occurring 56% of days at sites at 1,510 m (4,950 ft) and 75% of days at 1,960 m (6,430 ft) during the growing season (Berry et al 2014). Cloud immersion typically occurs in the morning and usually lasts for more than 2 h in the southern Appalachian Mountains (Berry and Smith 2012). Unfortunately, there are no published reports on cloud immersion for red spruce sites in the central Appalachians, however a meso-scale analysis of airport weather data shows cloud immersion is a common event along ridgelines in the mid-Appalachian Mountains (Kutta and Hubbart 2019).

Beyond the effects of cooling, when cloud immersion is accompanied by orographic winds, moisture deposited on spruce foliage subsequently drips to the ground, also called fog drip. This can increase water input by an estimated 50–100% of the rain gauge precipitation and results in an estimated 300–400 cm (120–160 in) per year of additional moisture at high elevations in both the northern and southern Appalachian Mountains (Lovett et al. 1982; Miller et al. 1993; Shubzda et al. 1995; Reinhardt and Smith 2008), influencing plant physiology (see Chap. 2). However, a more recent study found much lower amounts of fog interception in the southern Appalachians than previously reported (3–8% of rainfall), which may be related to foliar uptake (Praskievcz and Sigdel 2023).

1.3.3 Precipitation and Ice

Annual precipitation across the range of red spruce varies by location. Total annual precipitation ranges from 86 cm (34 in; northern interior) to 206 cm (81 in; southern Appalachian Mountains). However, localized amounts of precipitation vary depending on elevation. Winter also produces snow, which accounts for roughly 10% of the annual precipitation in the southern mountains, rising to 25% in the central Appalachians, 30% in the northern Appalachians, and 40% at the northern limits in Québec. Ice storms with freezing rain typically accumulating 2 cm (0.8 in) of ice on contact with branches occur sporadically (typically 2–3 times per decade), but cause major damage to the canopy, especially to deciduous trees just adjacent to spruce forest (Nicholas and Zedaker 1989; Rhoads et al. 2002; Boyce et al. 2003; Lafon 2004).

During cold winter temperatures, the conditions of cloud immersion, wind, and an intercepting surface combine to produce rime deposition. Supercooled cloud droplets entrained by the wind impact on objects such as foliage and instantly freeze as frost feathers growing into the wind. This is commonly observed as a *frost line* with white encased vegetation on mountain sides above the cloud base. In the Vermont mountains, the additional deposition of water contained in rime (60 cm [23.6 in]) in winter is roughly equal to the amount of water in the 60 cm (23.6 in) of snow deposited in the same period (Ryerson 1990). As in summer, the amount of water deposited in the high mountains in winter is up to double the amount measured as precipitation.

Red spruce is adapted to shedding heavy snow load, rime, or ice deposit, but tree height growth can be limited by broken branches or foliage damage in windy conditions at high elevations. More significant damage to red spruce and fir species are found with the regular wind-blown sleet or lofted ice (ice blast akin to sand blast) at high elevations, which abrades exposed branches and causes pruning of exposed foliage resulting in canopy flagging or stunted krummholz.

1.4 Summary

The last glacial maximum (approximately 20,00 YBP) influenced the genetics and biogeography of the modern red spruce species. Analysis of palaeobotanical data shows periods of increase and decrease in abundance through the Pleistocene and Holocene. While red spruce was pushed south by glaciation and moved north as the climate warmed, there were also refugia along the east coast. Red spruce occurs across a variety of latitudes and physiographic regions within its geographic distribution, with temperature, precipitation, cloud deposition, and elevation influencing occurrence throughout its range. Red spruce can be separated into five main zones: southern Appalachians, central Appalachians, northern Appalachians, coastal, and interior forests. Each region hosts distinct precipitation regimes that influence presence and unique disturbance regimes that influence stand dynamics. Each region also hosts different unique floral and faunal communities (see Chaps. 4 and 6 and Sidebar 1.3), with co-dominant tree species varying across the zones.

High moisture and cool growing season temperatures define red spruce sites for both past and present forests. This has implications for the impacts of contemporary climate change on these forests and for designing active restoration treatments. While we have evidence that red spruce is currently regaining ground lost to hardwood species after the exploitative pre-industrial logging period, this response could be impacted by changes in precipitation and summer air temperatures associated with anthropogenic climate change. Red spruce, as an evergreen, appears to benefit by cloud immersion compared to hardwood species, although potential climate change impacts to orographic cloud formation are difficult to estimate and quantify. Understanding the past of red spruce may help in understanding how climate change may impact this important tree species and the associated species found within its forests.

Sidebar 1.3 Ecological classification and mapping

In this book we use or reference several ecological community classification systems. The ecoregional boundary in Fig. 1.3 is based on the EPA's Ecoregions of North America, the NRCS system of Ecological Site Descriptions is described in Chap. 3, and the International Vegetation Classification system is used in Chap. 4 to describe the red spruce-associated communities. The overall goal of any classification system is to define, describe, and map repeated patterns, often in support of conservation and management efforts. Developing restoration actions using defined ecological units should aid in prioritizing where to act and allow for predictable results.

Classification systems should be hierarchical, that is the varying scale of ecological units should nest and build on each other with logical and consistent boundaries. Ecological communities are defined as the community of organisms and the physical environment. The physical environment defines the space the ecosystem occupies. This space can be described by independent variables of bedrock geology, geomorphic processes, and regional climate that influence the dependent variables of surface geology, landform, and local climate, which further combine to influence soil, site morphometry, and vegetation (Fig. 1.8). When relationships can be described between independent and dependent variables, independent variables may be useful for predicting dependent variables.

Ecological classification and mapping systems have shared characteristics, as all are developed to depict and describe the patterns and placement of ecosystems across multiple scales. The capacity of the land to support some action may also be included in these classification systems as many were first developed for specific management purposes, such as agricultural productivity or identification of geological hazards. The classification system used may be based on the preferences of individual managers or agencies and organizations. National Forest System lands are classified and mapped under the Terrestrial Ecological Unit Inventory system codified in the USDA Forest Service manual, handbook, and accompanying technical guide (Winthers et al. 2005). Other CASRI and SASRI partners may have different agency direction or no preference for a certain system.

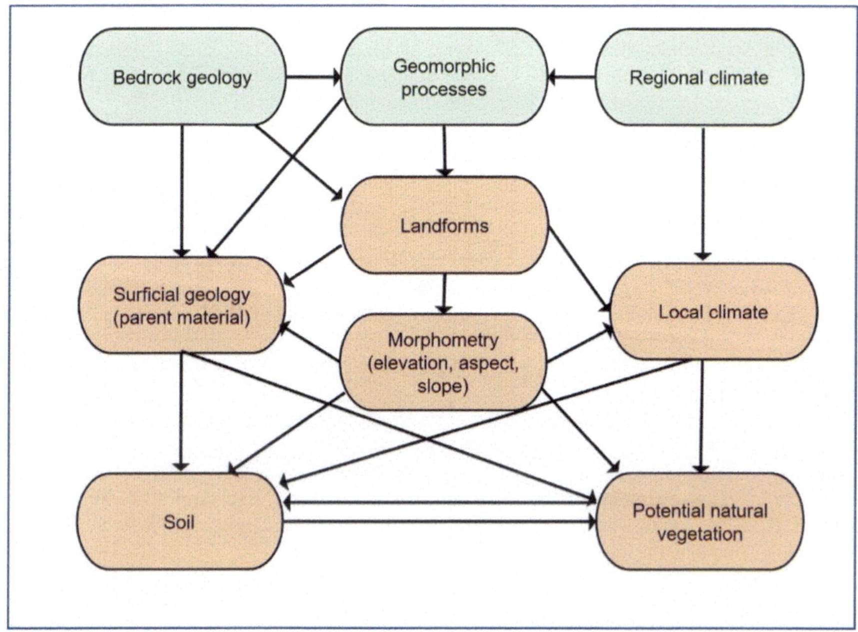

Fig. 1.8 A conceptual depiction of the general relationships between landscape elements. Elements are considered independent (light green shading) or dependent (light orange shading). Figure adapted from Winthers et al. (2005)

References

Abrams MD, Nowacki GJ (2015) Exploring the early Anthropocene burning hypothesis and climate-fire anomalies for the eastern US. J Sustain for 34:30–48

Abrams MD, Nowacki GJ (2019) Global change impacts on forest and fire dynamics using paleoecology and tree census data for eastern North America. Ann for Sci 76:8

Adams HS, Stephenson SL (1991) High elevation coniferous forests in Virginia. Va J Sci 42:391–399

Alley RB (2004) GISP2 ice core temperature and accumulation data (IGBPPAGES/World Data Center for Paleoclimatology Data Contribution Series #2004–013). Boulder, CO: NOAA/NGDC Paleoclimatology Program

Andrews C, Foster JR, Weiskittel A et al (2022) Integrating historical observations alters projections of eastern North American spruce–fir habitat under climate change. Ecosphere 13:e4016

Bailey CM, Ware S (1990) Red spruce forests of Highland County, Virginia: biogeographical considerations. Castanea 55:245–258

Bashaltkhanov S, Johnson JS, Rajora OP (2023) Postglacial phylogeography, admixture, and evolution of red spruce (*Picea rubens* Sarg.) in Eastern North American. Front Plant Sci 14:1272362

Beane NR, Rentch JS, Schuler TM (2013) Using maximum entropy modeling to identify and prioritize red spruce forest habitat in West Virginia. USDA Forest Service Northern Research Station Research Paper NRS-23, Newtown, Pennsylvania

Beckage B, Osborne B, Gavin DG et al (2008) A rapid upward shift of a forest ecotone during 40 years of warming in the Green Mountains of Vermont. Proc Natl Acad Sci 105:4197–4202

Berry ZC, Smith WK (2012) Cloud pattern and water relations in *Picea rubens* and *Abies fraseri*, southern Appalachian Mountains, USA. Agric for Meteorol 162–163:27–34
Berry ZC, Huges NM, Smith WK (2014) Cloud immersion: an important water source for spruce and fir saplings in the southern Appalachian Mountains. Oecologia 174:319–326
Bobola MS, Eckert RT, Klein AS et al (1996) Using nuclear and organelle DNA markers to discriminate among *Picea rubens*, *Picea mariana*, and their hybrids. Can J for Res 26:433–443
Boucher Y, Arseneault D, Sirois L (2006) Logging-induced change (1930–2002) of a preindustrial landscape at the northern range limit of northern hardwoods, eastern Canada. Can J for Res 36:505–517
Boyce RL, Friedland AJ, Vostral CB et al (2003) Effects of a major ice storm on the foliage of four New England conifers. Écoscience 10:342–350
Braun EL (1950) Deciduous forests of eastern North America. Hafner Publishing Company, New York
Brown JW (1941) Forest history of Mount Moosilauke. Thesis, Yale University, New Haven, Connecticut, New Hampshire
Busing RT, White PS, MacKenzie MD (1993) Gradient analysis of old spruce–fir forests of the Great Smoky Mountains circa 1935. Can J Bot 71:951–958
Capblancq T, Butnor JR, Deyoung S et al (2020) Whole-exome sequencing reveals a long-term decline in effective population size of red spruce (*Picea rubens*). Evol Appl 13:2190–2205
Clarkson R (1964) Tumult on the mountains: lumbering in West Virginia 1770–1920. McClain Printing Company, Parsons, West Virginia
Cogbill CV (2000) Vegetation of the presettlement forests of northern New England and New York. Rhodora 102:250–276
Cogbill CV, White PS (1991) The latitude-elevation relationship for spruce-fir forest and treeline along the Appalachian Mountain chain. Vegetatio 94:153–175
Collins B, Schuler TM, Ford WM et al (2010) Stand dynamics of relict red spruce in the Alarka Creek headwaters, North Carolina. In: Rentch, JS, Schuler, TM (eds) Proceedings from the conference on the ecology and management of high-elevation forests in the central and southern Appalachian Mountains. USDA Forest Service Northern Research Station General Technical Report NRS-P-64, Newtown Square, Pennsylvania, pp 22–27
Cox DD (1968) A late-glacial pollen record from the West Virginia-Maryland border. Castanea 33:137–149
Custer JF (1990) Early and middle archaic cultures of Virginia: culture change and continuity. In: Reinhart TR, Hodges MEN (eds) Early and middle archaic research in Virginia: a synthesis. Archaeological Society of Virginia Special Publication No. 22, Charles City, Virginia, pp 1–60
Danneyrolles V, Dupuis S, Fortin G et al (2019) Stronger influence of anthropogenic disturbance than climate change on century-scale compositional changes in northern forests. Nat Commun 10:1265
Davis JH Jr (1930) Vegetation of the Black Mountains of North Carolina: an ecological study. J Elisha Mitchell Sci Soc 45:291–318
Davis MB (1983) Holocene vegetation of the eastern United States. In: Wright HE (ed) Late-quaternary environments of the United States, vol 2. University of Minnesota Press. Minneapolis, Minnesota, pp 166–181
Davis MB (1981) Quaternary history and the stability of forest communities. In: West DC, Shugart HH, Botkin DB (eds) Forest succession: concepts and application. Springer-Verlag, New York, pp 132–153
Delcourt PA, Delcourt HR (1981) Vegetation maps for eastern North America: 40,000 yr B.P. to the present. In: Romans RC (ed) Geobotany II. Plenum Publishing Corporation, New York, New York, pp 123–165
Delcourt PA, Delcourt HR (1985) Quaternary palynology and vegetational history of the southeastern United States. In: Bryant VM, Holloway RG (eds) Pollen records of late-Quaternary North American settlements. American Association of Stratigraphic Palynologists' Foundation, Dallas, Texas, pp 1–37

DeRose RJ, Seymour RS (2012) Leaf area and structural changes after thinning in even-aged *Picea rubens* and *Abies balsamea* stands in Maine, USA. Int J For Res 2012:181057

Feranec RS, Christiansen M, Driver D et al (2021) "Man and the mastodon": revisiting the Northborough mastodon. Eastern Paleontologist 9:1–22

Foster JR, D'Amato AW (2015) Montane forest ecotones moved downslope in northeastern USA in spite of warming between 1984 and 2011. Glob Change Biol 21:4497–4507

Gaudreau DC (1988) The distribution of late Quaternary forest regions in the Northeast. In: Nicholas GP (ed) Holocene human ecology in northeastern North America. Plenum Publishing Company, New York, pp 215–256

Gordon AG (1976) The taxonomy and genetics of *Picea rubens* and its relationship to *P. mariana*. Can J Bot 54:781–813

Hayes M, Moody A, White PS et al (2007) The influence of logging and topography on the distribution of spruce-fir forests near their southern limits in Great Smoky Mountains National Park, USA. Plant Ecol 189:59–70

Henderson AG (1992) Dispelling the myth: seventeenth- and eighteenth-century Indian life in Kentucky. Regist Ky Hist Soc 90:1–25

Hopkins AD (1891) Black spruce. West Virginia Agricultural Experiment Station Bulletin No. 17, Morgantown, West Virginia

Hopkins AD (1899) Report on investigations to determine the cause of unhealthy conditions of the spruce and pine from 1880–1893. West Virginia Agricultural Experiment Station Bulletin No. 56, Morgantown, West Virginia

Iverson L, Prasad A, Matthews S (2008) Modeling potential climate change impacts on the trees of the northeastern United States. Mitig Adapt Strat Glob Change 13:487–516

Jackson ST, Weng C (1999) Late Quaternary extinction of a tree species in eastern North America. Proc Natl Acad Sci 96:13847–13852

Jackson ST, Whitehead DR (1991) Holocene vegetation patterns in the Adirondack Mountains. Ecology 72:641–653

Jagels R (1986) Acid fog, ozone and low elevation spruce decline. Int Assoc Wood Anat J 7:299–307

Jaramillo-Correa JP, Bousquet J (2003) New evidence from mitochondrial DNA of a progenitor-derivative species relationship between black spruce and red spruce (Pinaceae). Am J Bot 90:1801–1806

Jaramillo-Correa JP, Bousquet J (2005) Mitochondrial genome recombination in the zone of contact between two hybridizing conifers. Genetics 171:1951–1962

Koo KA, Madden M, Patten BC (2014) Projection of red spruce (*Picea rubens* Sargent) habitat suitability and distribution in the southern Appalachian Mountains, USA. Ecol Model 293:91–101

Korstian CF (1937) Perpetuation of spruce on cut-over and burned lands in the higher southern Appalachian Mountains. Ecol Monogr 7:125–167

Kutta E, Hubbart JA (2019) Observed mesoscale hydroclimate variability of North America's Allegheny Mountains at 40.2° N. Climate 7:91

Kutta E, Hubbart JA (2018) Changing climatic averages and variance: implications for mesophication at the eastern edge of North America's eastern deciduous forest. Forests 9:605

Lafon CW (2004) Ice-storm disturbance and long-term forest dynamics in the Adirondack Mountains. J Veg Sci 15:267–276

Lane L, Anderson DG (2001) Paleoindian occupations of the southern Appalachians. In: Sullivan LP, Prezzano SC (eds) Archaeology of the Appalachian highlands. University of Tennessee Press, Knoxville, Tennessee, pp 88–102

Lesser WH (1993) Prehistoric human settlement in the upland forest region. In: Stephenson SL (ed) Upland forests of West Virginia. McClain Printing Company, Parsons, West Virginia

Lindbladh M, Jacobson GL Jr, Schauffler M (2003) The postglacial history of three *Picea* species in New England, USA. Quatern Res 59:61–69

Little, EL Jr (1971). Atlas of United States trees. Volume 1, Conifers and important hardwoods. U.S. Department of Agriculture Miscellaneous Publication No. 1146, Washington, DC

Lovett GM, Reiners WA (1986) Canopy structure and cloud water deposition in subalpine coniferous forests. Tellus 38B:319–327

Lovett GM, Reiners WA, Olson RK (1982) Cloud droplet deposition in subalpine balsam fir forests: hydrological and chemical inputs. Science 218:1303–1304

Manley SAM (1972) The occurrence of hybrid swarms of red and black spruce in central New Brunswick. Can J for Res 2:381–391

Miller EK, Friedland AJ, Arons EA et al (1993) Atmospheric deposition to forests along an elevational gradient at Whiteface Mountain, NY, U.S.A. Atmos Environ 27A:2121–2136

Minckler LS (1940) Early planting experiments in the spruce-fir type of the southern Appalachians. J Forest 38:651–654

Minckler LS (1939) Spruce type problem analysis: analysis of problems in the reforestation of the spruce type of the southern Appalachians. USDA Forest Service Unpublished File Report

Mohnen VA (1992) Atmospheric deposition and pollutant exposure of eastern U.S. forests. In: Eagar C, Adams MB (eds) Ecology and decline of red spruce in the eastern United States. Springer-Verlag, New York, New York, pp 64–124

Mohnen VA, Vong RJ (1993) A climatology of cloud chemistry for the eastern United States derived from the mountain cloud chemistry project. Environ Rev 1:38–54

Morgenstern EK, Farrar JL (1964) Introgressive hybridization in red spruce and black spruce. University of Toronto Faculty of Forestry Technical Report 4, Toronto, Canada

Murphy LS (1917) The red spruce: its growth and management. U.S. Department of Agriculture Bulletin No. 544, Washington, DC

Nauman TW, Thompson JA, Teets J et al (2015) Pedoecological modeling to guide forest restoration using ecological site descriptions. Soil Sci Soc Am J 79:1406–1419

Nicholas NS, Zedaker SM (1989) Ice damage in spruce–fir forests of the Black Mountains, North Carolina. Can J for Res 19:1487–1491

Noss RF, LaRoe ET, Scott JM (1995) Endangered ecosystems of the United States: a preliminary assessment of loss and degradation. U.S. National Biological Service Biological Report No. 28, Washington, DC

Nowacki G, Wendt D (2010) The current distribution, predictive modeling, and restoration potential of red spruce in West Virginia. In: Rentch JS, Schuler TM (eds) Proceedings from the conference on the ecology and management of high-elevation forests in the central and southern Appalachian Mountains. USDA Forest Service Northern Research Station General Technical Report NRS-P-64, Newtown Square, Pennsylvania, pp 163–178

Nowacki G, Carr R, Van Dyck M (2010) The current status of red spruce in the eastern United States: distribution, population trends, and environmental drivers. In: Rentch JS, Schuler TM (eds) Proceedings from the conference on the ecology and management of high-elevation forests in the central and southern Appalachian Mountains. USDA Forest Service Northern Research Station General Technical Report NRS-P-64, Newtown Square, Pennsylvania, pp 140–162

Perron M, Bousquet J (2003) Natural hybridization between black spruce and red spruce. Mol Ecol 6:725–734

Pielke RA (1981) The distribution of spruce in west-central Virginia before lumbering. Castanea 46:201–216

Pinchot G (1898) The Adirondack spruce: a study of the forest in Ne-Ha-Sa-Ne Park. Critic Company, New York

Praskievicz S, Sigdel R (2023) Fog interception in spruce-fir and mixed northern hardwood forests of Great Smoky Mountains National Park, Southeast USA. Ecohydrol Hydrobiol 23:532–541

Pyle C, Schafale MP (1988) Land use history of three spruce-fir forest sites in southern Appalachia. J for Hist 32:4–21

Pyle C (1984) Pre-park disturbance in the spruce-fir forests of Great Smoky Mountains National Park. In: White PS (ed) The southern Appalachian spruce-fir ecosystem: its biology and threats. National Park Service Research/Resource Management Report SER-71, Atlanta, Georgia, pp 115–130

Reinhardt K, Smith WK (2008) Leaf gas exchange of understory spruce–fir saplings in relict cloud forests, southern Appalachian Mountains, USA. Tree Physiol 28:113–122

Rentch JS, Ford WM, Schuler TS et al (2016) Release of suppressed red spruce using canopy gap creation—ecological restoration in the central Appalachians. Nat Areas J 36:29–37

Rheinhardt RD, Ware SA (1984) The vegetation of the Balsam Mountains of southwest Virginia: a phytosociological study. Bull Torrey Bot Club 111:287–300

Rhoads AG, Hamburg SP, Fahey TJ et al (2002) Effects of an intense ice storm on the structure of a northern hardwood forest. Can J for Res 32:1763–1775

Richardson AD, Lee X, Friedland AJ (2004) Microclimatology of treeline spruce–fir forests in mountains of the northeastern United States. Agric for Meteorol 125:53–66

Roberti JA (2011) Monitoring interactions between cloud base and the subalpine-alpine environments of the Presidential Range. Plymouth State University, Plymouth, New Hampshire, Thesis

Rollins AW, Adams HS, Stephenson SL (2010) Changes in forest composition and structure across the red spruce-hardwood ecotone in the central Appalachians. Castanea 75:303–314

Rosencrance RL, Hirshman AJ (2021) Over the hills and far away: middle to late woodland archaeology and toolstone conveyance at Hyre Mound (46RD1), West Virginia. North American Archaeologist 42:140–176

Ryerson CC (1990) Atmospheric icing rates with elevation on northern New England mountains, U.S.A. Arct Alp Res 22:90–97

Schauffler M, Jacobson GL Jr (2002) Persistence of coastal spruce refugia during the Holocene in northern New England, USA, detected by stand-scale pollen stratigraphies. J Ecol 90:235–250

Seidel TM, Grant AN, Pszenny AAP et al (2007) Dewpoint and humidity measurements and trends at the summit of Mount Washington, New Hampshire, 1935–2004. J Clim 20:5629–5641

Seidel TM, Weihrauch DM, Kimball KD et al (2009) Evidence of climate change declines with elevation based on temperature and snow records from 1930s to 2006 on Mount Washington, New Hampshire, U.S.A. Arct Antarct Alp Res 41:362–372

Shubzda J, Lindberg SE, Garten CT et al (1995) Elevational trends in the fluxes of sulphur and nitrogen in throughfall in the southern Appalachian Mountains: some surprising results. Water Air Soil Pollut 85:2265–2270

Spear RW (1989) Late-Quaternary history of the high-elevation vegetation in the White Mountains of New Hampshire. Ecol Monogr 59:125–151

Stark JR, Fridley JD (2022) Microclimate-based species distribution models in complex forested terrain indicate widespread cryptic refugia under climate change. Glob Ecol Biogeogr 31:562–575

Stephenson SL (1993) Upland forests of West Virginia. McClain Printing Company, Parsons, West Virginia

Thomas-Van Gundy M, Strager M, Rentch J (2012) Site characteristics of red spruce witness tree locations in the uplands of West Virginia, USA. Journal of the Torrey Botanical Society 139:391–405

Thomas-Van Gundy MA, Strager MP (2012) European settlement-era vegetation of the Monongahela National Forest, West Virginia. USDA Forest Service Northern Research Station General Technical Report NRS-101, Newtown Square, Pennsylvania

USDA Forest Service (2006) Monongahela National Forest land and resource management plan. https://www.fs.usda.gov/Internet/FSE_DOCUMENTS/stelprdb5330420.pdf. Accessed 26 Jul 2024

Wason JW, Dovciak M (2017) Tree demography suggests multiple directions and drivers for species range shifts in mountains of northeastern United States. Glob Change Biol 23:3335–3347

Wason JW, Bevilacqua E, Dovciak M (2017) Climates on the move: implications of climate warming for species distributions in mountains of the northeastern United States. Agric for Meteorol 246:272–280

Watts WA (1979) Late Quaternary vegetation of central Appalachia and the New Jersey coastal plain. Ecol Monogr 49:427–469

Westveld M (1953) Ecology and silviculture of the spruce-fir forests of eastern North America. J Forest 51:422–430

Westveld M (1930) Suggestions for the management of spruce stands in the Northeast. U.S. Department of Agriculture Circular No. 134, Washington, DC

White PS, Cogbill CV (1992) Spruce-fir forests of eastern North America. In: Eagar C, Adams MB (eds) Ecology and decline of red spruce in the eastern United States. Springer-Verlag, New York, New York, p 3–39

Whittaker RH (1956) Vegetation of the Great Smoky Mountains. Ecol Monogr 26:1–80

Winthers E, Fallon, D, Haglund et al (2005) Terrestrial ecological unit inventory technical guide: landscape and land unit scales. USDA Forest Service General Technical Report WO-68, Washington, DC

Chapter 2
The Biology of Red Spruce

Jay E. Raymond, Sophan Chhin, Peter S. White, and William A. Whitter

2.1 Introduction

Red spruce (*Picea rubens*) is a medium sized, coniferous tree that grows in cool, moist climates, and acidic soils in eastern North America (Blum 1990; Fig. 2.1). It is a long-lived species with a slow growth rate due in part to its shade tolerance and subsistence in the understory for many years (Murphy 1917; Korstian 1937). Red spruce was first taxonomically described by the American botanist Charles Sprague Sargent in *Silva of North America—Vol. 12* (Sargent 1899). Modern taxonomy classifies red spruce as a gymnosperm in the family Pinaceae (International Plant Names Index 2022). Prior to the end of the nineteenth century, red spruce was common in many boreal and montane forests in eastern North America (Blum 1990; see Sect. 1.2.2 in Chap. 1). Extensive harvesting between the late 1800s and early 1900s dramatically reduced the area of red spruce occurrence (See Chap. 1). Today, red spruce communities are concentrated in three broad biogeographical regions—southern Appalachians, central Appalachians, and northern Appalachians (see Fig. 1.3 in Chap. 1).

J. E. Raymond (✉)
Virginia Western Community College, Integrated Environmental Studies, Roanoke, VA, USA
e-mail: jraymond@virginiawestern.edu

S. Chhin
West Virginia University, School of Natural Resources and the Environment, Morgantown, WV, USA

P. S. White
University of North Carolina at Chapel Hill, Department of Biology, Chapel Hill, NC, USA

W. A. Whitter
USDA Forest Service, National Forests in South Carolina, Columbia, SC, USA

D. J. Brown et al. (eds.), *Ecology and Restoration of Red Spruce Ecosystems in the Central and Southern Appalachians*, https://doi.org/10.1007/978-3-032-19616-3_2

Fig. 2.1 Red spruce (*Picea rubens*) stand, Spruce Knob, Monongahela NF, West Virginia. Photo by USDA Forest Service

2.2 Life History

2.2.1 Reproductive Structure, Seed Production, and Seed Dispersal

Red spruce is monoecious with both male (i.e., pollen cone) and female (i.e., seed cone; Fig. 2.2) reproductive structures on the same tree, each forming on different branches. Pollen and seed cones differentiate the year preceding fertilization in mid to late summer. Pollen cones (length: 1.5–2.5 cm [0.6–1.0 in]) develop on axillary apices part of the distal shoots or terminal apices of proximal shoots. Initial pollen cone colors range from purple to red and turn yellow to brown upon maturity. Seed cones (length: 2.5–6 cm [1.0–2.4 in]) form at the tip of stems (terminal) and are initially erect with a greenish, slight purple tinge. Similar to other coniferous species, upon maturity the color of seed cones changes to a reddish to purplish brown and the cones gradually become hanging (pendulous) on the branches. Although pollen is released from late April to May, seed cones are only receptive to pollen a few days during this period.

Seed cones mature the following year in late August to October (Hart 1959; Safford 1974; Blum 1990; Farjon 2017), with seeds falling throughout late fall into

Fig. 2.2 Red spruce (*Picea rubens*) seed cone, Spruce Knob, Monongahela NF, West Virginia. Photo by Elizabeth Byers

winter. Periods for seed dispersal can be highly variable, ranging from October to March (Heinselman 1965; Safford 1974). The lightweight, winged seeds of red spruce are well suited for wind dispersal. Silvicultural studies of red spruce seed dispersal into forest openings show an effective seedling dispersal range of 61 m (200 ft; Frank and Bjorkbom 1973), with the greatest concentration of seeds 25–30 m (82–98 ft; Hughes and Bechtel 1997) from the forest edge, but potentially viable regeneration up to 100 m (328 ft) away from the seed tree Randall 1974).

Although red spruce trees may produce seeds starting at age 15 (Frank and Bjorkbom 1973), viable, large seed crop production does not begin until the ages of 30–50 years (Heinselman 1965; Safford 1974). The largest seed crops occur when individual trees enter the canopy into full, direct sunlight (Korstian 1937). Under these conditions red spruce produces the largest seed crops every 3–8 years, with smaller, less viable crops annually that may continue until mortality (Hart 1965; Blum 1990). In the southern region, two studies found contradictory results related to the percentage of filled seeds and seed mass with increasing elevation. Whereas Nicholas et al. (1992) observed a decrease with increasing elevation, a more recent study by Butnor et al. (2019) found an increase.

2.2.2 Seed Germination

Red spruce germination is epigeal (i.e., occurring above the ground; Safford 1974; Farjon 2017). Most seedling germination occurs the year following fertilization from May to July but may also occur in October immediately after release (Blum 1990). Germination success varies among substrates. Most mineral soil substrates that have high soil moisture holding capacities with moderate temperatures (20–30 °C [68–86 °F]) have the highest rates of successful germination (Frank and Safford 1970). Substrates with extensive bryophyte (non-vascular mosses, hornworts, and liverworts) layers also have successful germination rates due to their high water holding capacity and nutrient content (Dibble et al. 1999). Germination is lower in areas with thicker organic horizons and on decomposing coarse woody debris due to rapid desiccation, resulting from highly variable temperatures. Germination will not occur below 20 °C (68 °F), and significant injury resulting in high mortality occurs when temperatures exceed 33 °C (91 °F; Heinselman 1965; Safford 1974; Blum 1990). In Maine, Greenwood and McConville (2002) found germination rates of 33–83% when there was no water stress, with optimal daily temperatures of 20 °C (68 °F) and nightly temperatures of 10 °C (50 °F).

2.2.3 Seedling Establishment and Growth

Dry growing mediums and frost damage are major causes of mortality for red spruce seedlings growing in areas where organic horizons exceed 5 cm (2 in) or on decaying logs in the first year after germination (Blum 1990). Because seedlings slowly develop a shallow, fibrous root system, young roots may not penetrate deep enough to avoid zones of droughty or freezing conditions, leading to mortality. The upper areas of the soil are also prone to higher soil surface temperatures especially in open canopy areas with hotter and drier microclimates, with significant mortality occurring when soil surface temperatures exceed 46 °C (115 °F) for short intervals (Heinselman 1965; Safford 1974; Blum 1990). Seedling establishment is most successful under canopies reducing direct sunlight to moderate forest floor temperatures and retain moisture (Fig. 2.3).

Red spruce seedlings are typically established when they reach a height of 15 cm (6 in) due to a fibrous root system extending below organic and soil surface horizons that can generally survive drought or frost heaving conditions (Heinselman 1965; Frank and Bjorkbom 1973; Safford 1974). Once established, seedling growth may be reduced by interspecific competition with seedlings of competing conifers (e.g., Fraser fir [*Abies fraseri*]), deciduous trees, ericaceous shrubs, and herbaceous species whose diversity and abundance depends on elevation, and the amount and type of localized disturbance.

Although low light levels slow seedling growth, red spruce is highly shade tolerant due to its foliar plasticity (Greenwood et al. 2008), increased resource allocation for

Fig. 2.3 Red spruce (*Picea rubens*) regeneration. Photo by USDA Forest Service

lateral versus height growth (Dumais and Prévost 2007), and its ability to photosynthesize with light levels found in sunflecks (Alexander et al. 1995). Given these factors, red spruce can survive extended periods in the understory and still respond to release treatments at 60–70 years (Nicholas et al. 1992), and even up to 150 years of age (Blum 1990). Since single-tree canopy disturbance is the most common disturbance type in red spruce and spruce-fir forests (see Chap. 5), this strategy is advantageous for a seedling to eventually reach the overstory.

Research across the entire red spruce range generally shows positive responses to silviculture prescriptions that mimic natural gap openings in the canopy that equate to changes in microenvironmental conditions. Specific silvicultural prescriptions relating to red spruce growth response are detailed in Chap. 8. Herein, we discuss seedling and sapling establishment in response to gap creation. Most silvicultural prescriptions aim to achieve an optimal light availability of 50% for red spruce seedlings (Seymour 1995; Moore et al. 2007; Dumais and Prévost 2014; Rentch et al. 2016). In the northern region in Québec, red spruce germination and establishment success depended on the type and size of gaps created in the canopy that influence the microenvironment (Dumais and Prévost 2007, 2008, 2014, 2016; Dumais et al.

2020). This research showed that although creating small (100m^2 [1076 ft^2]) gaps with a gradient of decaying wood and mosses initially appeared to be an optimum prescription, after 10 years seedling mortality appeared to increase compared to medium sized (100–300 m^2 [1076–3229 ft^2]) gaps, with larger (700 m^2 [7535 ft^2]) gaps having the lowest seedling survival.

In the central region, treatments with similar gap sizes were shown to increase red spruce seedling height and diameter growth by killing (but not removal) of more than 90% of the overstory competition with stem injection in gaps < 100 m^2 (1076 ft^2) and removal of 42% of overall basal area. Rollins et al. (2010) assessed seedling establishment across a plant community gradient and observed an increase in red spruce seedling establishment in the ecotone, a transitional zone of changing environmental conditions and species composition, between red spruce and deciduous communities. Red spruce seedling establishment also occurred in the red spruce-dominated study areas but was significantly reduced in the exclusively deciduous areas. Although creating small- to medium-sized gaps in the canopy is a viable silvicultural option to increase sapling red spruce growth and recruitment in the northern and central Appalachians, the size of gaps created in the southern Appalachians has not been well studied. Large gaps created by the death of hemlock (*Tsuga* spp.) and Fraser fir due to invasive insects (see Sect. 2.5.1) have caused overstory red spruce mortality due to wind throw (Busing 2004), indicating negative impacts of larger canopy gaps.

2.2.4 Forest Structure and Composition

Red spruce typically forms uneven-aged stands due to its persistance in the understory and is generally not released until gap formation in the canopy as domininant and co-dominant trees die (Blum 1990; Fraver and White 2005; Rentch et al. 2007; Rentch et al. 2016; see Fig. 2.1). Red spruce can also occur, although less frequently, as homogenous, single-aged stands due in part to historical clear-cut logging and subsequent fire activity (Korstian 1937; Allard and Leonard 1952) or large windthrow events (see Chap. 5). The high shade tolerance of red spruce allows seedlings and saplings to persist in the understory, potentially for decades (Blum 1990; D'Arrigo et al. 2012), which can make forest recovery slow after disturbances (Rollins et al. 2010).

Generally, red spruce growth rates are greater moving north to south due to a longer growing season in the more southern latitudes (Oosting and Billings 1951). The largest red spruce trees occur in old-growth forests, the majority of which occur in the Great Smoky Mountains in the southern Appalachians. For example, in the central and southern regions, stem diameters of larger trees may range from 54–61 cm (21–24 in; rarely exceeding 76 cm [30 in]), with maximum heights of 35 m (115 ft; Korstian 1937; Oostig and Billings 1951; Busing et al. 1988; Blum 1990; Soulé et al. 2012). However, the size and stature of red spruce in second growth stands following timber harvesting tends to be reduced compared to old-growth counterparts (Korstian 1937; Schuler et al. 2002; Rentch et al. 2010).

Red spruce's productivity is impacted by a variety of factors. One important growth factor for red spruce throughout its entire range is age. Red spruce is a long-lived tree species with maximal ages reaching up to 400 years (Hart 1965), as evidenced by dendrochronological studies. As red spruce ages, biomass productivity, photosynthesis rates, and foliar efficiency can decline (Day et al. 2001). Older red spruce trees can comparatively exhibit more symptoms of decline than their younger understory counterparts affected by similar stressors (Adams and Stephenson 1989; Goelz et al. 1999). The age of suppressed red spruce is important as older trees are generally less responsive to release compared to their younger counterparts (Seymour and Kenefic 2002), although two studies that indicated a possible response up to 145 years (Safford and Young 1968; Frank and Bjorkbom 1973). In addition to age, crown architecture can be important when considering red spruce productivity and efficiency. Studies have shown that dominant canopy red spruce, with the largest crown sizes, generally have decreased growth efficiency (Seymour and Kenefic 2002; DeRose and Seymour 2009).

One method to assess site and stand productivity for tree species is site index (Brown 1962). Site index is commonly expressed as the height of an overstory tree at a reference base age, usually 50 years in eastern North America, and is based on the measurement of free-growing canopy trees (Geyer and Lynch 1987; Mathiasen et al. 2006). Site index curves can then be developed that estimate tree growth and hence potential stand productivity. Although traditional site index relationships have assessed red spruce productivity (Carmean et al. 1989), they are not reliable estimates because true age is used to predict height of red spruce. Red spruce is shade tolerarant and has extended periods of suppressed, slow growth (Rentch et al. 2007) that introduces inaccuracies into the development of site index curves.

Although limited, there have been improvements to estimate red spruce site productivity using site index relationships in northern and central regions (Nicholas and Zedaker 1992; Seymour and Fajvan 2001; Yetter et al. 2021a). In the northern region, Seymour and Fajvan (2001) quantified free-to-grow conditions and alternate periods of tree suppression to adjust site index curves. Most recently, Yetter et al. (2021a) developed suppression-corrected age-height pair and DBH-height pair site index models in the central Appalachians (Fig. 2.4). Red spruce was sensitive to several climatic (e.g., number of frost-free days, extreme temperature over a 30-year period) and geographic (e.g., elevation) variables which can affect growth (Yetter et al. 2021a, 2021c). Appropriate site index curves for red spruce in the southern region still need to be developed.

2.2.5 Stand Dynamics

Forest dynamics in red spruce stands are typically controlled by wind (Busing and Pauley 1994; Busing 2004), ice (Nicholas and Zedaker 1989), and individual tree

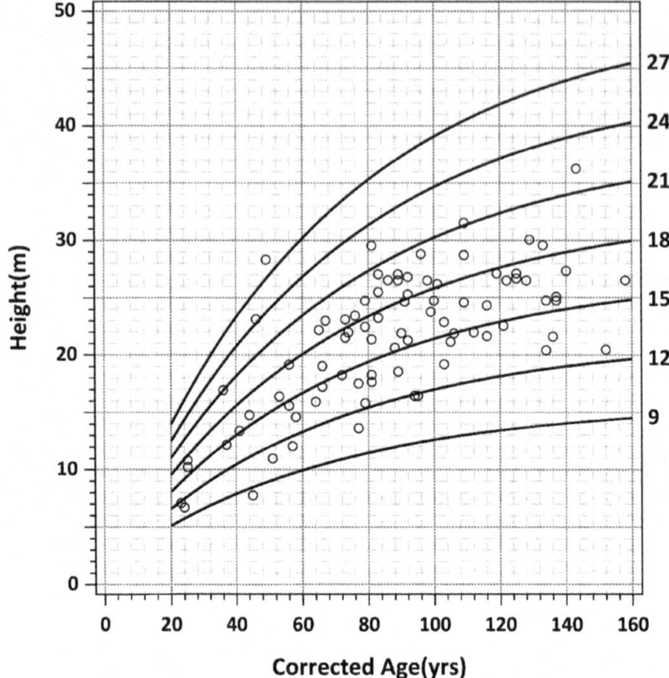

Fig. 2.4 Red spruce (*Picea rubens*) site index curves developed from a model of height- and suppression-corrected age. Individual lines are base age 50 site index curves. Open circles represent sample trees ($n = 83$). Adapted from Yetter et al. (2021a)

mortality (Cogbill and White 1991), although larger historical disturbance events caused by forest insects, large blowdown events, and landslides have been documented (Hopkins 1891, 1899; Murphy 1917; Korstian 1937; Cook and Johnson 1989; see Chap. 5). Gap dynamics of second-growth red spruce stands in the central region of West Virginia indicate that gaps are most frequently created by the death of American beech (*Fagus grandifolia*) due to beech bark disease (Rentch et al. 2010). Unfortunately, these typical gaps formed in the central Appalachians of West Virginia generally close before red spruce can reach the canopy, and restoration efforts require active management (i.e., thinning and release) to promote subsequent canopy openings to allow advanced red spruce regeneration to ascend to the overstory in hardwood dominant stands (Rentch et al. 2010; see Chap. 8). Gap phased dynamics in mixed stands of spruce-fir in the Great Smoky Mountains of North Carolina and Tennessee indicated a trend of replacement of canopy species, with Fraser fir prominent at higher elevations (White et al. 1985; Busing 1996).

In mixed-species stands of red spruce with a fir component, balsam woolly adelgid (*Adelges piceae*), a non-native insect, can lead to mortality of fir trees (Creed et al. 2004). In particular, Fraser fir is susceptible to balsam woolly adelgid and infestations can negatively impact fir regeneration while not impairing red spruce regeneration

(Stehn et al. 2013). Infestation with balsam woolly adelgid can in turn lead to an increased amount of coarse woody debris (Creed et al. 2004; Rose and Nicholas 2008). Past insect infestation of Fraser fir in Great Smoky Mountains National Park led to improved growth rates of red spruce in the stand, but mortality of red spruce increased due to increased wind exposure (Busing and Pauley 1994; Busing 2004).

Predation from herbivores has minimal impact on red spruce regeneration because the species has low palatability to many browsers, such as white-tailed deer (*Odocoileus virginianus*). Browse resistance provides a competitive advantage for red spruce regeneration over competing hardwood species, such as red maple and yellow birch (Griscom et al. 2011). In contrast, competition from other plant species limits spruce regeneration. For example, *Rhododendron* stems can grow into dense thickets and can negatively affect regeneration of red spruce (Adams and Stephenson 1989). In canopy gaps created by beech bark disease, American beech sprouts can grow fast and may end up filling canopy gaps with infected beech saplings, which takes up available space and resources for red spruce regeneration (Van Miegroet et al. 2007; Rentch et al. 2016).

2.3 Abiotic and Biotic Stressors

The interaction of abiotic and biotic stressors affects the reproduction, growth, and distribution of red spruce in the central and southern Appalachians (Harmon et al. 1984). Examples of abiotic stressors include extreme weather events such as ice storms (Nicholas and Zedaker 1989) and windstorms (Busing 2004), fire (Moore et al. 2002), and pollutant deposition, whereas biotic stressors include pests and pathogens. Stressors are also influenced by historical activities, including exploitative timber harvesting (Adams 1999) and climate change. For example, at the turn of the twentieth century extensive extractive harvest logging and subsequent fires caused significant soil erosion and nutrient loss from the ecosystem that reduced soil fertility and limited red spruce reproduction, growth, recovery, and potential restoration. More than a century after pre-industrial logging occurred, mountain slopes devoid of trees are still seen in Great Smoky Mountains National Park (Harmon et al. 1984).

2.3.1 *Abiotic Stressors*

Prior to 2000, most research on the abiotic stressors of red spruce forests in the central and southern Appalachians focused on understanding the effects of pollutant exposure and deposition on high elevation red spruce ecosystems (Eagar and Adams 1992; Rentch et al. 2010). Although these regional red spruce forests are confined to remote, high peaks, they are exposed to concentrated, high levels of pollutants originating from the midwestern and southern U.S. that are carried eastward on prevailing atmospheric patterns. Sources of this pollution include point sources (coal power plants emissions) and diffuse sources (automobile exhaust). Air masses are forced over the mountains, resulting in higher rainfall and cloud immersion at high elevations that concentrate pollutants such as ground level ozone, carbon monoxide, sulfur dioxide, and nitrogen dioxide. The levels of pollutant emissions from industrial sources increased from the 1950s through the 1990s but then declined due to the Clean Air Act, which focused on reducing pollution from point sources. Although passed in 1970, the Clean Air Act was strengthened by amendments in 1990 that helped reduce pollution levels (Mohnen 1992; Moore et al. 2002; Banks 2013; Lawrence and Baily 2021). As the successful impact of the Clean Air Act became increasingly apparent, research shifted to investigating questions relating to ecosystem recovery.

Research on the effects of pollutant deposition and exposure in Appalachian red spruce forests increased in the 1980s following initial observations that spruce populations were declining in the northern Appalachians and Europe (Eagar and Adams 1992). During this period the term *acid rain* was used to describe the acidic precipitation that formed when sulfur and nitrogen oxide pollutants combined with atmospheric water to become sulfuric and nitric acids. However, pollutant exposure includes both dry and wet forms. Clouds, mist, and fog may be more important pathways for pollutants as wet deposition in certain regions than rain. In addition, these high elevation forests may be even more vulnerable since the soils found in these environments are naturally acidic and low in buffering capacity due to low cation availability (Moore et al. 2002).

2.3.2 Biotic Stressors

When immature, red spruce is generally resistant to most native insects and diseases, and it is not until maturity that tree health may be affected (Blum 1990). Most research specific to insects and diseases is based on red spruce systems in the northern Appalachians and Canada, with limited studies specific to the central and southern regions. In the central Appalachians, research has focused on determining whether various spruce gall adelgids (e.g., eastern spruce gall adelgid [*Adelges abietis*] and *Cytospora* cankers [*Leucostoma kunzei*]) influence red spruce health (see Sect. 6.6.2 in Chap. 6). Two studies suggested that although spruce gall adelgids and *Cytospora* cankers were observed in several areas of West Virginia, they did not generally appear causing a decline in the overall health of individual red spruce trees (Audley and Skelly 1994; Audley et al. 1998). Spruce gall adelgid species, like aphids, form galls at the apex or base of new shoots depending on the species. Although unsightly and causing branch tips to weaken, infestations generally do not cause significant damage to individual trees and are mostly confined to single, scattered trees in a stand. Similarly, *Cytospora* cankers can damage branches and cause deformities, but they rarely translate to tree mortality. Similar results were also found for red spruce in the southern region. Bruck (1989) conducted a survey of diseases and insects on red spruce and Fraser fir in the Black Mountains of North Carolina and found the influence on individual tree health by the fungi and insects observed on red spruce was minor.

In contrast to minimally damaging biotic agents specific to red spruce in the central and southern Appalachians, other plant species associated with red spruce forests in these regions have been dramatically affected. Two specific examples of tree species include the eastern hemlock and Fraser fir. In the central Appalachians, hemlock woolly adelgid (*Adelges tsugae*) has caused significant mortality to eastern hemlock, occurring mixed stands with red spruce. In the southern Appalachians, Fraser fir occurs with red spruce at higher elevations and significant mortality has been caused by the balsam woolly adelgid (*Adelges piceae*, see Sect. 6.6.2.2 in Chap. 6) since its introduction and spread in the 1960s (Eagar and White 1984). Although it has been suggested that the deposition of atmospheric pollutants may have increased the vulnerability of the mature Fraser fir to balsam woolly adelgid infestation (McLaughlin et al. 1995), evidence has not been found to support this hypothesis (Lee et al. 1997). Although atmospheric deposition is a regional abiotic stressor, the timing and pattern of Fraser fir mortality in this region directly followed the geographic pattern of balsam woolly adelgid spread rather than a pattern associated with levels of pollutant deposition (White 1984).

The long-term persistence of Fraser fir in these ecosystems is uncertain. Fraser fir is a shade tolerant species that produces abundant regeneration (Peck 1990). Because the stems do not become vulnerable to balsam woolly adelgid attack until they are large enough to develop bark crevices, the initial waves of Fraser fir mortality left behind a high density of younger trees that enable recovery for Fraser fir populations (Smith and Nicholas 1998; Bowers and Bruck 2010; Lusk et al. 2010; McManamay et al. 2011; White et al. 2012; Kaylor et al. 2017). The health of young Fraser fir is further evidence that the balsam woolly adelgid was the principal agent of Fraser fir mortality rather than pollutant exposure. What is uncertain is whether the patches of existing healthy Fraser fir will survive long enough to provide seeds for future generations and open the possibility of long-term persistence of Fraser fir (Dale et al. 1991; Smith and Nichols 2000). Fraser fir develops bark fissures prior to cone production (Eagar and White 1984), suggesting Fraser fir may continue to decline despite the current recovery. It is possible that Fraser fir and balsam woolly adelgid continue to coexist in a dynamic patch system. In this system, balsam woolly adelgid would become locally scarce in young Fraser fir stands that had not developed bark crevices, possibly allowing Fraser fir reproduction through delayed recolonization of balsam woolly adelgids. The current patchiness of Fraser fir populations due to the previous balsam woolly adelgid infestation, combined with the geographically isolated Fraser fir forest populations, may also impact long-term viability of this tree species. Another possibility for long-term persistence of Fraser fir forests is the possibility that clones of Fraser fir are putatively resistant to balsam woolly adelgid produced sesquiterpenes after an adelgid feeding period of 20 weeks (Thomas et al. 2022). While it is not known whether these compounds are responsible for resistance to adelgids, Thomas et al. (2022) argued that they may be useful indicators of genetically resistant genotypes.

Although the balsam woolly adelgid does not attack red spruce itself, a sudden loss of the Fraser fir canopy in an old-growth forest in the Great Smoky Mountains caused compensatory mortality in red spruce by increasing spruce mortality to wind damage and windthrow (Busing and Pauley 1994). Another study showed that Fraser fir mortality from balsam woolly adelgid caused increased overstory red spruce mortality from 2 to 4% per year, with 70% of the dead canopy stems resulted from wind (Busing 2004). The increase in wind damage was especially marked on exposed topographic positions, such as ridgelines. As is shown in this example, the interaction between biotic and abiotic stressors can be difficult to separate.

Unlike the balsam woolly adelgid, the hemlock woolly adelgid feeds on the needles of all age classes of eastern hemlock, including younger stems, and there is no *window of escape* for this tree species as described for Fraser fir. Research has not focused on the indirect effects of the sudden loss of eastern hemlock co-dominant stems on red spruce forests in the central Appalachians, and it is uncertain what effect the loss of eastern hemlock will have on the lower elevation range of red spruce where these species co-occur. Attempts to identify and establish biocontrol agents for the hemlock woolly adelgid are now underway (Mayfield et al. 2020).

2.3.2.1 Acid Rain and Pollutant Deposition

Pollutant deposition increases with increasing elevation in mountainous regions, and studies have used this gradient to contrast red spruce response to various levels of exposure (Berry et al. 2014). The increase in pollutant deposition with elevation is due to increases in precipitation and cloud immersion. Cloud immersion contributes one-third to one-half of all water sources at upper elevation sites (Aneja et al. 1992; Reinhardt and Smith 2008b). In addition, cloud-intercepted water is more acidic (pH 2.5–4.5) than rainwater (pH 3.5–5.5; Aneja et al. 1992). Deposition of nitrogen oxides can be higher in cloud-intercepted water than in rainwater or dry deposition (Isil et al. 2022). Spruce-fir forests in the southern Appalachians are immersed in clouds 25–40% of the time and on 65% of all days annually (Aneja et al. 1992; Reinhardt and Smith 2008a, Berry et al. 2014; see Sect. 1.4.2 in Chap. 1). An important research need is to understand how these cloud immersion periods will change in duration, timing, and elevation as local climates change and what effect these changes will have on spruce-fir forests. Regardless of whether upwind pollutant emissions increase or decrease, the delivery of sulfuric and nitric acids will vary with precipitation and cloud exposure and may shift under contemporary rapid climate change.

As noted previously, pollutant emissions increased from the 1950s through the 1990s but then decreased with improved amendments to the Clean Air Act in 1990. Even areas with similar elevations and time periods may experience differences in pollutant exposure and deposition due to site specific microenvironmental differences. For example, no significant effect of acid deposition on spruce-fir forests in the central and southern Appalachians for elevations ranging from 980–1750 m (3215–5740 ft; Bryant et al. 1997). However, the Spruce-Fir Research Cooperative did find significant acidification effects in forests with elevations of 1800 m (5900 ft; Adams and Eagar 1992). Although the results of Bryant et al. (1997) did not show a significant effect of acidification, the systems sampled may have already begun recovery at the time of sampling due to the 1990 Clean Air Act amendments compared to the Spruce-Fir Research Cooperative data collected 5–10 years prior (Banks 2013). Pollutant exposure and deposition in high elevation forests has raised several lines of questions: (1) whether the deposition caused a short-term decline of red spruce; (2) whether the deposition caused a long-term acidification of soils; and (3) whether a reduction of soil nutrients due to the initial acidification would affect red spruce growth and aquatic ecosystems receiving runoff from the acidified terrestrial ecosystems. Although questions remain about the short- and long-term effects of acidic deposition, climate change may now be playing an outsized role in the resiliency of these spruce-fir forests and interacting with current pollution levels to impact mortality (Koo et al. 2014).

2.3.2.2 Pollutant Impacts on Red Spruce Growth and Mortality

Many studies have evaluated the impacts of pollutant deposition on red spruce growth and mortality in the Appalachians. In the northern region, evidence clearly showed pollutants slowing growth and increasing mortality starting in the 1960s (Adams and Eagar 1992). However, it is more difficult to ascertain obvious growth patterns for spruce-fir forests in the central and southern Appalachians. Some studies found growth declines in red spruce in the central and southern regions (Adams et al. 1985; McLaughlin et al. 1990, 1998; Adams and Eagar 1992; McLaughlin and Percy 1999; Elias et al. 2009; White et al. 2014). One study by Webster et al. (2004) found growth declined from the 1940s to the 1970s but that these declines reversed after the 1970s. Other studies found no growth rate declines in red spruce for these regions (Busing and Wu 1990; Cook and Zedaker 1992; Johnson et al. 1992; Goelz et al. 1999; Soulé 2022). Due to these mixed results, it is difficult to make definitive statements about the direct influence of pollutant deposition on red spruce mortality in the central and southern Appalachians (Reams et al. 1993; McLaughlin et al. 1994). Since 2000, most research has indicated no decline in red spruce growth rates (Bowers and Bruck 2010; Lusk et al. 2010), with some studies showing increases (Hornbeck and Kochenderfer 1998; Soulé 2022).

During the development of a forest, growth rates of trees change due in part to the competition for resources (e.g., light, nutrients) that may be further influenced by pollutants. At the start of a stem exclusion phase, the growth rate of red spruce declines due to increased stem crowding and hence increased competition for limited resources. During this phase, when there is a constant crown size and leaf area, even if biomass production is constant, tree-rings become narrower as tree circumference increases. In addition, red spruce is a late successional shade-tolerant species and therefore undergoes episodes of prolonged suppressed growth before it reaches the canopy. For example, growth declines in 25% of red spruce trees in the northern Appalachians after 1967, especially at higher elevations, but the growth decline was not greater than historical levels (LeBlanc et al. 1992). Yet, understanding specific influences of growth, and growth metrics in general, can be difficult to interpret (Reams et al. 1993; McLaughlin et al. 1994).

Understanding the influence of pollutants on red spruce growth rates is difficult to ascertain and most studies did not find an increase in mortality in these regions directly related to pollution (Bruck and Robarge 1988; Busing et al. 1988; McLaughlin et al. 1990, 1998; Cook and Zedaker 1992; Johnson et al. 1992; Busing and Pauley 1994; Smith and Nicholas 1999). Although Busing and Pauley (1994) observed an increase in red spruce mortality (2–4% per year) in an old-growth spruce-fir forest in the Great Smoky Mountains, the authors attributed this increase to the sudden loss of the co-dominant Fraser fir canopy trees and hence increased wind exposure to the remaining red spruce trees. Most studies after the 2000s have found no increase in red spruce mortality and potentially an increase in red spruce recruitment in certain systems (Nowacki et al. 2010; Mathias and Thomas 2018).

2.3.2.3 Acidification Effects on Red Spruce

Although it does not appear that pollutant deposition has created an ongoing decline in red spruce populations in the central and southern Appalachians, research has shown there is a more chronic, longer-term issue—ecosystem acidification. In areas of the Appalachians impacted by acid deposition, exposure to sulfuric and nitric acids in rainwater and cloud-intercepted water leaches cations from red spruce needles (Joslin et al. 1988; Robarge et al. 1989; McLaughlin et al. 1990, 1991; Thornton et al. 1990, 1992, 1994; Sayre and Fahey 1999; Crim et al. 2019). Foliar concentrations of calcium (Ca) and magnesium (Mg) in red spruce had decreased by more than 50% compared to the 1960s, while in Fraser fir concentrations had decreased from 25 to 50% (Shepard et al. 1995). A study showed significant levels of Ca and Mg leached from red spruce foliage when an acidic water solution (pH 3.1) was applied, especially in the spring when the outer cuticle on young needles may not have completely formed (Sayre and Fahey 1999). Evidence of foliar leaching, which can cause leaf mortality, is also supported by observed differences between water captured as throughfall compared to that in open areas, with throughfall containing twice the concentration of Ca, Mg, and other plant nutrients (Joslin et al. 1988; Johnson et al. 1991; Joslin and Wolfe 1992).

Soils with high concentrations of aluminum (Al) competitively inhibit calcium uptake, an important nutrient for structure in plants. Aluminum displaces certain cations from the surface of soil particles at low pH, causing reduced calcium plant uptake and increased likelihood of leaching. Ratios between certain nutrient cations, especially Ca and Al, have been proposed as indexes of acidification for both plant tissue and soil solutions (Shortle and Smith 1988). The risk of decline in red spruce increased as the Ca:Al ratio decreased, where the risk was 50% at a Ca:Al ratio of 1.0, 75% risk at 0.5, and near 100% risk at 0.2 (Cronan and Grigal 1995). In the southern Appalachians, Ca:Al ratios were among the lowest on record, suggesting the possibility of that aluminum toxicity could reduce root growth in the future (Robarge et al. 1989).

2.3.2.4 Acidification Effects on the Physiology of Red Spruce

Several physiological effects occur in red spruce due to acidification. One physiological effect is due to low Ca levels in cellular membranes of foliage. These low Ca levels are due to foliar and soil leaching, detailed above. Lower Ca levels in the membranes of cells decrease the cold hardiness of red spruce foliage, potentially translating to foliar damage during cold winter temperatures (Thornton et al. 1994; DeHayes et al. 1999; Schaberg et al. 2000). This connection between Ca deficiency in foliage and cold hardiness as a cause of spruce mortality in the northern Appalachians may help explain why central and southern red spruce populations did not show the same increased mortality as northern areas due to less severe winters (Johnson 1992; Johnson et al. 1992; McLaughlin et al. 1993). However, exposure to acid cloud water and ozone can still cause the loss of cold tolerance even in the southern Appalachians (Thornton et al. 1992, 1994). Another physiological effect from acidification is increased respiration and decreased photosynthesis (McLaughlin et al. 1990, 1991, 1993; Pier et al. 1992; McLaughlin and Percy 1999) that was correlated with foliar Ca and Ca:Al ratios. Red spruce trees with lower foliar Ca and Ca:Al ratios had less photosynthesis, higher respiration, hence indicating stressed conditions. Other physiological effects include cellular membrane disruption that causes foliar injury, including needle speckling (DeHayes et al. 1999) and stomatal cell functioning that can lead to a lessened response to water stress (Borer et al. 2005). Since acidification decreases Ca:Al ratios, those levels can become toxic to fine root growth, reducing the uptake of nutrients critical to growth (Shortle and Smith 1988; Robarge et al. 1989; Cronan and Grigal 1995). Because red spruce systems are generally Ca deficient and because Ca is critical to the physiology of red spruce, research has examined the effects of amending sites with Ca and other base cations to ameliorate ecosystem acidification. Adding Ca to red spruce sites in the southern Appalachians can improve foliar Ca at high elevation red spruce sites, but not sites at lower elevation (Van Miegroet et al. 1993; Joslin and Wolfe 1994).

2.3.2.5 Recent Reductions in Pollutant Deposition and Improvements in Forest Health

Studies have reported Ca, Mg, and Ca:Al ratios improved from the 1990s to the 2000s, suggesting the 1990 Clean Air Act amendments reduced atmospheric deposition of pollutants in the southern region of red spruce (Blintz and Butcher 2007; Rosenberg and Butcher 2010; Wilson and Butcher 2012). Recently, Schwartz et al. (2022) reported significant declines in acid deposition in the Great Smoky Mountains. Similar trends were observed in the northern Appalachians, with a 40–50% reduction in acidic deposition, improved soil pH, and a reduction in soil Al at certain locations (Lawrence et al. 2011, 2015; Lawrence and Bailey 2021). Although acidic deposition had decreased significantly, soil recovery is probably at an early stage and additional research is required to understand linkages between terrestrial and aquatic systems as this relates to forest health (Lawrence and Bailey 2021). Improved forest health in red spruce forests was also observed in Canada following air quality regulations (Kosiba et al. 2018).

2.3.2.6 Ozone Exposure Effects on Red Spruce

One pollutant receiving more attention in recent years is ground-level ozone. This pollutant can form in remote areas when sunlight provides the energy for a chemical reaction between nitrogen oxides (NO_X), released during fossil fuel combustion, and volatile organic compounds, a broad term that includes several classes of airborne compounds that derive from both human and natural sources. Examples of volatile organic compound sources include the burning of fossil fuels, forest fires, and repellants or toxins that reduce insect attack released by tree species (Montero-Montoya et al. 2018).

Ground-level ozone can also be transported from urbanized areas by atmospheric patterns, like other pollutants previously discussed. Ground-level ozone is considered a pollutant because it can cause significant damage to the foliage of plants. Like acid deposition in the central and southern regions of red spruce, ground-level ozone has been found to increase with elevation, especially during the summer months (Aneja et al. 1992; Neufield et al. 2019).

Although ground-level ozone has been shown to cause significant foliar damage in certain plant species, most research to date generally shows minimal direct effects on red spruce foliage, physiology, and growth (Laurence et al. 1989; Kohut et al. 1990; Thorton et al. 1990, 1992, 1994; Pier et al. 1992; Rebbeck et al. 1993). The earliest studies looked at ground-level ozone effects on seedling height, branching, mass of stem, needle and roots, biomass production, and rate of photosynthesis. No direct or interactive effects were observed when ozone was combined with acidic wet deposition. Even studies using elevated levels of ozone (0.5–2.0 times ambient levels) combined with simulated acid precipitation have found no significant effects of either pollutant, alone or in combination (Laurence et al. 1989; Kohut et al. 1990).

A few studies showed differences in red spruce response to ozone, mainly when combined with water from clouds. One such study showed elevated ozone effected red spruce physiology (Rebbeck et al. 1993). Differences were not observed with ambient ozone levels, but there was a 29–40% reduction in photosynthesis in young red spruce when ozone was elevated to 75 and 150 ppb above ambient. Stomatal conductance was also higher in juvenile stems, indicating that ozone uptake may be as much as twice as high as on older stems and that seedlings will have different responses than canopy branches of established trees. Another study in the central and southern Appalachians found a significant effect of combining ozone with cloud water. Thornton et al. (1992) found lower foliar Ca and Mg concentrations and higher respiration rates for both seedlings and mature trees of red spruce compared to ozone only treatments. They also observed increased height growth for seedlings with the ozone and cloud water treatment, possibly due to a fertilization effect from nitrogen and sulfur in the cloud water. It is clear from the mixed results of these studies that additional research is needed to understand the effects that ozone, including its interactions with atmospheric precipitation, have on red spruce ecosystems.

2.4 Climate Change

The spruce-fir forests of southeastern North America occur at the highest elevations in Appalachian Mountains and depend on the cool, wet conditions that occur at these elevations. As the climate changes, environmental conditions supporting these montane ecosystems is also likely to change, possibly creating upslope migration of these forests. Among the trees of eastern North America, red spruce has been classified one of the most vulnerable tree species to climate change (Rogers et al. 2016). Similar to other long-lived trees, red spruce is vulnerable to changing climatic conditions because of its long generation time, inability to migrate northward at the southern parts of its range, and low genetic diversity (Potter et al. 2010; see Chap. 7). Analyzing changes in mean temperatures and the seasonal progression of temperatures are also important metrics to understand the effects of climate change in these ecosystems (Day 2000). Red spruce seedlings, saplings, and the radial growth of mature trees are sensitive to hot and dry summers (Yetter et al. 2021c) and some future environmental changes might be favorable to red spruce productivity. Higher atmospheric CO_2 levels (if water and nitrogen are not limiting), longer growing seasons, and the moderation of winter temperatures, may also increase red spruce productivity (Yetter et al. 2021c). However, it is also possible that an earlier release from winter dormancy may increase vulnerability to late frosts (Koo et al. 2014; Soulé 2022), possibly offsetting productivity gains.

Cloud immersion and precipitation can have mixed effects on red spruce populations (Yetter et al. 2021c). Although these environmental variables are responsible for most of the pollutant deposition in these forests, red spruce is sensitive to summer heat and drought and depends on high elevation habitats with high humidity and water supply. Cloudy days have been found to have both positive (Johnson and Smith 2006; Reinhardt and Smith 2008a; Berry and Smith 2012, 2013) and negative (Yetter et al. 2021c; Soulé 2022) effects on red spruce photosynthesis. Red spruce growth rates in the central Appalachians decreased with hot summer temperatures, but increased with fall temperatures, indicating that warming climates would have a net negative effect on red spruce growth (Yetter et al. 2021b). In the northern Appalachians, a tree ring analysis between 1925 and 2012 found that higher temperatures outside the traditional growing season were related to increased growth (Kosiba et al. 2018). In Massachusetts, red spruce populations were sensitive to both temperature and precipitation, with a negative growth response to the regional warming that had occurred over the last 100 years (Ribbons 2014). Although the effect of climate change on precipitation and cloud immersion is critical to spruce-fir forests (Berry et al. 2014), it is hard to predict (Berry and Smith 2013). Some have asserted that the average cloud ceiling sets the lower boundary of spruce-fir forests, and this cloud ceiling is moving up in elevation in some areas of the Appalachians, as in the north, or stable or decreasing in other areas, such as in the south (Richardson et al. 2003). Cloud immersion may also affect microclimatic conditions during regional warm periods, wherein low elevations experience higher heat effects, but high elevation sites experienced mitigated temperatures due to low cloud cover (Fridley 2009). Therefore, high-elevation spruce-fir forests may have buffering effects during extreme regional heat events due to cloud immersion, providing suitable microclimates for long-term persistence despite regional warming. See Sect. 7.2.2 in Chap. 7 for additional information on how climatic variation and extreme weather events influence the physiology of red spruce.

2.5 Genetics

Understanding genetic diversity is an important variable to consider when assessing long-term species viability in a region. Populations may develop regional adaptations to local abiotic and biotic stressors. Although these adaptations may be well-suited for conditions in one region, they may not translate favorably to another. Yet these regional adaptations are important for maintaining a high level of genetic diversity that allow populations to possibly endure anomalous events, such as rapid environmental changes. They may also assist species in dealing with longer term climatic changes as are occurring in our modern world. Populations that become increasingly isolated have reduced gene flow (i.e., bottleneck effect), which increases genetic inbreeding and equates to decreased genetic diversity. Isolated populations experiencing this cycle of reduced genetic diversity are at a higher risk of not developing positive adaptations in rapidly changing conditions, potentially contributing to their disappearance from the landscape.

A mostly continuous *genetic core* of red spruce ranges from the Canadian Maritimes west into east-central Canada, south through New England into the Adirondack and Catskill Mountains of New York, with small disjunct populations occurring further west in Algonquin Park, Ontario (Major et al. 2015). Although there are small populations in Pennsylvania, the two largest disjunct populations occur in the central (West Virginia) and southern (Virginia, North Carolina, and Tennessee) Appalachian Mountains (see Fig. 1.3 in Chap. 1). The gene flow for these disjunct populations in the central and southern regions was initially constrained by the last glaciation at the end of the Pleistocene, but exacerbated by the destructive logging at the end of the nineteenth century, creating isolated fragmented populations as *sky islands* of spruce-fir in the southernmost part of its range (see Chap. 1). In addition to these genetically isolated regions, research has shown genetic variation decreasing significantly from the *geneticcore* to trailing edge populations (Hawley and DeHayes 1994; Prakash et al. 2021), with higher instances of inbreeding in the central and southern Appalachians (Capblancq et al. 2021), thus highlighting the isolation and lack of gene flow between disjunct populations. Due to these combined genetic attributes, it is uncertain whether the central and southern regions contain sufficient genetic variation to survive the increased warming expected with climate change

(Capblancq et al. 2020; Prakash et al. 2021). See Chaps. 7 and 9 for more information on climate change impacts and adaptation solutions employing ecological genomics.

Some of the earliest red spruce genetics research, although limited, was done in Canada (Fowler et al. 1988; Blum 1990). In the early 1990s, research began looking at how genetic diversity might help explain red spruce declines across the eastern U.S. The earliest studies suggested that because genetic diversity of red spruce was relatively uniform across the entire range, this species was poorly suited to adjust to sudden environmental changes, which could contribute to its decline in certain areas (DeHayes and Hawley 1992). The authors noted that populations with increased genetic variability had higher vigor compared to populations with less genetic variability, suggesting these populations may be more resilient and capable of maintaining viable populations in the future. Although old-growth forests are thought to have high genetic diversity due to their stand age and dynamics, evidence suggests they may suffer similar genetic diversity limitations if they have small population sizes (Mosseler et al. 2003). Despite potential limitations of genetic diversity in smaller populations, it is still important to characterize phenotypic variations of populations, regardless of their size, due to the potential for selecting seed sources for restoration efforts. For example, phenotypic variation was observed on climate-associated traits, with seeds collected along an elevational gradient on Kuwohi Mountain (a.k.a., Clingmans Dome) in Tennessee (Butnor et al. 2019) suggesting that seed sources might be able to be transferred from lower to higher elevation populations to achieve successful restoration of red spruce in the southeastern U.S. Interestingly, the same relationship was not observed on material collected along an elevational gradient on Mt. Mitchell in North Carolina (Butnor et al. 2019). It is possible that an unknown seed source was used to artificially regenerate the red spruce populations in the area that was replanted on Mt. Mitchell after extensive resource exploitation in the early to mid-twentieth century (Davis 1930; Frothingham 1931; Minckler 1940).

Understanding the unique regional phenotypic and genotypic variation that may improve the chances of future survival in the context of changing abiotic and biotic stressors may be key in successfully maintaining and expanding existing red spruce populations (Beane 2010; Thomas et al. 2014; Koo et al. 2015; Andrews 2016). The acceptance and use of climate change adaptation tools like assisted species migration has increased in recent years as an option to maintain or restore ecosystems (Vitt et al. 2010; Thomas et al. 2014). This concept tries to match the viability of specific phenotypic and/or genotypic variation to future modeled climatic scenarios. For plants, this usually requires determining the genetic variation of species in a specific area, assessing adaptations in relation to biotic and abiotic stressors, and moving genetic sources to areas predicted to experience similar conditions under future climate models. For red spruce, this area of research is in its initial stages (Butnor et al. 2019; Capblancq et al. 2020).

2.6 Conclusion

Red spruce has been an integral part of eastern forest landscapes due to its natural resiliency as a shade tolerant, slow-growing, medium-sized conifer with the ability to thrive under harsh environmental conditions. Although historical human influences, such as exploitive logging and fires, dramatically reduced the area occupied by red spruce in the eastern forests, the species is gradually expanding due to natural processes and significant restoration efforts led by groups such as CASRI and SASRI. These groups are using silvicultural prescriptions to increase gap dynamics to improve red spruce growth in existing forests, and planting seedlings in areas with the genetics to match future climatic model predictions that will hopefully improve resiliency in specific areas. However, there are still potential obstacles for the future of red spruce. The interaction of pollution and climate change on species survival, especially in the southern region, is uncertain. Climate change may also influence cloud and fog development, which is critical to red spruce in the southern Appalachians. Future pests and pathogens are unknown, as are invasive species. Despite these potential headwinds, much progress continues to be made in expanding the range of red spruce throughout the entire Appalachian Mountains.

References

Adams MB (1999) Acidic deposition and sustainable forest management in the central Appalachians, USA. For Ecol Manage 122:17–28

Adams MB, Eagar C (1992) Impacts of acidic deposition on high-elevation spruce-fir forests: results from the spruce-fir research cooperative. For Ecol Manage 51:195–205

Adams HS, Stephenson SL (1989) Old-growth red spruce communities in the mid-Appalachians. Vegetatio 85:45–56

Adams HS, Stephenson SL, Blasing TJ et al (1985) Growth-trend declines of spruce and fir in mid-Appalachian subalpine forests. Environ Exp Bot 25:315–325

Alexander JD, Donnelly SR, Shane JB (1995) Photosynthetic and transpirational responses of red spruce understory trees to light and temperature. Tree Physiol 15:393–398

Allard HA, Leonard EC (1952) The Canaan and the Stony River valleys of West Virginia, their former magnificent spruce forests, their vegetation and floristics today. Castanea 17:1–60

Andrews C (2016) Modeling and forecasting the influence of current and future climate on eastern North American spruce-fir (*Picea-Abies*) forests. University of Maine, Orono, Maine, Thesis

Aneja VP, Robarge WP, Claiborn CS et al (1992) Chemical climatology of high elevation spruce-fir forests in the southern Appalachian Mountains. Environ Pollut 75:89–96

Audley DE, Skelly JM (1994) A *Phomopsis* species associated with nonlethal adelgid galls on upper crown branchlets of red spruce in West Virginia. Plant Dis 78:569–571

Audley DE, Skelly JM, McCormick LH et al (1998) Crown condition and nutrient status of red spruce (*Picea Rubens* Sarg.) in West Virginia. Water Air Soil Pollut 102:177–199

Banks SA (2013) Forest response to the U.S. 1990 Clean Air Act: the southern spruce-fir ecosystem. Thesis, North Carolina State University, Raleigh, North Carolina

Beane NR (2010) Using environmental and site-specific variables to model current and future distribution of red spruce (*Picea rubens* Sarg.) forest habitat in West Virginia. Dissertation, West Virginia University, Morgantown, West Virginia

Berry ZC, Smith WK (2012) Cloud pattern and water relations in *Picea rubens* and *Abies fraseri*, southern Appalachian Mountains, USA. Agric For Meteorol 162–163:27–34
Berry ZC, Smith WK (2013) Ecophysiological importance of cloud immersion in a relic spruce–fir forest at elevational limits, southern Appalachian Mountains, USA. Oecologia 173:637–648
Berry ZC, Hughes NM, Smith WK (2014) Cloud immersion: an important water source for spruce and fir saplings in the southern Appalachian Mountains. Oecologia 174:319–326
Blintz WW, Butcher DJ (2007) Characterization of the health of southern Appalachian red spruce (*Picea rubens*) through determination of calcium, magnesium, and aluminum concentrations in foliage and soil. Microchem J 87:170–174
Blum BM (1990) Red spruce *Picea rubens* Sarg. In: Burns RM, Honkala BH (eds) Silvics of North America, vol 1. Conifers. USDA Forest Service Handbook 654, Washington, DC, pp 250–259
Borer CH, Schaberg PG, DeHayes DH (2005) Acidic mist reduces foliar membrane-associated calcium and impairs stomatal responsiveness in red spruce. Tree Physiol 25:673–680
Bowers TA, Bruck RI (2010) Evidence of montane spruce-fir forest recovery on the high peaks and ridges of the Black Mountains, North Carolina: recent trends, 1986–2003. In: Rentch JS, Schuler TM (eds) Proceedings from the conference on the ecology and management of high-elevation forests in the central and southern Appalachian Mountains. USDA Forest Service Northern Research Station General Technical Report NRS-P-64, Newtown Square, Pennsylvania, p 203
Brown JH (1962) Success of tree planting on strip-mined areas in West Virginia. West Virginia Agricultural and Forestry Experiment Station Bulletin No. 473. Morgantown, West Virginia
Bruck RI (1989) Survey of diseases and insects of Fraser fir and red spruce in the southern Appalachian Mountains. Eur J for Pathol 19:389–398
Bruck RI, Robarge WP (1988) Change in forest structure in the boreal montane ecosystem of Mount Mitchell, North Carolina. Eur J for Pathol 18:357–366
Bryant KN, Fowlkes AJ, Mustafa SF et al (1997) Determination of aluminum, calcium, and magnesium in Fraser fir, balsam fir, and red spruce foliage and soil from the southern and middle Appalachians. Microchem J 56:382–392
Busing RT (1996) Estimation of tree replacement patterns in an Appalachian *Picea-Abies* forest. J Veg Sci 7:685–694
Busing RT (2004) Red spruce dynamics in an old southern Appalachian forest. J Torrey Bot Soc 131:337–342
Busing RT, Pauley EF (1994) Mortality trends in a southern Appalachian red spruce population. For Ecol Manage 64:41–45
Busing RT, Wu X (1990) Size-specific mortality, growth, and structure of a Great Smoky Mountains red spruce population. Can J for Res 20:206–210
Busing RT, Clebsch EEC, Eagar CC et al (1988) Two decades of change in a Great Smoky Mountains spruce-fir forest. Bull Torrey Bot Club 115:25–31
Butnor JR, Verrico BM, Johnsen KH et al (2019) Phenotypic variation in climate-associated traits of red spruce (*Picea rubens* Sarg.) along elevational gradients in the southern Appalachian Mountains. Castanea 84:128–143
Capblancq T, Butnor JR, Deyoung S et al (2020) Whole-exome sequencing reveals a long-term decline in effective population size of red spruce (*Picea rubens*). Evol Appl 13:2190–2205
Capblancq T, Munson HL, Butnor JR et al (2021) Genomic drivers of early-life fitness in *Picea rubens*. Conserv Genet 22:963–976
Carmean WH, Hahn JT, Jacobs RD (1989) Site index curves for forest tree species in the eastern United States. USDA Forest Service General Technical Report NC-128, St. Paul, Minnesota
Cogbill CV, White PS (1991) The latitude-elevation relationship for spruce-fir forest and treeline along the Appalachian Mountain chain. Vegetatio 94:153–175
Cook ER, Zedaker SM (1992) The dendroecology of red spruce decline. In: Eagar C, Adams MB (eds) Ecology and decline of red spruce in the eastern United States. Springer-Verlag, New York, New York, pp 192–231
Cook ER, Johnson AH (1989) Climate change and forest decline: a review of the red spruce case. Water Air Soil Pollut 48:127–140

Creed IF, Morrison DL, Nicholas NS (2004) Is coarse woody debris a net sink or source of nitrogen in the red spruce—Fraser fir forest of the southern Appalachians, U.S.A.? Can J for Res 34:716–727
Crim PM, McDonald LM, Cumming JR (2019) Soil and tree nutrient status of high elevation mixed red spruce (*Picea rubens* Sarg.) and broadleaf deciduous forests. Soil Syst 3:80
Cronan CS, Grigal DF (1995) Use of calcium/aluminum ratios as indicators of stress in forest ecosystems. J Environ Qual 24:209–226
D'Arrigo R, Anchukaitis KJ, Bukley B et al (2012) Regional climatic and North Atlantic oscillation signatures in West Virginia red cedar over the past millennium. Global Planet Change 84–85:8–13
Dale VH, Gardner RH, DeAngelis DL et al (1991) Elevation-mediated effects of balsam woolly adelgid on southern Appalachian spruce–fir forests. Can J For Res 21:1639–1648
Davis JH Jr (1930) Vegetation of the Black Mountains of North Carolina: an ecological study. J Elisha Mitchell Sci Soc 45:291–318
Day ME (2000) Influence of temperature and leaf-to-air vapor pressure deficit on net photosynthesis and stomatal conductance in red spruce (*Picea Rubens*). Tree Physiol 20:57–63
Day ME, Greenwood MS, White AS (2001) Age-related changes in foliar morphology and physiology in red spruce and their influence on declining photosynthetic rates and productivity with tree age. Tree Physiol 21:1195–1204
DeHayes DH, Hawley GJ (1992) Genetic implications in the decline of red spruce. Water Air Soil Pollut 62:233–248
DeHayes DH, Shaberg PG, Hawley GJ et al (1999) Acid rain impacts on calcium nutrition and forest health: alteration of membrane-associated calcium leads to membrane destabilization and foliar injury in red spruce. Bioscience 49:789–800
DeRose RJ, Seymour RS (2009) The effect of site quality on growth efficiency of upper crown class *Picea rubens* and *Abies balsamea* in Maine, USA. Can J For Res 39:777–784
Dibble AC, Brissette JC, Hunter ML Jr (1999) Putting community data to work: some understory plants indicate red spruce regeneration habitat. For Ecol Manage 114:257–291
Dumais D, Prévost M (2007) Management for red spruce conservation in Québec: the importance of some physiological and ecological characteristics – a review. For Chron 83:378–392
Dumais D, Prévost M (2008) Ecophysiology and growth of advance red spruce and balsam fir regeneration after partial cutting in yellow birch–conifer stands. Tree Physiol 28:1221–1229
Dumais D, Prévost M (2014) Physiology and growth of advance *Picea rubens* and *Abies balsamea* regeneration following different canopy openings. Tree Physiol 34:194–204
Dumais D, Prévost M (2016) Germination and establishment of natural red spruce (*Picea rubens*) seedlings in silvicultural gaps of different sizes. For Chron 92:90–100
Dumais D, Raymond P, Prévost M (2020) Eight-year ecophysiology and growth dynamics of *Picea rubens* seedlings planted in harvest gaps of partially cut stands. For Ecol Manage 478:118514
Eagar C, White PS (1984) Review of the biology and ecology of the balsam woolly aphid in southern Appalachian spruce–fir forests. In: White PS (ed) The southern Appalachian spruce–fir ecosystem: its biology and threats. National Park Service Southeast Region Research/Resources Management Report SER-71, Atlanta, Georgia, pp 36–50
Elias PE, Burger JA, Adams MB (2009) Acid deposition effects on forest composition and growth on the Monongahela National Forest, West Virginia. For Ecol Manage 258:2175–2182
Farjon A (2017) A handbook of the world's conifers (2 vol), revised and updated edition. Brill, Leiden, Netherlands
Fowler DP, Park YS, Goren AG (1988) Genetic variation of red spruce in the Maritimes. Can J For Res 18:703–709
Frank RM, Bjorkbom JC (1973) A silvicultural guide for spruce-fir in the Northeast. USDA Forest Service Northeastern Forest Experiment Station General Technical Report NE-6, Upper Darby, Pennsylvania
Frank RM, Safford LO (1970) Lack of viable seed in the forest floor after clearcutting. J Forest 68:776–778

Fraver S, White AS (2005) Disturbance history of old-growth red spruce stands in northern Maine: linking tree-ring and stem-mapped data. In: Kenefic LS, Twery MJ (eds) Changing forests—challenging times: proceedings of the New England Society of American Foresters 85th Winter Meeting, Portland, Maine, 16–18 Mar 2005

Fridley J (2009) Downscaling climate over complex terrain: high finescale (<1000 m) spatial variation of near-ground temperatures in a montane forested landscape (Great Smoky Mountains). J Appl Meteorol Climatol 48:1033–1049

Frothingham EH (1931) Timber growing and logging practices in the southern Appalachian region. USDA Forest Service Appalachian Forest Experiment Station Technical Bulletin No. 250, Washington, DC

Geyer WA, Lynch KD (1987) Use of site index as a forestry management tool. Trans Kans Acad Sci 90:46–51

Goelz JCG, Burk TE, Zedaker SM (1999) Long term growth trends of red spruce and Fraser fir at Mt. Rogers, Virginia and Mt. Mitchell, North Carolina. For Ecol Manag 115:49–59

Greenwood M, McConville D (2002) Factors affecting the regeneration of red spruce and balsam fir. University of Maine Cooperative Forestry Research Unit Annual Report 2001–2002:37–40

Greenwood MS, O'Brien CL, Schatz JD et al (2008) Is early life cycle success a determinant of the abundance of red spruce and balsam fir? Can J for Res 38:2295–2305

Griscom BG, Griscom HP, Deacon S (2011) Species-specific barriers to tree regeneration in high elevation habitats of West Virginia. Restor Ecol 19:660–670

Harmon ME, Bratton SP, White PS (1984) Disturbance and vegetation response in relation to environmental gradients in the Great Smoky Mountains. Vegetatio 55:129–139

Hart AC (1959) Silvical characteristics of red spruce (*Picea rubens*). USDA Forest Service Northeastern Forest Experiment Station Paper 124, Upper Darby, Pennsylvania

Hart AC (1965) Red spruce (*Picea rubens* Sarg.). In: Fowells HA (ed) Silvics of forest trees of the United States. U.S. Department of Agriculture Handbook No. 271, Washington, DC, p 305–310

Hawley GJ, DeHayes DH (1994) Genetic diversity and population structure of red spruce (*Picea rubens*). Can J Bot 72:1778–1786

Heinselman ML (1965) Black spruce (*Picea mariana* (Mill.) B. S. P.). In: Fowells HA (ed) Silvics of forest trees of the United States. U.S. Department of Agriculture Handbook No. 271, Washington, DC, pp 288–298

Hopkins AD (1891) Black spruce. West Virginia Agricultural Experiment State Bulletin No. 17, Morgantown, West Virginia, USA

Hopkins AD (1899) Report on investigations to determine the cause of unhealthy conditions of the spruce and pine from 1880–1893. West Virginia Agricultural Experiment State Bulletin No. 56, Morgantown, West Virginia, USA

Hornbeck JW, Kochenderfer JN (1998) Growth trends and management implications for West Virginia's red spruce forests. North J Appl for 15:197–202

Hughes JW, Bechtel DA (1997) Effect of distance from forest edge on regeneration of red spruce and balsam fir in clearcuts. Can J For Res 27:2088–2096

International Plant Names Index (2022) *Picea rubens* Sarg. The Royal Botanic Gardens, Kew, Harvard University Herbaria and Libraries, and Australian National Botanic Gardens. http://www.ipni.org. Accessed 16 May 2022

Isil SMS, Collett J Jr, Lynch J et al (2022) Cloud and fog deposition: monitoring in high elevation and coastal ecosystems. The past, present, and future. Atmospheric Environment 274:118997

Johnson AH (1992) The role of abiotic stresses in the decline of red spruce in high elevation forests of the eastern United States. Annu Rev Phytopathol 30:349–367

Johnson DM, Smith WK (2006) Low clouds and cloud immersion enhance photosynthesis in understory species of a southern Appalachian spruce–fir forest (USA). Am J Bot 93:1625–1632

Johnson DW, Van Miegroet H, Lindberg SE et al (1991) Nutrient cycling in red spruce forests of the Great Smoky Mountains. Can J for Res 21:769–787

Johnson AH, McLaughlin SB, Adams MB et al (1992) Why are red spruce declining at high elevations? Synthesis and conclusions from epidemiological and mechanistic studies of red spruce decline. In: Eagar C, Adams MB (eds) Ecology and decline of red spruce in the eastern United States. Springer-Verlag, New York, New York, pp 385–412

Joslin JD, Wolfe MH (1992) Red spruce soil solution chemistry and root distribution across a cloud water deposition gradient. Can J for Res 22:893–904

Joslin JD, Wolfe MH (1994) Foliar deficiencies of mature southern Appalachian red spruce determined from fertilizer trials. Soil Sci Soc Am J 58:1572–1579

Joslin JD, McDuffie CM, Brewer PF (1988) Acidic cloud water and cation loss from red spruce foliage. Water Air Soil Pollut 39:355–363

Kaylor SD, Hughes MJ, Franklin JA (2017) Recovery trends and predictions of Fraser fir (*Abies fraseri*) dynamics in the southern Appalachian Mountains. Can J for Res 47:125–133

Kohut RL, Laurence JA, Amundson RG et al (1990) Effects of ozone and acidic precipitation on the growth and photosynthesis of red spruce after two years of exposure. Water Air Soil Pollut 51:277–286

Koo KA, Patten BC, Teskey RO et al (2014) Climate change effects on red spruce decline mitigated by reduction in air pollution within its shrinking habitat range. Ecol Model 293:81–90

Koo KA, Patten BC, Madden M (2015) Predicting effects of climate change on habitat suitability of red spruce (*Picea rubens* Sarg.) in the southern Appalachian Mountains of the USA: understanding complex systems mechanisms through modeling. Forests 6:1208–1226

Korstian CF (1937) Perpetuation of spruce on cut-over and burned lands in the higher southern Appalachian Mountains. Ecol Monogr 7:125–167

Kosiba AM, Schaberg PG, Rayback SA et al (2018) The surprising recovery of red spruce growth shows links to decreased acid deposition and elevated temperature. Sci Total Environ 637–638:1480–1491

Laurence JA, Kohut RJ, Amundson RG (1989) Response of red spruce seedlings exposed to ozone and simulated acidic precipitation in the field. Arch Environ Contam Toxicol 18:285–290

Lawrence GB, Bailey SW (2021) Recovery processes of acidic soils experiencing decreased acidic deposition. Soil Systems 5:36

Lawrence GB, Shortle WC, David MB (2011) Early indications of soil recovery from acidic deposition in U.S. red spruce forests. Soil Sci Soc Am J 76:1407–1417

Lawrence GB, Hazlett PW, Fernandez IJ et al (2015) Declining acidic deposition begins reversal of forest-soil acidification in the northeastern U.S. and eastern Canada. Environ Sci Technol 49:13103–13111

LeBlanc DC, Nicholas NS, Zedaker SM (1992) Prevalence of individual-tree growth decline in red spruce populations of the southern Appalachian Mountains. Can J for Res 22:905–914

Lee CE, Cox JM, Foster DM et al (1997) Determination of aluminum, calcium, and magnesium in Fraser fir (*Abies fraseri*) foliage from five native sites by atomic absorption spectrometry: the effect of elevation upon nutritional status. Microchem J 56:236–246

Lusk L, Mutel M, Walker E et al (2010) Forest change in high-elevation forests of Mt. Mitchell, North Carolina: re-census and analysis of data collected over 40 years. In: Rentch JS, Schuler TM (eds) Proceedings from the conference on the ecology and management of high-elevation forests in the central and southern Appalachian Mountains. USDA Forest Service Northern Research Station General Technical Report NRS-P-64, Newtown Square, Pennsylvania, pp 104–112

Major JE, Mosseler A, Johnsen KH et al (2015) Growth and allocation of *Picea rubens*, *Picea mariana*, and their hybrids under ambient and elevated CO_2. Can J for Res 45:877–887

Mathias JM, Thomas RB (2018) Disentangling the effects of acidic air pollution, atmospheric CO_2, and climate change on recent growth of red spruce trees in the central Appalachian Mountains. Glob Change Biol 24:3938–3953

Mathiasen RL, Olsen WK, Edminster CB (2006) Site index curves for white fir in the southwestern United States developed using a guide curve method. West J Appl for 21:87–93

Mayfield AE III, Salom SM, Sumpter K et al (2020) Integrating chemical and biological control of the hemlock woolly adelgid: a resource manager's guide. USDA Forest Service Forest Health Assessment and Applied Sciences Team Report FHAAST-2018–04, Morgantown, West Virginia

McLaughlin S, Percy K (1999) Forest health in North America: some perspectives on actual and potential roles of climate and air pollution. Water Air Soil Pollut 116:151–197

McLaughlin SB, Andersen CP, Edwards NT et al (1990) Seasonal patterns of photosynthesis and respiration of red spruce saplings from two elevations in declining southern Appalachian stands. Can J for Res 20:485–495

McLaughlin SB, Andersen CP, Hanson PJ et al (1991) Increased dark respiration and calcium deficiency of red spruce in relation to acidic deposition at high-elevation southern Appalachian Mountain sites. Can J for Res 21:1234–1244

McLaughlin SB, Tjoelker MG, Roy WK (1993) Acid deposition alters red spruce physiology: laboratory studies support field observations. Can J for Res 23:380–386

McLaughlin SB, Blasing TJ, Downing DJ (1994) Two hundred year variation of southern red spruce radial growth as estimated by spectral analysis: comment. Can J for Res 24:2299–2304

McLaughlin SB, Wullschleger S, Stone A et al (1995) Effects of acid deposition on calcium nutrition and health of southern Appalachian spruce fir forests. Paper presented at the International Union of Forest Resource Organizations, New Brunswick, Canada, 7 Sep 1994. https://www.osti.gov/biblio/32571/. Accessed 31 Jul 2024

McLaughlin SB, Joslin JD, Robarge W et al (1998) The impacts of acidic deposition and global change on high elevation southern Appalachian spruce-fir forests. In: Mickler RA, Fox S (eds) The productivity and sustainability of southern forest ecosystems in a changing environment. Springer-Verlag, New York, New York, pp 255–277

McManamay RH, Resler LM, Campbell JB et al (2011) Assessing the impacts of balsam woolly adelgid (*Adelges piceae* Ratz.) and anthropogenic disturbance on the stand structure and mortality of Fraser fir [*Abies fraseri* (Pursh) Poir.] in the Black Mountains. North Carolina. Castanea 76:1–19

Minckler LS (1940) Early planting experiments in the spruce-fir type of the southern Appalachians. J Forest 38:651–654

Mohnen VA (1992) Atmospheric deposition and pollutant exposure of eastern U.S. forests. In: Eagar C, Adams MB (eds) Ecology and decline of red spruce in the eastern United States. Springer-Verlag, New York, New York, pp 64–124

Montero-Montoya R, López-Vargas R, Arellano-Aguilar O (2018) Volatile organic compounds in air: sources, distribution, exposure and associated illness in children. Ann Glob Health 84:225–238

Moore PT, Van Miegroet H, Nicholas NS (2007) Relative role of understory and overstory in carbon and nitrogen cycling in a southern Appalachian spruce-fir forest. Can J for Res 37:2689–2700

Moore JA, Bartlett JG, Boggs JL et al (2002) Abiotic factors. In: Wear DN, Greis JG (eds) Southern forest resource assessment. USDA Forest Service Southern Research Station General Technical Report SRS-53, Asheville, North Carolina, pp 429–452

Mosseler A, Major JE, Rajora OP (2003) Old-growth red spruce forests as reservoirs of genetic diversity and reproductive fitness. Theor Appl Genet 106:931–937

Murphy LS. 1917. The red spruce: its growth and management. U.S. Department of Agriculture Bulletin No. 544, Washington, DC

Neufeld HS, Sullins A, Sive BC et al (2019) Spatial and temporal patterns of ozone at Great Smoky Mountains National Park and implications for plant responses. Atmos Environ: X 2:100023

Nicholas N, Zedaker SM (1989) Ice damage in spruce–fir forests of the Black Mountains, North Carolina. Can J for Res 19:1487–1491

Nicholas NS, Zedaker SM (1992) Expected stand behavior: site quality estimation for southern Appalachian red spruce. For Ecol Manage 47:39–50

Nicholas NS, Zedaker SM, Eagar C (1992) A comparison of overstory community structure in three southern Appalachian spruce-fir forests. Bull Torrey Bot Club 119:316–332

Nowacki G, Carr R, Van Dyck M (2010) The current status of red spruce in the eastern United States: distribution, population trends, and environmental drivers. In: Rentch JS, Schuler TM (eds) Proceedings from the conference on the ecology and management of high-elevation forests in the central and southern Appalachian Mountains. USDA Forest Service Northern Research Station General Technical Report NRS-P-64, Newtown Square, Pennsylvania, pp 140–162

Oostig HJ, Billings WD (1951) A comparison of virgin spruce-fir forest in the northern and southern Appalachian system. Ecology 32:84–103

Peck, DE (1990) Frasier fir *Abies fraseri* (Pursh) Poir. In: Burns RM, Honkala BH (eds) Silvics of North America, vol 1. Conifers. USDA Forest Service Handbook 654, Washington, DC, pp 47–51

Pier PA, Thornton FC, McDuffie C Jr et al (1992) CO_2 exchange rates of red spruce during the second season of exposure to ozone and acidic cloud deposition. Environ Exp Bot 32:115–124

Potter KM, Hargrove WW, Koch FH (2010) Predicting climate change extirpation risk for central and southern Appalachian forest tree species. In: Rentch JS, Schuler TM (eds) Proceedings from the conference on the ecology and management of high-elevation forests in the central and southern Appalachian Mountains. USDA Forest Service Northern Research Station General Technical Report NRS-P-64, Newtown Square, Pennsylvania, pp 179–189

Prakash A, DeYoung S, Lachmuth S et al (2021) Genotypic variation and plasticity in climate-adaptive traits after range expansion and fragmentation of red spruce (*Picea rubens* Sarg.). Philosophical Transactions of the Royal Society B 377:20210008

Randall AG (1974) Seed dispersal into two spruce-fir clearcuts in eastern Maine. University of Maine Life Sciences and Agriculture Experiment Station Research in the Life Sciences 21:15

Reams GA, Nichols NS, Zedaker SM (1993) Two hundred year variation of southern red spruce radial growth as estimated by spectral analysis. Can J for Res 23:291–301

Rebbeck J, Jensen KF, Greenwood MS (1993) Ozone effects on grafted mature and juvenile red spruce: photosynthesis, stomatal conductance, and chlorophyll concentration. Can J for Res 23:450–456

Reinhardt K, Smith WK (2008a) Impacts of cloud immersion on microclimate, photosynthesis and water relations of *Abies fraseri* (Pursh.) Poiret in a temperate mountain cloud forest. Oecologia 158:229–238

Reinhardt K, Smith WK (2008b) Leaf gas exchange of understory spruce–fir saplings in relict cloud forests, southern Appalachian Mountains, USA. Tree Physiol 28:113–122

Rentch JS, Schuler TM, Ford WM et al (2007) Red spruce stand dynamics, simulations, and restoration opportunities in the central Appalachians. Restor Ecol 15:440–452

Rentch JS, Schuler TM, Nowacki GJ et al (2010) Canopy gap dynamics of second-growth red spruce-northern hardwood stands in West Virginia. For Ecol Manag 260:1921–1929

Rentch JS, Ford FM, Schuler TS et al (2016) Release of suppressed red spruce using canopy gap creation—ecological restoration in the central Appalachians. Nat Areas J 36:29–37

Ribbons RR (2014) Disturbance and climatic effects on red spruce community dynamics at its southern continuous range margin. PeerJ 2:e293

Richardson AD, Denny EG, Siccama TG et al (2003) Evidence for a rising cloud ceiling in eastern North America. J Clim 16:2093–2098

Robarge WP, Pye JM, Bruck RI (1989) Foliar elemental composition of spruce-fir in the southern Blue Ridge province. Plant Soil 114:19–34

Rogers BM, Jantz P, Goetz SJ et al (2016) Vulnerability of tree species to climate change in the Appalachian Landscape Conservation Cooperative. In: Hansen AJ, Monahan WB, Olliff ST et al (eds) Climate change in wildlands. Island Press, Washington, DC

Rollins AW, Adams HS, Stephenson SL (2010) Changes in forest composition and structure across the red spruce-hardwood ecotone in the central Appalachians. Castanea 75:303–314

Rose AK, Nicholas NS (2008) Coarse woody debris in a southern Appalachian spruce-fir forest of the Great Smoky Mountains National Park. Nat Areas J 28:342–355

Rosenberg MB, Butcher DJ (2010) Investigation of acid deposition effects on southern Appalachian red spruce (*Picea rubens*) by determination of calcium, magnesium, and aluminum in foliage and surrounding soil using ICP-EOS. Instrum Sci Technol 38:341–358

Safford LO, Young HE (1968) Nutrient content of the current foliage of red spruce growing on three soils in Maine. University of Maine Research in the Life Sciences, April Issue, Orono, Maine

Safford LO (1974) *Picea* A. Dietr. Spruce. In: Schopmeyer CS (ed) Seeds of woody plants in the United States. U.S. Department of Agriculture Handbook No. 450, Washington, DC, pp 587–597

Sargent CS (1899) The silva of North America: a description of the trees which grow naturally in North America exclusive of Mexico, vol 12. Houghton, Mifflin and Company, Boston, Massachusetts

Sayre RG, Fahey TJ (1999) Effects of rainfall acidity and ozone on foliar leaching in red spruce (*Picea rubens*). Can J for Res 29:487–496

Schaberg PG, DeHayes DH, Hawley GJ et al (2000) Acid mist and soil Ca and Al alter the mineral nutrition and physiology of red spruce. Tree Physiol 20:73–85

Schuler TM, Ford WM, Collins RJ (2002) Successional dynamics and restoration implications of a montane coniferous forest in the central Appalachians, USA. Nat Areas J 22:88–98

Schwartz JS, Veeneman A, Kulp MA et al (2022) Throughfall deposition chemistry in the Great Smoky Mountains National Park: landscape and seasonal effects. Water Air Soil Pollut 233:107

Seymour RS (1995) The northeastern region. In: Barrett JW (ed) Regional silviculture of the United States, 3rd edn. Wiley, New York, pp 31–79

Seymour RS, Fajvan MA (2001) Influence of prior growth suppression and soil on red spruce site index. North J Appl for 18:55–62

Seymour RS, Kenefic LS (2002) Influence of age on growth efficiency of *Tsuga canadensis* and *Picea rubens* trees in mixed-species, multiaged northern conifer stands. Can J for Res 32:2032–2042

Shepard MR, Lee CE, Woosley RS et al (1995) Determination of calcium and magnesium by flame atomic absorption spectrometry in Fraser fir (*Abies fraseri*) and red spruce (*Picea rubens*) foliage from Richland Balsam Mountain, North Carolina. Microchem J 52:118–126

Shortle WC, Smith KT (1988) Aluminum-induced calcium deficiency syndrome in declining red spruce. Science 240:1017–1018

Smith GF, Nicholas NS (1998) Patterns of overstory composition in the fir and fir-spruce forests of the Great Smoky Mountains after balsam woolly adelgid infestation. Am Midl Nat 139:340–352

Smith GF, Nicholas NS (1999) Post-disturbance spruce-fir stand dynamics at seven disjunct sites. Castanea 64:175–186

Smith GF, Nicholas NS (2000) Size- and age-class distributions of Fraser fir following balsam woolly adelgid infestation. Can J for Res 30:948–957

Soulé PT (2022) Changing climate, atmospheric composition, and radial tree growth in a spruce-fir ecosystem on Grandfather Mountain, North Carolina. Nat Areas J 31:65–74

Soulé PT, White PB, van de Gevel SL (2012) Succession and disturbance in an endangered red spruce–Fraser fir forest in the southern Appalachian Mountains, North Carolina, USA. Endangered Species Research 18:17–25

Stehn SE, Jenkins MA, Webster CR et al (2013) Regeneration responses to exogenous disturbance gradients in southern Appalachian *Picea-Abies* forests. For Ecol Manage 289:98–105

Thomas E, Jalonen R, Loo J et al (2014) Genetic considerations in ecosystem restoration using native tree species. For Ecol Manage 333:66–75

Thomas A, Tilotta DC, Frampton J et al (2022) Sesquiterpene induction by the balsam woolly adelgid (*Adelges piceae*) in putatively resistant Fraser fir (*Abies fraseri*). Forests 13:716

Thornton FC, Pier PA, McDuffie C Jr (1990) Response of growth, photosynthesis, and mineral nutrition of red spruce seedlings to ozone and acidic cloud deposition. Environ Exp Bot 30:313–323

Thornton FC, Pier PA, McDuffie C Jr (1992) Red spruce response to ozone and cloudwater after three years exposure. J Environ Qual 21:196–222

Thornton FC, Joslin JD, Pier PA et al (1994) Cloudwater and ozone effects upon high elevation red spruce: a summary of study results from Whitetop Mountain, Virginia. J Environ Qual 23:1158–1167

Van Miegroet H, Johnson DW, Todd DE (1993) Foliar response of red spruce saplings to fertilization with Ca and Mg in the Great Smoky Mountains National Park. Can J for Res 23:89–95

Van Miegroet H, Moore PT, Tewksbury CE et al (2007) Carbon sources and sinks in high-elevation spruce-fir forests of the southeastern US. For Ecol Manage 238:249–260
Vitt P, Havens K, Kramer AT et al (2010) Assisted migration of plants: changes in latitudes, changes in attitudes. Biol Cons 143:18–27
Webster KL, Creed IF, Nicholas NS et al (2004) Exploring interactions between pollutant emissions and climatic variability in growth of red spruce in the Great Smoky Mountains National Park. Water Air Soil Pollut 159:225–248
White PS, MacKenzie MD, Busing RT (1985) Natural disturbance and gap phase dynamics in southern Appalachian spruce–fir forests. Can J for Res 15:233–240
White PB, Soulé P, van de Gevel S (2014) Impacts of human disturbance on the temporal stability of climate-growth relationships in a red spruce forest, southern Appalachian Mountains, USA. Dendrochronologia 32:71–77
White PS (ed) (1984) The southern Appalachian spruce–fir ecosystem: its biology and threats. National Park Service Southeast Region Research/Resources Management Report SER-71, Atlanta, Georgia
White PB, van de Gevel SL, Soulé PT (2012) Succession and disturbance in an endangered red spruce–Fraser fir forest in the southern Appalachian Mountains, North Carolina, USA. Endanger Species Res 18(1):69–82
Wilson LE, Butcher DJ (2012) Determination of aluminum, calcium, and magnesium in Fraser fir (*Abies fraseri*) foliage and surrounding soil in the southern Appalachians. Instrum Sci Technol 40:457–467
Yetter E, Brown J, Chhin S (2021a) Anamorphic site index curves for central Appalachian red spruce in West Virginia, USA. Forests 12:94
Yetter E, Chhin S, Brown JP (2021b) Dendroclimatic analysis of central Appalachian red spruce in West Virginia. Can J for Res 51:1607–1620
Yetter E, Chhin S, Brown JP (2021c) Sustainable management of central Appalachian red spruce. Sustainability 13:10871

Chapter 3
Soils

James A. Thompson, James E. Leonard, and S. Jason Teets

3.1 Introduction

Soils are dynamic, natural bodies that serve as the foundation of most terrestrial ecosystems (Whalen and Sampedro 2010; Binkley and Fisher 2019). Soils play a crucial role in ecosystem function, serving as a non-renewable source of mineral nutrients and organic matter in addition to being the growth medium for vegetation (Urbanska et al. 1997). Consequently, considering soil information during restoration planning efforts has been shown to improve the success of ecological restoration activities (Callaham et al. 2008; Heneghan et al. 2008; Stanturf et al. 2021).

Recent research in the red spruce (*Picea rubens*) forests of the central Appalachians demonstrates the importance and value of incorporating soils information into restoration planning. These investigations showed that current soil properties can be used to determine if a site was historically dominated by red spruce forests (Nauman et al. 2015a). High-elevation red spruce and spruce-fir ecosystems in the southern Appalachians share many similarities with red spruce ecosystems of the central Appalachians (land use history, climate, topography), but less research on the links between soil properties and site history has been conducted in the southern Appalachians (Wolfe 1967; Busing 1985; Lietzke and McGuire 1987; Feldman 1989; Feldman et al. 1991a, b). This chapter begins with a discussion of some of the

J. A. Thompson (✉)
West Virginia University, School of Natural Resources and the Environment, Morgantown, WV, USA
e-mail: james.thompson@mail.wvu.edu

J. E. Leonard
USDA Natural Resources Conservation Service, Morgantown, WV, USA
e-mail: james.leonard@usda.gov

S. J. Teets
USDA Forest Service, Monongahela National Forest, Morgantown, WV, USA
e-mail: jason.teets@usda.gov

D. J. Brown et al. (eds.), *Ecology and Restoration of Red Spruce Ecosystems in the Central and Southern Appalachians*, https://doi.org/10.1007/978-3-032-19616-3_3

basic properties of soils and the ecosystem services they support before examining the notable characteristics of soils that underlie red spruce forests in central and southern Appalachia. The chapter concludes with a discussion of the impacts of past management on red spruce soils and the critical ways that soils inform and direct the restoration of red spruce ecosystems.

3.2 Soils 101

Given the importance of soil properties and functions to red spruce ecosystems, it is beneficial to begin with an introduction to soils. Soils are the unconsolidated material on the immediate surface of the Earth that serve as a medium for plant growth. The role of soils in sustaining ecosystem functions derives from soils being a dynamic mixture of organic and mineral materials, water (with its associated dissolved components), and various gases. Importantly, soils also provide habitat to a biologically diverse collection of organisms, ranging from bacteria and fungi to macroinvertebrates and small mammals.

The soil properties we observe and measure are a result of complex and ongoing interactions among climate, organisms (plants, animals, fungi, bacteria, etc.), topography, and parent material (bedrock or other sediments) over time (Dokuchaev 1899; Jenny 1941). Climate influences many factors of soil development, driving processes like weathering rates, leaching, and biotic influences (Binkley and Fisher 2019). At any given location, the development of soil properties is influenced by the topography, chemical and physical characteristics of the parent material, ecological disturbances, and land use history (Binkley and Fisher 2019).

The interconnected nature of these soil forming factors results in a great deal of variability in soil properties across even short distances. Soil scientists characterize differences in soil properties in a variety of ways (Soil Survey Staff 2024), and ultimately use that information to classify soils using the U.S. system of Soil Taxonomy (Soil Survey Staff 1999). For a detailed description of soil profiles and soil horizons, see Sidebar 3.1. For a discussion of soil classification, see Sidebar 3.2.

Sidebar 3.1 Soil Profiles and Soil Horizonation Soil properties change with depth below the soil surface. Many soils develop distinct layers that have formed in place within the soil profile. These soil horizons differ chemically, physically, or biologically from the layers above and below. To soil scientists, the soil horizons are the context for interpreting the stories a soil has to tell about the environmental conditions that influenced the formation of the soil over thousands of years, the effects of more recent land use and management practices, and the future potential for land use and management practices. The types and sequence of horizons at a given site are also used to classify the soil (see Sidebar 3.2).

When investigating soils, the observed soil horizons are given alphanumeric labels as shorthand names that reflect their dominant properties (e.g., color, texture, structure). At a minimum, each horizon is labeled with a capital letter that designates the master horizon. There are six master soil horizons commonly used to name the naturally occurring layers within the soil in central and southern Appalachia (see Table 3.1 and Fig. 3.1). Additional properties of each master horizon are often denoted by adding lowercase letters that serve as subordinate distinctions to the master horizons (see Table 3.1 for subordinate distinctions common to red spruce ecosystems). The resulting names for soil horizons (e.g., Ap, Bhs, Bt, Cr) help communicate the key properties of soil horizons. Other conventions for labeling soil horizons include the use of transitional horizons share the properties of two adjacent master horizons (e.g., AB, EB). Numbers can be added as prefixes to indicate changes in parent materials within the soil profile (e.g., 2Bw) or as suffixes to indicate a sequence of horizons with the same letter designations (e.g., Bs1-Bs2-Bs3).

Of particular importance in the red spruce (*Picea rubens*) ecosystem are organic horizons (O, Oi, Oe, Oa) and spodic horizons (Bs, Bh, Bhs). O horizons may have one of three subordinate distinctions to indicate the degree of decomposition of the organic soil materials in the horizon, with Oi horizons the least decomposed, Oa the most decomposed, and Oe representing intermediate levels of decomposition. Spodic horizons are often darker and/or redder in color than adjacent horizons and exhibit smeary consistence due to the poorly crystalline organically-bound Fe and Al minerals that have accumulated through the podzolization process (Soil Survey Staff 1999, 2024). In soils with well-expressed spodic materials, an E horizon usually separates the O horizons from the spodic materials. For more information on horizon nomenclature, refer to the Keys to Soil Taxonomy (Soil Survey Staff 2022, p 377–384), the Field Book for Describing and Sampling Soils (Soil Survey Staff 2024), or a soil science textbook.

Table 3.1 Soil master horizons and commonly associated subordinate distinctions in the central and southern Appalachians (Soil Survey Staff 2022)

Master horizon	Description	Subordinate distinctions commonly used with each master horizon
O	Comprised of organic material from dead plant and animal residues that generally accumulates at the surface above the mineral soil	i. Slightly decomposed e. Intermediate decomposition a. Highly decomposed
A	Topmost mineral horizon that generally contains enough partially decomposed organic matter to give the soil a darker color than the horizons below it	p. Indicates disturbance by tillage, pasturing, or similar agricultural activities
E	Layer dominated by leaching and/or the eluviation of clay particles, organic matter, and/or iron and aluminum oxides, resulting in a gray or white layer dominated by sand and silt sized particles	
B	Layer that forms below O, A and/or E horizons that has undergone soil development such that the original structure of the parent material (bedrock) is no longer discernable	h. Indicates an illuvial accumulation of organic matter evidenced by a smeary consistence and dark soil colors s. Indicates illuvial Fe and Al oxides evidenced by soil colors that are redder or more orange than adjacent layers w. Indicates a distinct color or structure change from horizons above or below without clay accumulation t. Indicates the accumulation of clay that has weathered in the horizon or been translocated to the horizon through leaching and illuviation g. Indicates the layer is saturated with water, which results in gray (low chroma) colors from reduced forms of iron x. Indicates a fragipan or fragic properties evidenced by firm and brittle consistence; increased soil densities of this layer may impede water flow or root penetration
C	Layer that forms below the A, E and/or B horizons that has not been sufficiently altered by soil genesis to qualify as a B horizon	r. Indicates weathered, soft, or diggable bedrock
R	Hard consolidated bedrock with little evidence of weathering	

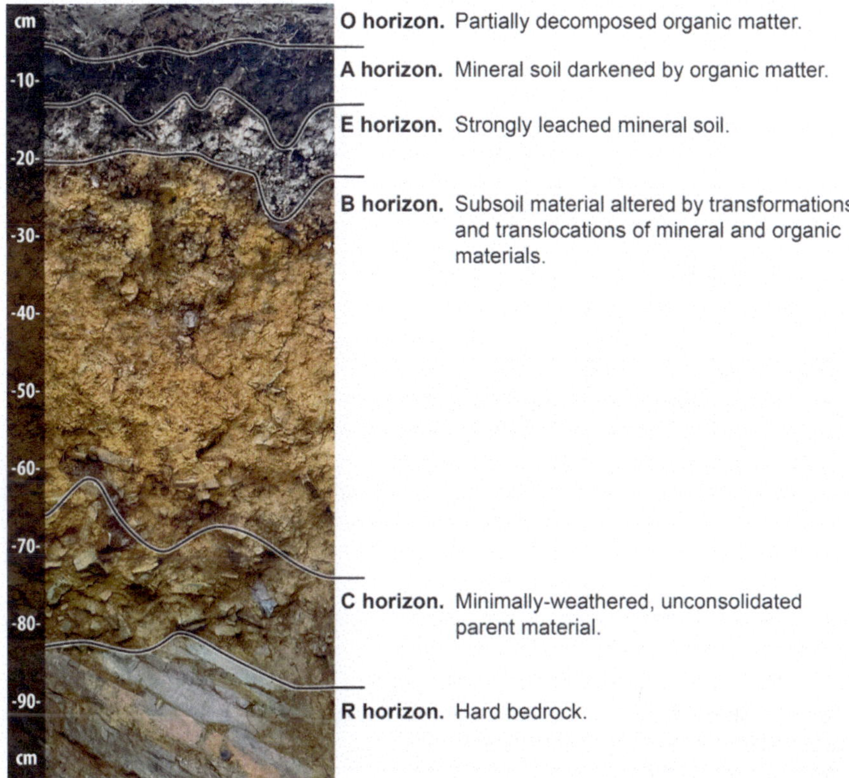

Fig. 3.1 A typical sequence of master horizons frequently observed in high-elevation red spruce (*Picea rubens*) ecosystems (photo by James Leonard)

Sidebar 3.2 Soil Classification We classify soils to improve communication by conveying information about the important characteristics of soils, particularly those properties that we use to make decisions about land use and management. Classification also helps us organize our knowledge about soils by recognizing relationships among the individuals we classify, and it facilitates our inventory and mapping of soil resources. Much like biological classification, Soil Taxonomy is a hierarchical system of classification with soil orders at the highest and most general level, and soil series at the lowest and most specific level. There are currently 12 soil orders recognized in Soil Taxonomy (Soil Survey Staff 2022). Differences among the soil orders are commonly recognized to be associated with differences in climate, vegetation, topography, geology, and/or age. Soil horizons and other diagnostic soil characteristics (e.g., soil moisture regime, soil temperature regime) are used to place a soil into a soil order.

Soil Taxonomy has a unique nomenclature that is used to connote the major characteristics of a soil. Furthermore, unlike biological taxonomy, Soil Taxonomy uses a hierarchical naming structure that helps identify the relationships among soils using unique formative elements to identify soil orders, suborders, great groups, and subgroups. Consequently, names used in Soil Taxonomy include the root elements of the soil order plus formative elements that suggest other distinctive properties of the soil (Table 3.2). As a result, the names of lower-level categories in Soil Taxonomy provide increased specificity about the suitability and limitations of each soil.

Soil orders commonly found in the high-elevation red spruce (*Picea rubens*) ecosystems of the central and southern Appalachians include Spodosols, Inceptisols, Histosols, Ultisols, and Alfisols (Table 3.3). The primary characteristic of Spodosols is that they have well-expressed spodic materials in the subsoil (Soil Survey Staff 2022). Both Ultisols and Alfisols have subsoil horizons with accumulated clays, but Alfisols are inherently more fertile (Soil Survey Staff 2022). Inceptisols show less evidence of soil development than Ultisols and Alfisols (Soil Survey Staff 2022). Additionally, soils that have spodic materials but do not meet the specific requirements of a spodic horizon are often classified as Inceptisols, such as Spodic Dystrudepts (Nauman et al. 2015a, 2015b; Soil Survey Staff 2022). Histosols are dominated by organic soil materials, and are commonly found in wetlands; however, in the high elevations of the central and southern Appalachians where colder temperatures can slow the rates of organic matter decomposition, thick organic layers can accumulate at the surface, and these soils can be classified as Histosols. For more information on soil classification and Soil Taxonomy, refer to the Keys to Soil Taxonomy (Soil Survey Staff 2022) or a soil science textbook.

Soils are described, analyzed, and classified to better understand their inherent properties, limitations, and capabilities to aid in sustainable land management. In this way it is important to recognize the association between soil properties and soil functions. Soil properties, such as pH and porosity, support soil functions, such as biomass production and water storage, which promote ecosystem services, such as supporting food production and regulating the local hydrologic cycle (Adhikari and Hartemink 2016). In fact, soil contributes to multiple provisioning services, regulating services, cultural services, and supporting services, as described in the Millennium Ecosystem Assessment (https://www.millenniumassessment.org/en/Index-2.html).

Specific ecosystem services provided by soils include carbon sequestration, water retention, water purification, flood control, raw materials, habitat, recreation, and aesthetics (Daily et al. 1997; Robinson et al. 2012; Adhikari and Hartemink 2016). Because management and land use actions directly and indirectly influence soil properties, soil functions, and ecosystem services, restoring forests to regain the interconnected functions of an ecosystem, such as with red spruce, is a goal for many managers. More broadly, stewardship of the soil in these and other landscapes

Table 3.2 An example of the hierarchical nature of soil taxonomy using the Mandy soil series (loamy-skeletal, mixed, active, frigid Spodic Dystrudepts)

Level in soil taxonomy	Example taxa	Connotation
Order	Inceptisol	Soils with a weakly developed subsoil with little to no accumulation of clay or other materials
Suborder	Udept	Inceptisol (-ept) in a udic soil moisture regime (ud-)
Great group	Dystrudept	Udept with a relatively acidic pH (dystr-)
Subgroup	Spodic Dystrudept	Dystrudept with a weak expression of spodic materials
Family	Loamy-skeletal, mixed, active, frigid Spodic dystrudept	Spodic Dystrudept where the soil texture of the subsoil is not particularly clayey or sandy (loamy) but has more than 35% rock fragments (skeletal), the mineral content is not dominated by a specific mineral type (mixed), the cation exchange capacity is moderately high (active), and the mean annual soil temperature is between 8 and 15 °C (46 and 59 °F; frigid)
Series	Mandy	Soil series are named after a town or other geographic feature near where the soil was first described
Phase	Mandy channery silt loam, 8–15% slopes	Mandy soils where the texture of the surface horizon is silt loam with more than 15% channers, located on moderately steep slopes

Table 3.3 Soil orders that are commonly found across high elevations of the central and southern Appalachians (Soil Survey Staff 2022)

Soil order	Description/characteristics
Alfisol	Relatively fertile soils formed in humid to subhumid climates; soils have an accumulation of clay (Bt horizons) in the subsoil; commonly found in forested environments
Histosols	Soils comprised primarily of organic materials; often found in cool and/or wet environments where decomposition rates are slow, which allows organic matter to accumulate
Inceptisols	Soils with weakly developed subsoil horizons with little to no accumulation of clay or other materials (Bw horizons); lack of development is commonly attributed to relatively young age or the presence of environmental conditions that inhibit mineral weathering
Spodosols	Soils formed in cool, moist climates under confer forest vegetation; soils have an accumulation of organic matter complexed with Al and/or Fe oxides (Bh, Bs, Bhs horizons)
Ultisols	Relatively infertile soils formed in humid climates; soils have an accumulation of clay (Bt horizons); commonly found in forested environments

contributes to improving the security of the soil resource base (McBratney et al. 2014). An understanding of how the soils in these ecosystems develop and function is needed to effectively incorporate soils information into restoration planning.

3.3 Soils of Red Spruce Ecosystems

Soils develop and evolve in response to climate, vegetation, topography, geology, and age. In the higher elevations of the central and southern Appalachians, the cool, moist climates and acidic parent materials that promote and sustain the characteristic conifer forest vegetation also lead to the development of locally unique soils. For example, the type of vegetation controls the amount and location of organic matter additions to the soil. Red spruce forests produce inherently acidic leaf litter and needles, which decompose more slowly than hardwood leaf litter (Phillips and FitzPatrick 1999; Hobbie et al. 2007). Root secretions of organic acids from conifer vegetation further acidify the soil (Jongmans et al. 1997; Lundström et al. 2000a). This, coupled with the cool moist climates found at higher elevations, often fosters the development of thick organic layers at the soil surface (USDA NRCS 2016a). These organic horizons can range in thickness from <5 cm (2 in) to <1 m (3.3 ft). The needle litter that accumulates in conifer dominated ecosystems accentuates the acidic soil conditions, acting as a positive feedback loop for red spruce regeneration (USDA NRCS 2016a, 2016b).

The thick organic soil layers in red spruce ecosystems produce organic acids that drive a process known as podzolization by which organic compounds bound with iron and aluminum are transported downward in the soil profile and accumulate in the subsoil (Lundström et al. 2000a, 2000b). The accumulation of organic compounds bound with iron and aluminum are what soil scientists recognize as spodic materials. When spodic materials are strongly expressed in the subsoil, that soil is normally classified as a Spodosol in the U.S. system of Soil Taxonomy (Soil Survey Staff 2022), while a soil with weaker expression of spodic materials is recognized as a spodic intergrade of other taxa (see Sidebar 3.2). As a result of the podzolization process, Spodosols are distinguished by thick, dark-colored organic surface horizons, white to gray horizons from which most of the organic matter and secondary minerals have been stripped, and dark red to black spodic horizons that indicate the accumulations of organic matter and/or aluminum and iron compounds in the subsoil (Lundström et al. 2000a). Within the U.S., soils with spodic materials (Spodosols and spodic intergrades) are most associated with the colder and wetter climates of the Pacific Northwest, northern Great Lakes, northern New England, and southern Alaska (Bockheim 2014), although wet Spodosols are also found in coastal zones from New Jersey to Florida. Within the central and southern Appalachians, Spodosols and spodic intergrades are minor components of the landscape (West et al. 2016), relegated to the coldest and wettest areas.

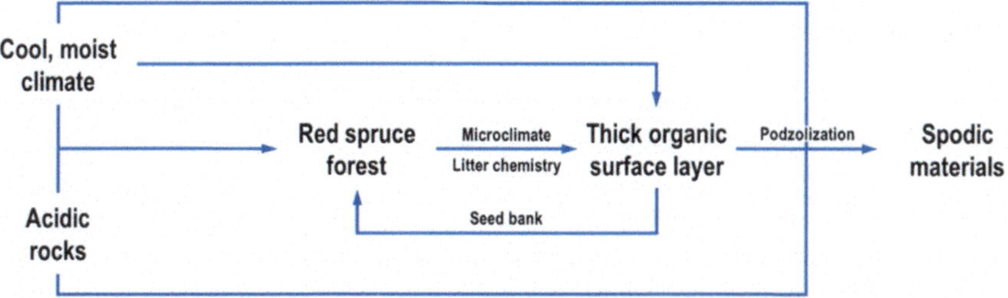

Fig. 3.2 Conceptual diagram showing connections between red spruce (*Picea rubens*) ecosystems and climate, vegetation, geology, and soils. The cool, moist climate and acidic parent materials at high elevations of the central and southern Appalachians favor red spruce over other tree species. The properties of the red spruce litter—along with the cool, moist climatic conditions—promotes the accumulation of thick organic layers at the soil surface. The organic acids released by the slow decomposition of the organic soil materials—in conjunction with the cool, moist climate, and acid parent materials—drive the podzolization process that leads to the development of the organic-rich spodic materials in the subsoil. Those organic soil materials in the forest floor also contain the majority of the red spruce seed bank that enables continued red spruce regeneration

Podzolization is a complex process involving a suite of biogeochemical interactions. The local climate, vegetation, and parent materials all contribute to the development of spodic materials (Lundström et al. 2000a). Consequently, the red spruce ecosystem is intricately connected through the interactions of climate, vegetation, geology, and soils (Fig. 3.2). Variability in site conditions and soil forming processes influences how strongly or weakly spodic materials are expressed. For instance, a high-elevation ecosystem with base-poor acidic geology would be more conducive to podzolization than a site with alkaline geology. Similarly, changes or shifts in the soil forming factors that drive podzolization (such as a change from conifer to hardwood vegetation) may result in the degradation of spodic materials.

As a result of variations in environmental conditions and soil-forming processes, high-elevation soils across central and southern Appalachia demonstrate varied degrees of expression of spodic materials and other features associated with Spodosols (Fig. 3.3). As the environmental conditions promote more podzolization, the thickness of the organic surface layer increases and the expression of spodic materials increases. Increased podzolization can also lead to the development of a leached layer above the spodic materials (Lundström et al. 2000a, 2000b; Sauer et al. 2007).

Many of the soil functions and associated ecosystem services attributed to the soils in red spruce ecosystems are directly or indirectly derived from the accumulation of organic materials at the soil surface and in the subsoil. Soil organic matter content is a dynamic soil property (Tugel et al. 2005, 2008), which is strongly influenced by both natural processes and human management practices (Lal 2005; Stockman et al. 2013; Ramesh et al 2019). The interactions among soil organic matter and other soil properties and processes are complex (Fig. 3.4). Increasing the amount of organic matter in the soil tends to improve soil structure, decrease bulk density, increase cation exchange capacity, and promote nutrient cycling. Changes to the physical

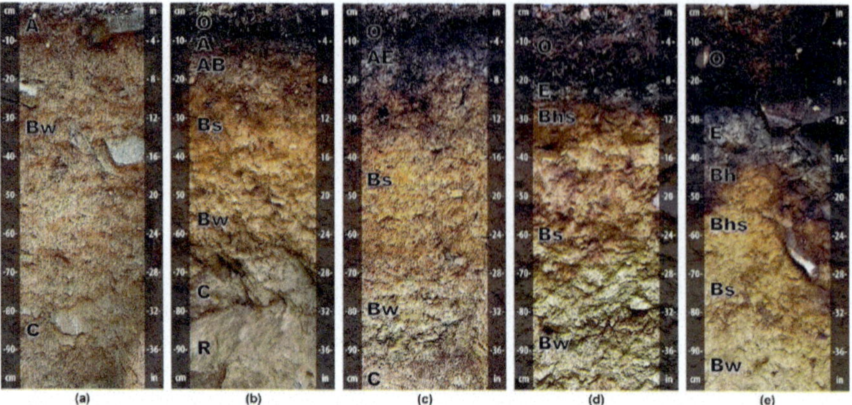

Fig. 3.3 Five soil profiles that depict varying degrees of expression of spodic materials and other features associated with Spodosols in red spruce (*Picea rubens*) forests. From left to right, the thickness of the organic surface layer (O) increases from none to over 20 cm (8 in). Similarly, spodic materials increase from none (Bw) to moderate (Bs) to strong (Bh, Bhs). The development of a leached layer (E) above the spodic materials also increases from left to right. Profiles **b** and **c** are classified as spodic intergrades while profiles **d** and **e** are Spodosols. See Sidebars 3.1 and 3.2 for more information on soil horizon nomenclature and soil classification. Photos by James Leonard

properties in turn increase porosity, increase infiltration, and increase water holding capacity. This can lead to reduced peak flows in streams and rivers, reducing the risk for downstream flooding. Increased organic matter also improves aggregate stability, which reduces soil erodibility and minimizes sediment delivery to surface water. The decomposition of organic matter increases the availability of multiple plant nutrients, particularly nitrogen and phosphorus. Globally, soils contain more carbon than the atmosphere and vegetation combined (Schlesinger and Bernhardt 2013). In central and southern Appalachia, the capacity for red spruce soils to sequester carbon through podzolization makes these forests ideal locations for increasing and maintaining terrestrial carbon stocks while simultaneously improving the various other dynamic soil properties that are influenced by soil organic matter. It also contributes to the region's high biodiversity and makes these soils and these ecosystems more resilient to drought, extreme weather events, or other disturbances.

In both the central and southern Appalachians, the cool, moist climates where precipitation exceeds evapotranspiration favor conifer and ericaceous vegetation and are conducive to the podzolization process. The presence of soils with spodic materials has long been recognized in both the central Appalachians (Williams and Fridley 1931; Coile 1938; Losche and Beverage 1967) and southern Appalachians (Coile 1938; Springer and Elder 1980; Springer 1984; Feldman et al. 1991a, 1991b). However, differences in bedrock geology create different soil-forming environments in the two regions. In the central Appalachians, the soils have formed from sedimentary rocks, particularly acid sandstones and shales. In the southern Appalachians, metamorphic and igneous rocks are the most common soil parent materials, which

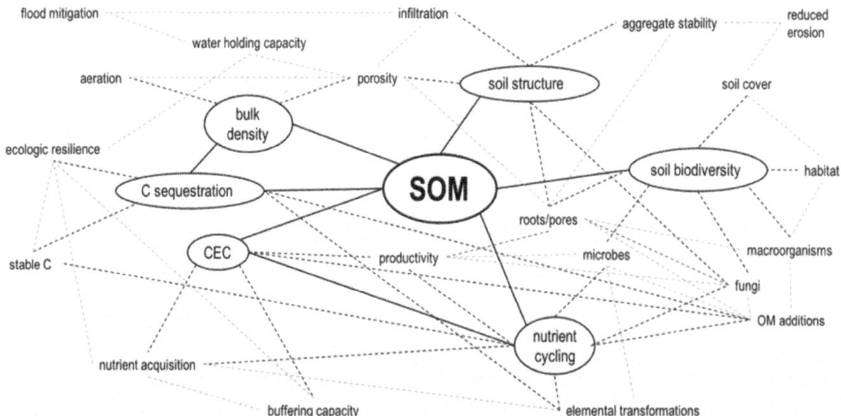

Fig. 3.4 Conceptual diagram illustrating how increasing soil organic matter (SOM) improves water storage, increases nutrient cycling, and enhances ecosystem health, which leads to more resilient landscapes. Solid lines represent primary relationships among soil properties, while dashed lines represent secondary and tertiary relationships

produce soils that are less acidic (Minckler 1939). In the central Appalachians, the most strongly expressed spodic materials are found in sandy and other coarser-textured soils derived from sandstone bedrock, which allow for ample leaching and the movement of the organic acids into the subsoil (Lundström et al. 2000a, 2000b; Sauer et al. 2007; Nauman et al. 2015a, 2015b). While finer-textured soils (i.e., less sand, more silt and clay), such as the soils derived from metamorphic and igneous rocks in the southern Appalachians, may limit podzolization and the expression of spodic materials, recent investigations in the central Appalachians has demonstrated that Spodosols and spodic intergrades can also form in geologies with finer-textured parent materials such as shale geologies (Nauman et al. 2015a).

Another difference in the soil-forming environments between the central and southern Appalachians that may affect the amount of podzolization and, therefore, the strength of expression of spodic materials is the amount of snowfall received and the annual depth of snowpack. Schaetzl et al. (2015) reported that the strength of podzolization in Spodosols in Michigan was related to the thickness of the snowpack and, consequently, the volume of water that enters the soil over a relatively short period of time during spring snowmelt. Compared to the central Appalachians, the highest elevations in the southern Appalachians receive less snowfall and maintain less of a snowpack (Feldman et al. 1991a).

Spodosols and other soils with spodic materials are not the only soils found in red spruce ecosystems of central and southern Appalachia. Bogs, fens, and other wetlands are common across these high elevation ecosystems. The cool temperatures and saturated soil conditions create an environment where organic matter accumulation exceeds decomposition. In some places, these soils are comprised almost entirely of organic materials with little to no mineral soil materials. The high organic matter content of the soils contributes to multiple ecosystem services such as flood control, water purification, and habitat for endemic plants and animals. Other soils commonly found in red spruce ecosystems include finer-textured soils with accumulations of clay in the subsoil, and more immature soils with minimal soil development. Some of these soils may be wetter, with relatively shallow water tables.

3.4 Impacts of Disturbances on Soils in Red Spruce Ecosystems

The occurrence of red spruce in the central and southern Appalachians has been greatly reduced relative to pre-European settlement (see Chap. 1). In the central Appalachians, up to 99% of the historical range for red spruce was disturbed in some manner between the mid-1800s and early-1900s (Byers et al. 2010). Broad-scale intensive logging and subsequent wildfires in red spruce forests resulted in dramatic changes in tree community composition (Nowacki et al. 2010). Logging of red spruce forests was less prevalent in the southern Appalachians, but atmospheric acid deposition contributed to declines in the 1900s (McLaughlin et al. 1998).

Soil properties and processes were severely impacted by the disturbances, which reduced the ecosystem services the soils could provide. Intensive logging removed the tree canopy, eliminating the organic matter additions that produced and maintained the thick O horizons characteristic of red spruce soils. The loss of forest cover also allowed more solar radiation to reach the soil surface, resulting in warmer temperatures and drier site conditions, which likely facilitated faster decomposition of organic horizons at the soil surface (Allard and Leonard 1952; Pauley 2008; Boggs and McNulty 2010; Fig. 3.5). In some locations, wildfires consumed most of the organic soil materials down to bare rock or the mineral soil surface (Korstian 1937, Allard and Leonard 1952, Clarkson 1964, Lewis 1998; see Sect. 1.2.3 in Chap. 1), resulting in massive CO_2 emissions and the loss of ecosystem services tied to soil organic matter. Landscapes were denuded of vegetation and organic matter after these disturbances, greatly reducing soil water holding capacity across the headwaters of major rivers (Clarkson 1964). The burning of the O horizons also destroyed much of the red spruce seed bank and greatly restricted red spruce regeneration in the burned landscapes (Rentch and Schuler 2018). With no vegetation and no organic horizons to protect the soil from rain-drop impact across steep topography, many areas are believed to have lost some or all topsoil (Korstian 1937; Allard and Leonard 1952). Logging and associated disturbances also modified the physical, chemical, and biological properties of the soil, which contributed to decreased fertility and productivity. More recently, atmospheric acid deposition has further altered the chemistry and fertility of these forest soils (McLaughlin et al. 1998; Elias et al. 2009). Over time, the disturbances had cascading effects that ultimately favored hardwood regeneration (Rentch et al. 2007; Nowacki and Wendt 2010).

Changes in microclimate, specifically increased solar radiation and temperatures, resulted in decreased soil moisture and less leaching. The change from conifer to

Fig. 3.5 Podzolized soil in a former virgin red spruce (*Picea rubens*) forest on Gaudineer Knob, Pocahontas County, West Virginia. The area was logged 20 years prior to this photograph (taken in July 1938) but was not burned. The original O horizon was rapidly decomposed after logging such that only a thin layer (<15 cm [6 in]) of highly decomposed organic material remains at the surface. The leached E horizons and the spodic B horizons that formed under the original red spruce cover are still present. The scale is 30 cm (12 in). See Minckler (1939) for additional information (photo by USDA Forest Service)

hardwood-dominated overstories eliminated the input of organic acids that contribute to the development and maintenance of spodic materials and Spodosols. The degradation of spodic materials, sometimes called depodzolization (Barrett and Schaetzl 1998), became a dominant soil process in the ecosystems where conifer cover was lost. While evidence of podzolization and spodic soil properties has been documented throughout the southern Appalachians, Spodosols do not appear on soil maps, likely because the areas are not extensive enough. Less intensive podzolization may be due to the differences in climate and parent materials, significant depodzolization after the loss of spruce cover, or a combination of both. It may also be due to mapping protocols that did not emphasize or prioritize the identification and delineation of soils with spodic materials. Further investigation is needed to look for possible signatures in the soil of the historic presence of red spruce in the southern Appalachians. This would improve the existing soil maps in these areas and could add to the inventory of recognized ecological sites in the southern Appalachians.

3.5 Red Spruce Ecological Sites

The characteristic soils and landscapes that support the distinguishing plant and animal communities of the high-elevation red spruce forests represent specific ecological sites (ES) across the central and southern Appalachians. Corresponding ES descriptions (ESD) have been developed to better understand the complex relationship between the vegetation, climate, geology, soils, management, and disturbance history of red spruce ecosystems (Fig. 3.2). For a detailed description of ES and the associated ESDs, see Sidebar 3.3. For more on how to use red spruce ESDs to guide management toward reference conditions see Sect. 3.6 and Sect. 8.2.1 in Chap. 8.

Sidebar 3.3 Ecological Site Descriptions Ecological sites (ES) are defined as "a distinctive kind of land based on recurring soil, landform, geological, and climate characteristics that differs from other kinds of land and its ability to produce distinctive kinds and amounts of vegetation and its ability to respond similarly to management actions and natural disturbances" (USDA NRCS 2017). An ecological site description (ESD) is a means for characterizing an ES, and is a tool used to implement practices to meet restoration and management goals for a landscape (Bestelmeyer and Brown 2010). An ES represents a combination of recurring soil, landform, geologic, and climatic influences that produce a unique group of ecological states, which are further comprised of specific community phases.

An important aspect of the utility of ESDs is, in part, a product of correlating one or more soil map unit components to an ES based on observable changes in soil-plant relationships, soil processes, and community composition (Duniway et al. 2010; USDA NRCS 2017). Often, differences between ES and individual ecological states are due to changes in soil properties (Duniway et al. 2010), which affect the potential range of vegetation communities for that site. In practice, an ES is tied to one or more soil map unit components that produce a unique vegetation type (USDA NRCS 2017). These soil map unit components include a soil name as well as other information about the land such as soil texture, slope steepness, or surface rock fragments designators. The relationship between a soil and its associated ESD is very important—if the soil correlated to the ESD is not present, then neither is the ES. One soil map unit component cannot be correlated to more than one ESD, although, a single ESD can have multiple correlated map unit components.

Ecological states are temporally related plant communities and dynamic soil properties that provide recurring structural and functional attributes to a given ecosystem and make up the ES (Bestelmeyer et al. 2009). Of the states comprising an ES, the most important is the reference state. A reference state can be defined as the state that can support the largest number of ecosystem services, and from which every other state in an ES can be derived

(Bestelmeyer et al. 2009; Bestelmeyer and Brown 2010). Often the reference state is accepted as the pre-European conditions that comprise a vegetation community. Throughout much of the eastern U.S., reference states are no longer easily observable due to past anthropogenic disturbances (Drohan and Ireland 2016). Within an ecological state there may be community phases. Community phases are specific plant communities and dynamic soil properties that reoccur in an observable pattern within a state (Briske et al. 2008; Bestelmeyer et al. 2009; Bestelmeyer and Brown 2010). In the reference state there is an associated reference phase community, which imparts structural and functional qualities that determine the reference state's resilience and integrity (Briske et al. 2008; Bestelmeyer and Brown 2010).

Linking ecological states within an ES are what are referred to as restoration and transitional pathways. These pathways represent a suite of restorative management practices that guide the development, or degradational events or processes, that result in a new ecological state. Restoration pathways represent management practices that are used to restore an alternative ecological state back to the reference state condition. A transitional pathway is a degradational pathway attributed to some disturbance. Drought, disease, logging, fires, and many other disturbances affect the trajectory of an ecological community. When extreme disturbances have created highly degraded conditions, restoration back to the reference state of the ES may not be feasible.

The component of an ESD that is most important to management is the state-and-transition model (STM). A STM is a broad conceptual framework used to group together vegetation and soil dynamics inherent to an ES, organized to easily implement restoration and management practices (Briske et al. 2006; Bestelmeyer et al. 2009). A STM provides a way to organize and portray key boundaries that comprise ecological states within an ES, and act as a map for management practices that can be applied to reach a desired ecological result. A STM is composed of ecological states, community phases, and transitional pathways to provide a comprehensive guide for developing management plans specific to a given ES. Triggers, drivers, and disturbance mechanisms that cause transition between states are also conceptually outlined by a STM (Briske et al. 2008; Bestelmeyer et al. 2009). Figure 8.1 in Chap. 8 shows an example of a red spruce ES STM, with ecological states and transitional pathways.

Currently there are two official ESDs for red spruce ecosystems in West Virginia (USDA NRCS 2016a, 2016b). Both ESDs were created for soils developed on shale geologies interbedded with sandstone that produce low pH soils and contain similar vegetation communities; however, both ESDs have now been associated with soils across multiple parent materials derived from sandstone, siltstone, and shale. The primary difference between the two ESs is the vegetation composition and the degree of podzolization specific to the soil associated with each ESD. These ESs are referred to as the Spodic Shale Upland Conifer Forest, which is associated with the Wildell (classified as loamy-skeletal, mixed, superactive, frigid Typic Haplorthods), Gauley (classified as loamy-skeletal, siliceous, superactive, frigid Typic Haplorthods), and Gaudineer (loamy-skeletal, siliceous, superactive, frigid Typic Haplorthods) soils (see Sidebar 3.2) and the Spodic Intergrade Shale Upland Hardwood and Conifer Forest, which is associated with the Mandy series (loamy-skeletal, mixed, active, frigid Spodic Dystrudepts). A third red spruce ESD, the Rubbly Upland Conifer Forest, has been drafted but it is still in its provisional stages of development. This ESD occurs on the Pottsville sandstone formation and is characterized by a high degree of surface rock fragments. The Rubbly Upland Conifer Forest is correlated mainly to the Gauley, Gaudineer, Blandburg (loamy-skeletal, siliceous, semiactive, frigid Typic Haplorthods), and Leetonia (sandy-skeletal, siliceous, mesic Entic Haplorthods) soils with 15% or more of the soil surface covered with stones or boulders. These three ESs represent a small portion of the range for red spruce within the central and southern Appalachians. Looking ahead, we expect that more ESDs will be developed to further characterize the diversity of red spruce ecosystems in the two regions. These and all other recognized ESDs are publicly available and can be accessed using the Ecosystem Dynamics Interpretive Tool (https://edit.jornada.nmsu.edu).

3.6 Soil Maps as Tools for Decision Making

An inventory of the soil resources of the U.S. has been created by the NRCS as part of the National Cooperative Soil Survey (NCSS). The NRCS is responsible for the management and maintenance of soils information for land use planning across the U.S. This best available data about soil science is continuously updated to reflect new knowledge to ensure that the maps, data, and interpretations are complete, consistent, correct, comprehensive, and current. These resources are publicly available and can be accessed through the Web Soil Survey (https://websoilsurvey.nrcs.usda.gov/).

The Web Soil Survey contains information about specific soil properties and qualities, which can be displayed for user-selected depth ranges, as well as information about how those soil properties may affect various types of management (called interpretations). The tools available within the Web Soil Survey emphasize the critical role that soils have for broad-scale land management and planning. While the information contained in Web Soil Survey can be used to predict or estimate the potentials or limitations of soils for a variety of uses, it is important to recognize that soils can be highly variable over short distances, such that the scale of the soil maps in the Web Soil Survey is not detailed enough to provide site-specific information. Consequently, current soil survey products are most suited for broad land management planning. For these reasons, it is imperative that soil survey information be used alongside any other applicable information such as on-site investigations when making site-specific decisions.

The Web Soil Survey has many uses to support land use and management decision-making. Specific to red spruce restoration efforts, work is underway to tie specific soils to vegetative communities through the development of ESDs. With the incorporation of ES information, the Web Soil Survey can also be used as a tool to support landscape-scale restoration. A demonstration of the value of the Web Soil Survey for informing red spruce restoration is shown in Fig. 3.6. The upper map (a) shows the base soil map for the area, which is roughly 60 km^2 (23 mi^2; 6 km by 10 km [3.7 mi by 6.2 mi]). The orange lines represent the delineations of the individual soil map units. A map unit may represent one or more dominant soils that are specified in the map unit name (e.g., Mandy-Wildell complex, 35–55% slopes, very stony; MfwF), but also will include smaller areas of other soils. The minor components may or may not influence the use and management interpretations for the areas covered by a given map unit. The middle map (b) depicts the dominant soil order in each map unit delineation. The teal colors on the west side of the map are Spodosols, while the red colors on the east side of the map are Inceptisols (mainly Spodic Dystrudepts). The lower map (c) displays the ESDs associated with the dominant component in each map unit delineation. The dark blue areas, which correspond to the Spodosols, are within the Spodic Shale Upland Conifer Forest ES. The teal areas, which correspond to the Spodic Dystrudepts, are within the Spodic Intergrade Shale Upland Hardwood and Conifer Forest. The purple areas on the west side of the map are within the provisional Rubbly Upland Conifer Forest ES. If the goal was to restore red spruce within this area of interest, the ESDs associated with these three ESs, particularly their STMs, could be used to guide landscape-scale prescriptions for red spruce planting and/or release to guide the current vegetation communities toward the desired state.

3.7 Future of Red Spruce Soils

Red spruce forests within the Appalachians are considered to be one of the most endangered forest types in the U.S. (Noss et al. 1995; Christensen et al. 1996). Interest in red spruce restoration stems from multiple ecological, recreational, biodiversity, and carbon mitigation interests. Red spruce forests provide habitat for threatened, endangered, and sensitive species such as the Cheat Mountain salamander (*Plethodon nettingi*; Pauley 2008) and Virginia northern flying squirrel (*Glaucomys sabrinus fuscus*; Menzel et al. 2004). The soils that form as part of red spruce ecosystems also provide valuable ecosystem services including increased water holding capacity and filtration (Sauer et al. 2007) and soil carbon buildup both at the soil surface and in the subsoil (Herbauts and De Buyl 1981; Miles 1985; Sohet et al. 1988; Tarnocai et al. 2009; Averill et al. 2014). Restoration of red spruce forests across the central and southern Appalachians is essential to building ecological resiliency across the landscape. Research suggests that red spruce restoration efforts can sequester significant soil organic carbon at the soil surface in less than 100 years (Nauman et al. 2015b), and is also expected to increase soil organic carbon in the spodic materials in the subsoil (Barrett and Schaetzl 1998; Lundström et al. 2000a, 2000b) (Fig. 3.7). Available ESDs developed for red spruce ecosystems (e.g., USDA NRCS 2016a, 2016b) can be used to inform potential carbon sequestration during forest restoration

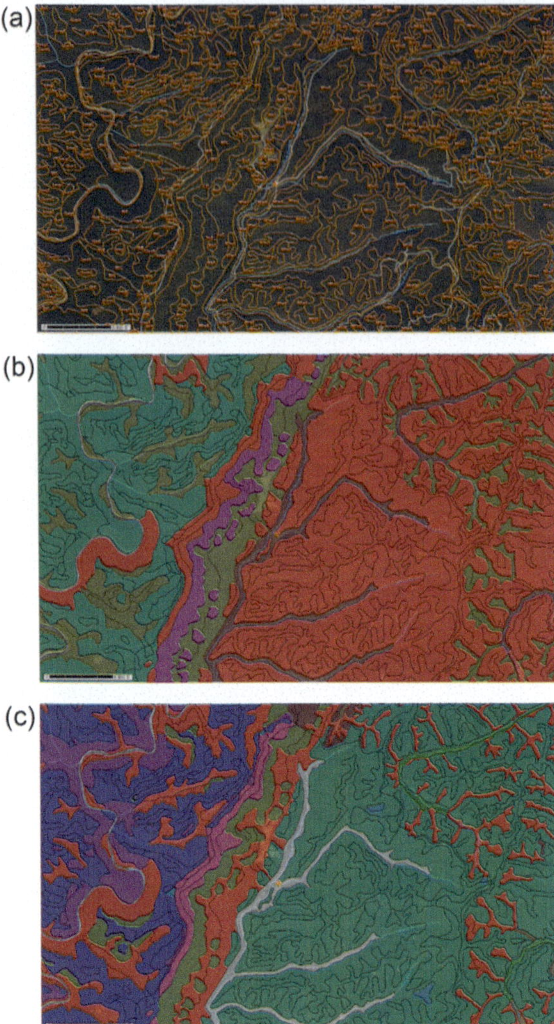

Fig. 3.6 Example of the value of the web soil survey for informing red spruce (*Picea rubens*) restoration, using a 60 km^2 (23 mi^2) area near Wildell, West Virginia. Soil map units, **a** represent one or more dominant soils that are specified in the map unit name. Soil orders, **b** depict the dominant soil in each map unit delineation (teal: Spodosols; red: Inceptisols; green: Ultisols; purple: Alfisols; gray: Entisols). Ecological sites, **c** indicate the ecological site description (ESD) associated with the dominant component in each map unit delineation (blue: Spodic Shale Upland Conifer Forest; teal: Spodic Intergrade Shale Upland Hardwood and Conifer Forest; red: Convergent Uplands; purple: Rubbly Upland Conifer Forest; green: Divergent Uplands; brown: Acidic Shale Upland Oak/Heath, magenta: Frigid High Elevation Uplands). By associating soil map units with specific ESDs (and their associated STMs), landowners and managers can use soil maps for a given area of interest to develop management plans to meet restoration or conservation goals

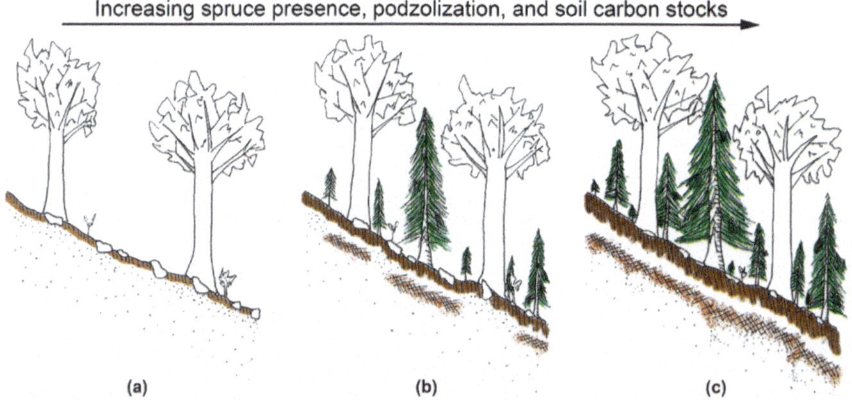

Fig. 3.7 Illustration showing that as the red spruce (*Picea rubens*) component of ecosystems increases, the soil organic carbon stocks (indicated by dark shading) also increase at the soil surface and in the subsoil. The greater presence of red spruce leads to increased thickness of the forest floor. The organic acids that leach from the forest floor promote more podzolization and greater accumulation of organic carbon in the subsurface through the formation of spodic horizons

activities, with initial investigations showing that soils of the Spodic Shale Upland Conifer Forest contain greater soil organic carbon stocks than the Spodic Intergrade Shale Upland Hardwood and Conifer Forest, but greater opportunities for increasing carbon stocks in the Spodic Intergrade Shale Upland Hardwood and Conifer Forest (Leonard 2022).

In general, environments that alter the types and amounts of processes that occur in soil can create persistent features that record the history of the environment in which they formed, even after those environmental conditions have changed. Consequently, every soil has a story to tell through the horizons and the associated colors, textures, structures, and other properties we observe (Targulian and Goryachkin 2010; Lin 2011; Monger and Rachal 2013). In the central and southern Appalachians, the presence or absence of spodic materials, including how strongly or weakly they are expressed, serve as a record of historical site conditions, referred to as pedomemory (Nauman et al. 2015a). This information can be observed directly in the field when investigating possible sites for restoration efforts, and it can be inferred when examining soil maps from across the region. A Spodosol with well-expressed spodic morphology indicates that red spruce or other conifers were dominant historically, even if conifers are a limited or absent component in the stand today. The absence of spodic properties, however, may not necessarily indicate that red spruce was never there historically; it may indicate that other conditions were not conducive to podzolization or that site conditions promoted more rapid depodzolization. The lack of spodic materials provides useful information too because if a site did not support red spruce in the past, it is likely not conducive to contemporary red spruce restoration efforts. To summarize, spodic properties and the degree to which they are expressed is a valuable land management tool in planning and prioritizing red spruce restoration across the central and southern Appalachians. Soil maps, with their associated ESDs, are particularly useful for targeting restoration efforts at the scale of landscapes to regions. For site-specific assessments, or where existing soil maps lack either the necessary detail or the necessary ESDs, an investigation of the soils can reveal the presence or absence of spodic materials.

Today, red spruce ecosystems only occupy a fraction of their historical range (see Chap. 1). Furthermore, forest impact models predict that habitat quality will decrease for red spruce in the central Appalachians over the next few decades (Butler et al. 2015; see Chap. 7). Thus, the challenge of restoring red spruce ecosystems in the central and southern Appalachians is a daunting one. Deciding where to focus red spruce restoration efforts within the current geographic distribution of the species is even more challenging. Although long-gone, historical red spruce forests left signatures in the soil. This pedomemory, when utilized in conjunction with topographic conditions and climatic projections, can serve as an effective tool for land managers when prioritizing where to focus red spruce restoration across the landscape.

References

Adhikari K, Hartemink AE (2016) Linking soils to ecosystem services—A global review. Geoderma 262:101–111

Allard HA, Leonard EC (1952) The Canaan and Stony River valleys of West Virginia, their former magnificent spruce forests, their vegetation and floristics today. Castanea 17:1–60

Averill C, Turner BL, Finzi AC (2014) Mycorrhiza-mediated competition between plants and decomposers drives soil carbon storage. Nature 505:543–545

Barrett LR, Schaetzl RJ (1998) Regressive pedogenesis following a century of deforestation: evidence for depodzolization. Soil Sci 163:482–497

Bestelmeyer BT, Tugel AJ, Peacock GL Jr et al (2009) State-and-transition models for heterogeneous landscapes: a strategy for development and application. Rangel Ecol Manag 62:1–15

Bestelmeyer BT, Brown JR (2010) An introduction to the special issue on ecological sites. Rangelands 32:3–4

Binkley D, Fisher RF (2019) Ecology and management of forest soils. Wiley, Hoboken, New Jersey

Bockheim JG (2014) Soil geography of the USA: a diagnostic horizon approach. Springer, New York

Boggs JL, McNulty SG (2010) Changes in canopy cover alter surface air and forest floor temperature in a high-elevation red spruce (*Picea rubens* Sarg.) forest. In: Rentch JS, Schuler TM (eds) Proceedings from the conference on the ecology and management of high-elevation forests in the central and southern Appalachian Mountains. USDA Forest Service Northern Research Station General Technical Report NRS-P-64, Newtown Square, Pennsylvania, pp 13–21

Briske DD, Fuhlendorf SD, Smeins FE (2006) A unified framework for assessment and application of ecological thresholds. Rangel Ecol Manage 59:225–236

Briske DD, Bestelmeyer BT, Stringham TK et al (2008) Recommendations for development of resilience-based state-and-transition models. Rangel Ecol Manag 61:359–367

Busing RT (1985) Gap and stand dynamics of southern Appalachian spruce fir forests. Dissertation, University of Tennessee, Knoxville, Tennessee

Butler PR, Iverson LR, Thompson FR III et al (2015) Central Appalachians forest ecosystem vulnerability assessment and synthesis: a report from the central Appalachians climate change response framework project. U.S. Department of Agriculture Forest Service Northern Research Station General Technical Report NRS-146, Newtown Square, Pennsylvania

Byers EA, Vanderhorst JP, Streets BP (2010) Classification and conservation assessment of upland red spruce communities in West Virginia. West Virginia Division of Natural Resources Report, Elkins, West Virginia

Callaham MA Jr, Rhoades CC, Heneghan L (2008) A striking profile: soil ecological knowledge in restoration management and science. Restor Ecol 16:604–607

Christensen NL, Bartuska AM, Brown JH et al (1996) The report of the ecological society for America committee on the scientific basis for ecosystem management. Ecol Appl 6:665–691

Clarkson RB (1964) Tumult on the mountains: lumbering in West Virginia, 1770–1920. McClain Printing Company, Parsons, West Virginia

Coile TS (1938) Podzol soils in the southern Appalachian Mountains. Soil Sci Soc Am J 3:274–279

Daily GC, Matson PA, Vitousek PM (1997) Ecosystem services supplied by soil. In: Daily GC (ed) Nature services: societal dependence on natural ecosystems. Island Press, Washington, DC, pp 113–132

Dokuchaev VV (1899) On the doctrine of nature zones: horizontal and vertical soil zones. St. Petersburg City Administration, St. Petersburg, Russia

Drohan PJ, Ireland AW (2016) Provisional, forested ecological sites in the northern Appalachians and their state-and-transition models. Rangelands 38:350–356

Duniway MC, Bestelmeyer BT, Tugel A (2010) Soil processes and properties that distinguish ecologic sites and states. Rangelands 32:9–15

Elias PE, Burger JA, Adams MB (2009) Acid deposition effects on forest composition and growth on the Monongahela National Forest, West Virginia. For Ecol Manag 258:2175–2182

Feldman SB (1989) Taxonomy, genesis, and parent material distribution of high elevation forest soils in the southern Appalachians. Dissertation, Virginia Polytechnic Institute and State University, Blacksburg, Virginia

Feldman SB, Zelazny LW, Baker JC (1991a) High-elevation forest soils of the southern Appalachians: I. Distribution of parent materials and soil-landscape relationships. Soil Sci Soc Am J 55:1629–1637

Feldman SB, Zelazny LW, Baker JC (1991b) High-elevation forest soils of the southern Appalachians: II. Geomorphology, pedogenesis, and clay mineralogy. Soil Sci Soc Am J 55:1782–1791

Heneghan L, Miller SP, Baer S et al (2008) Integrating soil ecological knowledge into restoration management. Restor Ecol 16:608–617

Herbauts J, De Buyl E (1981) The relation between spruce monoculture and incipient podzolisation in ochreous brown earths of the Belgian Ardennes. Plant Soil 59:33–49

Hobbie SE, Ogdalh M, Chorover J et al (2007) Tree species effects on soil organic matter dynamics: the role of soil cation composition. Ecosystems 10:999–1018

Jenny H (1941) Factors of soil formation: a system of quantitative pedology. McGraw-Hill Book Company, New York

Jongmans AG, van Breemen N, Lundström U et al (1997) Rock-eating fungi. Nature 389:682–683

Korstian CF (1937) Perpetuation of spruce on cut-over and burned lands in the higher southern Appalachian Mountains. Ecol Monogr 7:125–167

Lal R (2005) Forest soils and carbon sequestration. For Ecol Manag 220:242–258

Leonard JL (2022) Red spruce ecological sites, ecological states, and restoration pathways quantified through soil organic carbon. Thesis, West Virginia University, Morgantown, West Virginia

Lewis RL (1998) Transforming the Appalachian countryside: railroads, deforestation, and social change in West Virginia, 1880–1920. University of North Carolina Press, Chapel Hill, North Carolina

Lietzke DA, McGuire GA (1987) Characterization and classification of soils with spodic morphology in the southern Appalachians. Soil Sci Soc Am J 51:165–170

Lin H (2011) Three principles of soil change and pedogenesis in time and space. Soil Sci Soc Am J 75:2049–2070

Losche CK, Beverage WW (1967) Soil survey of Tucker County and part of northern Randolph County, West Virginia. USDA Soil Conservation Service and Forest Service, Washington, DC

Lundström US, van Breeman N, Bain D (2000a) The podzolization process. A review. Geoderma 94:91–107
Lundström US, van Breeman N, Bain DC et al (2000b) Advances in understanding the podzolization process resulting from a multidisciplinary study of three coniferous forest soils in the Nordic countries. Geoderma 94:335–353
McBratney A, Field DJ, Koch A (2014) The dimensions of soil security. Geoderma 213:203–213
McLaughlin SB, Joslin JD, Robarge W et al (1998) The impacts of acidic deposition and global change on high elevation southern Appalachian spruce-fir forests. In: Mickler RA, Fox S (eds) The productivity and sustainability of southern forest ecosystems in a changing environment. Springer-Verlag, New York, pp 255–277
Menzel JM, Ford WM, Edwards JW et al (2004) Nest tree use by the endangered Virginia northern flying squirrel in the central Appalachian Mountains. Am Midl Nat 151:355–368
Miles J (1985) The pedogenic effects of different species and vegetation types and the implications of succession. J Soil Sci 36:571–584
Minckler LS (1939) Spruce type problem analysis: analysis of problems in the reforestation of the spruce type of the southern Appalachians. USDA Forest Service Unpublished File Report
Monger C, Rachal DM (2013) Soil and landscape memory of climate change—How sensitive, how connected? In: Driese SG, Nordt LC (eds) New frontiers in paleopedology and terrestrial paleoclimatology: paleosols and soil surface analog systems. SEPM Society for Sedimentary Geology, pp 63–70
Nauman TW, Thompson JA, Teets SJ et al (2015a) Ghosts of the forest: mapping pedomemory to guide restoration. Geoderma 247:51–64
Nauman TW, Thompson JA, Teets J et al (2015b) Pedoecological modeling to guide forest restoration using ecological site descriptions. Soil Sci Soc Am J 79:1406–1419
Noss RF, LaRoe ET III, Scott JM (1995) Endangered ecosystems of the United States: a preliminary assessment of loss and degradation. U.S. National Biological Service Biological Report 28, Washington, DC
Nowacki G, Wendt D (2010) The current distribution, predictive modeling, and restoration potential of red spruce in West Virginia. In: Rentch JS, Schuler TM (eds) Proceedings from the conference on the ecology and management of high-elevation forests in the central and southern Appalachian Mountains. USDA Forest Service Northern Research Station General Technical Report NRS-P-64, Newtown Square, Pennsylvania, pp 163–178
Nowacki G, Carr R, Van Dyck M (2010) The current status of red spruce in the eastern United States: distribution, population trends, and environmental drivers. In: Rentch JS, Schuler TM (eds) Proceedings from the conference on the ecology and management of high-elevation forests in the central and southern Appalachian Mountains. USDA Forest Service Northern Research Station General Technical Report NRS-P-64, Newtown Square, Pennsylvania, p 140–162
Pauley TK (2008) The Appalachian inferno: historical causes for the disjunct distribution of *Plethodon nettingi* (Cheat Mountain salamander). Northeast Nat 15:595–606
Phillips DH, FitzPatrick EA (1999) Biological influences on the morphology and micromorphology of selected Podzols (Spodosols) and Cambisols (Inceptisols) from the eastern United States and north-east Scotland. Geoderma 90:327–364
Ramesh T, Bolan NS, Kirkham MB et al (2019) Soil organic carbon dynamics: impact of land use changes and management practices: a review. Adv Agron 156:1–107
Rentch JS, Schuler TM (2018) Early red spruce restoration research by the Appalachian Forest Experiment Station, 1922–1954. J Forest 116:192–196
Rentch JS, Schuler TM, Ford WM et al (2007) Red spruce stand dynamics, simulations, and restoration opportunities in the central Appalachians. Restor Ecol 15:440–452
Robinson DA, Emmett BA, Reynolds B et al (2012) Soil natural capital and ecosystem service delivery in a world of global soil change. In: Hester RE, Harrison RM (eds) Soils and food security. Royal Society of Chemistry Publishing, Cambridge, England, pp 41–68
Sauer D, Sponagel H, Sommer M et al (2007) Podzol: soil of the year 2007. A review on its genesis, occurrence, and functions. J Plant Nutr Soil Sci 170:581–597

Schaetzl RJ, Luehmann MD, Rothstein D (2015) Pulses of podzolization: the relative importance of spring snowmelt, summer storms, and fall rains on Spodosol development. Soil Sci Soc Am J 79:117–131

Schlesinger WH, Bernhardt ES (2013) Biogeochemistry: an analysis of global change. Elsevier, San Diego, California

Sohet K, Herbauts J, Gruber W (1988) Changes caused by Norway spruce in an ochreous brown earth, assessed by the isoquartz method. J Soil Sci 39:549–561

Soil Survey Staff (1999) Soil taxonomy: a basis system of soil classification for making and interpreting soil surveys. In: USDA Natural resources conservation service agriculture handbook, vol 436, Washington, DC

Soil Survey Staff (2022) Keys to soil taxonomy, 13th edn. USDA Natural Resources Conservation Service, Washington, DC

Soil Survey Staff (2024) Field book for describing and sampling soils, version 4.0. USDA Natural Resources Conservation Service, Washington, DC

Springer ME (1984) Soils in the spruce-fir region of the Great Smoky mountains. In: White PS (ed) The southern Appalachian spruce-fir ecosystem: its biology and threats. National Park Service Southeastern Region Research/Resources Management Report SER-71, Atlanta, Georgia, pp 201–210

Springer ME, Elder JA (1980) Soils of Tennessee. In: University of Tennessee agricultural experiment station bulletin, vol 596, Knoxville, Tennessee

Stanturf JA, Callaham MA, Madsen P (2021) Soils are fundamental to landscape restoration. In: Stanturf JA, Callaham MA (eds) Soils and landscape restoration. Academic Press, New York, pp 1–37

Stockmann U, Adams MA, Crawford JW et al (2013) The knowns, known unknowns and unknowns of sequestration of soil organic carbon. Agr Ecosyst Environ 164:80–99

Targulian VO, Goryachkin SV (2010) Soil memory and environmental reconstructions. Eurasian Soil Sci 44:464–465

Tarnocai C, Canadell JG, Schuur, EAG et al (2009) Soil organic carbon pools in the northern circumpolar permafrost region. Glob Biogeochem Cycl 23:GB2023

Tugel AJ, Herrick JE, Brown JR et al (2005) Soil change, soil survey, and natural resources decision making: a blueprint for action. Soil Sci Soc Am J 69:738–747

Tugel AJ, Wills SA, Herrick JE (2008) Soil change guide: procedures for soil survey and resource inventory, ver 1.1. USDA Natural Resources Conservation Service, Lincoln, Nebraska

Urbanska KM, Webb NR, Edwards PJ (eds) (1997) Restoration ecology and sustainable development. Cambridge University Press, Cambridge, England

[USDA NRCS] USDA Natural Resources Conservation Service (2016a) Ecologic site F127XY001WV: spodic shale upland conifer forest. Available via the Ecosystem Dynamics Interpretive Tool. https://edit.jornada.nmsu.edu. Accessed 17 Jun 2024

[USDA NRCS] USDA Natural Resources Conservation Service (2016b) Ecologic site F127XY002WV: spodic intergrade shale upland hardwood and conifer forest. Available via the Ecosystem Dynamics Interpretive Tool. https://edit.jornada.nmsu.edu. Accessed 17 Jun 2024

[USDA NRCS] USDA Natural Resources Conservation Service (2017) National ecological site handbook. https://www.nrcs.usda.gov/resources/guides-and-instructions/national-ecological-site-handbook. Accessed 17 Jun 2024

West LT, Singer MJ, Hartemink AE (eds) (2016) The soils of the USA. Springer, Cham, Switzerland

Whalen JK, Sampedro L (2010) Fundamental properties of the soil ecosystem. In: Whalen JK, Sampedro L (eds) Soil ecology and management. Cambridge University Press, Cambridge, England, pp 3–26

Williams BH, Fridley HM (1931) Soil survey of Randolph County, West Virginia. USDA Bureau of Chemistry and Soils, Washington, DC

Wolfe JA (1967) Forest soil characteristics as influenced by vegetation and bedrock in the spruce-fir zone of the Great Smoky Mountains. Dissertation, University of Tennessee, Knoxville, Tennessee

BY

Chapter 4
Plant Communities

Elizabeth A. Byers and Michael P. Schafale

4.1 Introduction

Plant communities are assemblages of plant species that share specific habitat conditions along similar environmental gradients, with species interacting where they occur together. Plant communities have a characteristic range of stand physiognomy and suite of ecological processes, including natural disturbance regimes (Grossman et al. 1998; Jennings et al. 2002, 2003). They also are believed to support distinctive assemblages of microbial and animal species, especially among smaller and less mobile organisms. Terrestrial ecological communities are classified based on vegetation because plants are the most persistent and observable life form in these systems.

Plant communities are an important component of biodiversity, crucial for the maintenance of ecological processes and food pathways necessary for survival of many interdependent species. Because communities provide habitat for a multitude of common and poorly known organisms, conservation of their diversity offers a means to protect the many species that are not the focus of individual species conservation efforts. Even where we expect climate change to lead to different species assemblages, conserving the present diversity of communities protects the diversity of site conditions or habitat niches that will make up future communities as well as the pool of species that will compose future communities.

The plant communities in red spruce ecosystems across the central and southern Appalachians share many common features. Nevertheless, each community has its

E. A. Byers (✉)
West Virginia Division of Natural Resources (retired), Elkins, WV, USA
e-mail: Appalachian.Ecology@gmail.com

M. P. Schafale
North Carolina Department of Natural and Cultural Resources, North Carolina Natural Heritage Program, Raleigh, NC, USA
e-mail: michael.schafale@dncr.nc.gov

D. J. Brown et al. (eds.), *Ecology and Restoration of Red Spruce Ecosystems in the Central and Southern Appalachians*, https://doi.org/10.1007/978-3-032-19616-3_4

own distinct diagnostic species, substrate, topographic position, moisture regime, and disturbance regime. For some aspects of management and study, they may be addressed collectively. For others, the distinctions among different communities are crucial, and recognition of them may be essential for conservation success.

Communities in which red spruce is dominant or co-dominant often share a number of associated plant species. Co-dominant tree canopy species include Fraser fir (*Abies fraseri*), balsam fir (*Abies balsamea*), yellow birch (*Betula alleghaniensis*), eastern hemlock (*Tsuga canadensis*), yellow buckeye (*Aesculus flava*), red maple (*Acer rubrum*), American beech (*Fagus grandifolia*), black cherry (*Prunus serotina*), Fraser magnolia (*Magnolia fraseri*), and pitch pine (*Pinus rigida*). These communities range from closed canopy forests to woodlands (i.e., <60% canopy cover).

Shrubs may vary from a sparse to dense layer of lower-stature woody vegetation. Great laurel (*Rhododendron maximum*) is the most commonly associated shrub. Other frequently associated shrub species are common winterberry (*Ilex verticillata*), mountain laurel (*Kalmia latifolia*), Catawba rhododendron (*Rhododendron catawbiense*), southern mountain cranberry (*Vaccinium erythrocarpum*), minniebush (*Menziesia pilosa*), skunk currant (*Ribes glandulosum*), alderleaf viburnum (*Viburnum lantanoides*), and northern wild raisin (*Viburnum nudum* var. *cassinoides*).

Herbaceous layers are sparse to dense in spruce forests. Ferns are often an important ground cover in upland sites. Common fern associates are mountain woodfern (*Dryopteris campyloptera*), evergreen woodfern (*Dryopteris intermedia*), and fern-ally clubmosses (*Lycopodium* spp.). A small set of broadleaf herbs and graminoids is common in upland sites, including white woodsorrel (*Oxalis montana*), Canada mayflower (*Maianthemum canadense*), clustered goldenrod (*Solidago glomerata*), Fernald's bladder sedge (*Carex intumescens* var. *fernaldii*), and flattened oatgrass (*Danthonia compressa*). In wetlands, graminoids such as tawny cotton-grass (*Eriophorum virginicum*) and three-seeded sedge (*Carex trisperma* var. *trisperma*) may be characteristic.

Mosses, liverworts, and lichens often grow densely on live tree trunks, fallen logs, and the forest floor, giving these forests a distinctive carpeted appearance. Significant diversity in the red spruce ecosystem is contained within the bryophyte layer. More than 20 species of peatmoss (*Sphagnum* spp.) are associated with red spruce wetland types. Common bazzania liverwort (*Bazzania trilobata*) is typical in both upland and wetland communities of the central Appalachians, while several species of haircap mosses (*Polytrichum* spp.) and stairstep moss (*Hylocomium splendens*) are abundant in both regions.

4.1.1 Vegetation Patterns

Among spruce-dominated communities, the distinction between uplands and wetlands is a fundamental division leading to very different floristic and soil characteristics. Upland communities occur in a broad range of exposed and sheltered topographic settings at high elevations. Wetland communities tend to occur in frost pockets and along small headwater streams. They are especially abundant in the central Appalachians where the broad ridgetops of the Allegheny Highlands are dissected into gently sloping headwater basins. Much of the diversity of the red spruce ecosystem in the central Appalachians is found in its wetland plant communities. Spruce wetlands are rare and very limited in extent in the more heavily dissected southern Appalachians, but are an important component of community diversity.

Substrate structure, rock type, and soil type can be important factors differentiating communities. Shallow soils over bedrock and relict talus and periglacial boulder fields each support very different communities from deep soils. Soil chemistry matters as well. Additionally, wetlands with peat soils support unique communities. In the central Appalachians, most upland red spruce communities occur on acidic, nutrient-poor soils that develop on ridgetop-forming sandstone. Limestone or calcareous shale support more nutrient-demanding plant communities, with spruce-fir communities in wetland areas and richer northern hardwood components in upland areas. In the southern Appalachians, spruce and spruce-fir communities occur on a wide variety of soils formed from igneous and metamorphic rocks (see Chap. 3). Most are on highly acidic rocks such as quartzite, granitic gneiss, and felsic schist; no limestone occurs at high elevation in the southern Appalachians. Where examples also occur on amphibolite or calc-silicate, rocks which are associated with distinctive vegetation at lower elevations, the vegetation does not appear to be notably different. In the central Appalachians, most upland red spruce communities occur on Spodosols or soils that show evidence of spodic horizon development. Soils in the southern Appalachians appear to be a heterogeneous mix of primarily Entisols and Inceptisols, along with organic soils on bedrock. Extensive Spodosols have not been reported (see Chap. 3).

While descriptions of plant communities such as the International Vegetation Classification (IVC) generally focus on the best examples of natural vegetation presently available, disturbance history plays a large role in determining the vegetation growing on the landscape today. Anthropogenic disturbances, particularly logging and burning, have resulted in many successional communities such as mixed red spruce – northern hardwood forests and mixed red spruce – pitch pine forests (see Sect. 1.2.3 in Chap. 1). The local invasion of young red spruce into northern hardwood understories on middle-slope positions in the central Appalachians suggests that red spruce associations once occupied, and will in the future occupy, a wider variety of environments in the region (Fleming and Moorhead 1996). In the southern Appalachians, large expanses of spruce-fir forests that were logged and burned still do not support any forest after more than 100 years (see Sect. 1.2.3 in Chap. 1). Logging of swamps and subsequent burning of the peat substrate have left the central Appalachians with large expanses of successional shrubland areas where red spruce – hemlock – yellow birch swamps once existed. Natural disturbances also play a significant role in determining plant community composition, especially in central Appalachian wetland areas where periodic flooding and beaver activity support cyclic patterns of plant communities.

An interesting aspect of vegetation patterns in the southern Appalachians is the existence of several mountain ranges that extend to elevations that support red spruce in nearby ranges, around 1,680 m (5,510 ft), but which lack any red spruce. Fraser fir is also absent from these ranges, but some other tree, shrub, and herb species primarily found in red spruce forests are sometimes present. The regional climate of these ranges is not apparently different, but all have their highest peaks somewhat lower than the highest peaks in ranges that contain red spruce, generally below 1,800 m (6,000 ft). It is believed that these ranges contained red spruce and possibly Fraser fir forests at the end of the Pleistocene, but that during the warmer, drier Hypsithermal period between the Pleistocene and the present, they were not cool enough to support the species (see Sect. 1.2.2 in Chap. 1). As community zones shifted uphill, red spruce as a species and as a community was "pushed off the top of the mountain" (Schafale 2024). This hypothesis assumes that the dispersal ability of red spruce is limited enough that it has not been able to disperse to these ranges in the thousands of years since that time.

4.1.2 Plant Communities and the International Vegetation Classification

The plant community descriptions in this chapter are based on the IVC, supplemented by studies focused on red spruce communities in West Virginia (Byers et al. 2007, 2010), and the state classification of North Carolina (Schafale 2024), the two states with most of the red spruce ecosystems in the region. The IVC (Jennings et al. 2004; Federal Geographic Data Committee 2008; Faber-Langendoen et al. 2018; NatureServe 2022) is a hierarchical standardized classification of plant communities which is intended to document biodiversity, rank conservation priorities, and provide a structure to detect ecosystem responses to threats such as invasive species, land use changes, and climate change. Individual state Natural Heritage Programs have often produced classifications to address natural community diversity within their jurisdiction, which match the IVC to varying degrees. West Virginia uses the same associations and names as the IVC. North Carolina uses different names but all communities in the red spruce ecosystem have a one-to-one correspondence with associations in the IVC. Both names are used in the description of these communities.

Thirty-two red spruce associations are documented in the IVC, including 20 upland red spruce associations and 12 wetland red spruce associations ranging from southeastern Canada southward to North Carolina and Tennessee. The central and southern Appalachians are home to 11 upland associations and 8 wetland associations. All central and southern Appalachian red spruce associations are at risk of extinction or elimination due to their restricted range, relatively few occurrences, recent and widespread declines, and threats such as introduced insects, acid deposition, and climate change.

Conservation status ranks for IVC associations have been developed and applied by NatureServe in cooperation with its network members including states, provinces, and tribes. "G" ranks indicate global (range-wide) conservation status and "S" ranks indicate subnational, e.g. state-level, conservation status. Conservation status ranks are defined in Table 4.1.

Table 4.1 NatureServeconservation status ranks for plant communities (NatureServe 2022)

Rank	Definition
GX, SX	Presumed Collapsed—Collapsed throughout its range (GX), or extirpated in the jurisdiction (SX), due to loss of key dominant and characteristic taxa and/or elimination of the sites and ecological processes on which the type depends. Not located despite intensive searches of historical sites and other appropriate habitat, and virtually no likelihood that it will be rediscovered
GH, SH	Possibly Collapsed—Known from only historical occurrences but still some hope of rediscovery range-wide (GH) or from the jurisdiction (SH). The association has been searched for unsuccessfully, but not enough to state with certainty that it is not present
G1, S1	Critically Imperiled—At very high risk of collapse (G1) or extirpation in the jurisdiction (S1) due to very restricted range, very few occurrences, very steep declines, very severe threats, or other factors
G2, S2	Imperiled—At high risk of collapse (G2) or extirpation in the jurisdiction (S2) due to restricted range, few occurrences, steep declines, severe threats, or other factors
G3, S3	Vulnerable—At moderate risk of collapse (G3) or extirpation in the jurisdiction (S3) due to a fairly restricted range, relatively few occurrences, recent and widespread declines, threats, or other factors
G4, S4	Apparently Secure—At fairly low risk of collapse (G4) or extirpation in the jurisdiction (S4) due to an extensive range and/or many occurrences, but with possible cause for some concern as a result of local recent declines, threats, or other factors
G5, S5	Secure—At very low or no risk of collapse (G5) or extirpation in the jurisdiction (S5) due to a very extensive range, abundant occurrences, and little to no concern from declines or threats
GNR, SNR	Unranked—Global or subnational rank not yet assessed
G#G#, S#S#	Range Rank—A numeric range rank (e.g., G1G2, S2S3). A question mark (?) is used to indicate uncertainty about the exact status of an association

4.2 Central Appalachians

Four upland red spruce forest associations and one upland woodland association occur in the central Appalachians. All five of the associations have high global and state conservation priority.

4.2.1 Upland Red Spruce Communities

Upland red spruce forests and woodlands occupy the highest, coolest, and wettest environments in the central Appalachians, topping the ridges and knobs of the Allegheny Mountains at elevations above 1,000 m (3,300 ft) and two knobs on North Fork Mountain in the Ridge and Valley above 1,190 m (3,900 ft). The Red Spruce/Southern Mountain Cranberry Forest association is found in the highest and wettest topographic positions, while the Red Spruce/Heath Rocky Woodland association may be just as high in elevation but is restricted to drier summits along the Allegheny Front. Throughout the middle and upper elevations, Red Spruce–Yellow Birch Forest is the most common association. The Red Spruce/Rhododendron Forest association is generally found in sheltered locations at middle and lower elevations within the red spruce zone. At lower elevations and grading into northern hardwood forests, the Red Spruce – Hemlock – Beech Forest association may extend to elevations as low as 875 m (2,870 ft). In the central Appalachians, red spruce forest and woodland associations are characterized by relatively gentle slopes and less disturbed landscapes, as compared to other forest associations.

Upland red spruce habitats are characterized by a short growing season, low temperatures, high precipitation, high relative humidity, and low evapotranspiration rates. Solar radiation is high. The coldest temperatures and highest precipitation are found in the high elevation Red Spruce/Southern Mountain Cranberry Forest association. The warmest temperatures are experienced by the lower elevation Red Spruce – Hemlock – Beech Forest association. Temperature extremes, including frost and summer heat, are greatest in the Red Spruce/Heath Rocky Woodland association that occurs on the exposed summits of the Allegheny Front.

Dominant species in upland red spruce communities are red spruce, yellow birch, and common bazzania liverwort. Other important species are red maple, eastern hemlock, mountain holly (*Ilex montana*), great laurel, southern mountain cranberry, evergreen woodfern, pellucid plait moss (*Hypnum imponens*), and common broom moss (*Dicranum scoparium*).

The flora of central Appalachian upland red spruce communities is characterized by a generally northern affiliation. A few of the species whose largest distributions occur to the north include the foundation species red spruce as well as white woodsorrel, painted trillium (*Trillidium undulatum*), and late lowbush blueberry (*Vaccinium angustifolium*). Mixed with the northern-affiliated species, and giving a unique character to the ecosystem, are several characteristic central and southern

Appalachian species such as Blue Ridge St. Johns-wort (*Hypericum mitchellianum*), minniebush, and southern mountain cranberry. One hundred sixty-six vascular plant species, 45 bryophyte species, and 105 fungi have been documented within upland red spruce vegetation plots in West Virginia (Byers et al. 2010).

Myxomycetes, or slime molds, have been documented in the red spruce ecosystem in West Virginia by Steven L. Stephenson (Appendix D in Byers et al. 2010). He noted that five species are possibly restricted to spruce and spruce-fir forests. One of these, *Diderma simplex*, is also a rare species. *Colloderma oculatum* and *Diderma roanense* are occasional, and *Barbeyella minutissima* and *Lepidoderma tigrinum* are abundant.

4.2.1.1 Red Spruce – Yellow Birch Forest

Association CEGL008501. *Picea rubens/Betula alleghaniensis/Bazzania trilobata* forest (Fig. 4.1). NatureServe Conservation Status: G2; VA: S1, WV: S2.

This is the typical forest encountered in the heart of red spruce habitat in the central Appalachians, with widespread distribution in the middle and upper elevations (1,070–1,400 m [3,500–4,600 ft]) of the red spruce zone in West Virginia. The community occurs on both gentle slopes bordering high-elevation valley floors and on more exposed ridge crests and rocky summits. The canopy typically has strong

Fig. 4.1 Example of the Red Spruce – Yellow Birch association. Photo by Elizabeth Byers

dominance by red spruce, with yellow birch next in importance. Shrub cover is sparse to moderate, herbs are sparse, and the forest floor often has high cover by bryophytes. In West Virginia, two common floristic variants of this community occur. In one variant, shrubs and herbs are almost entirely absent, and the dense canopy shades a luxuriant green carpet of the common bazzania liverwort. Another variant is characterized by post-fire increases in the shrub mountain laurel, which locally dominates the shrub layer. The local invasion of young red spruce into northern hardwood understories on middle-slope positions suggests that this association once occupied, or will in the future occupy, a wider variety of environments in the region (Fleming and Moorhead 1996).

This community is geographically and environmentally restricted, occurring primarily in the Allegheny Mountains of West Virginia (Greenbrier, Pendleton, Pocahontas, Randolph, and Tucker counties) and at two sites in the Ridge and Valley region of Pendleton County. It also occurs in scattered stands in Virginia (Highland and Rockingham counties; Stevens 1969) in the Alleghenies and adjacent Ridge and Valley region. Its former extent has been reduced by logging and subsequent fires (Allard and Leonard 1952; Clarkson 1964; Pielke 1981; Stephenson and Clovis 1983). This type has good short-term viability, healthy regeneration, and protected status of many stands, but is restricted to the highest elevations of the central Appalachians and only occupies a small portion of its former range.

Within the narrowly defined ecology of the red spruce zone in the central Appalachians, this community has a relatively broad ecological amplitude and a central position, and as a result it has no indicator species that differentiate it from other upland red spruce communities. Instead, the other four upland red spruce communities differentiate themselves from this community through their diagnostic species, which reflect their own particular niches along environmental gradients within the red spruce zone.

The canopy is strongly dominated by red spruce, with lower cover of yellow birch and red maple. Other trees that occasionally occur with low cover in the canopy include eastern hemlock, sweet birch (*Betula lenta*), striped maple (*Acer pensylvanicum*), eastern white pine (*Pinus strobus*), American mountain-ash (*Sorbus americana*), Allegheny serviceberry (*Amelanchier laevis*), and northern red oak (*Quercus rubra*). The subcanopy is very similar in dominance and composition to the canopy, with the occasional addition of Fraser magnolia, mountain maple (*Acer spicatum*), downy serviceberry (*Amelanchier arborea* var. *arborea*), and American beech.

The tall shrub stratum is strongly dominated by saplings of the regenerating canopy species, especially red spruce and yellow birch. Mountain holly, mountain laurel, and great laurel occur frequently in the tall shrub layer. Rarely, a small amount of American witch-hazel (*Hamamelis virginiana*), minniebush, or catberry (*Nemopanthus mucronatus*) may be present. The short shrub stratum is also dominated by regenerating canopy species, with mountain holly, southern mountain cranberry, mountain laurel, and great laurel present. Occasionally very low cover of minniebush, northern lowbush blueberry, velvetleaf blueberry (*Vaccinium myrtilloides*), black huckleberry (*Gaylussacia baccata*), common greenbrier (*Smilax rotundifolia*), and alderleaf viburnum may be present.

The herbaceous stratum often includes evergreen woodfern, white woodsorrel, and eastern hay-scented fern (*Dennstaedtia punctilobula*). Less common are painted trillium, Canada mayflower, mountain woodfern, tree clubmoss (*Lycopodium obscurum*), ghost pipe (*Monotropa uniflora*), white-edge sedge (*Carex debilis* var. *rudgei*), partridge-berry (*Mitchella repens*), and whorled aster (*Oclemena acuminata*). The non-vascular stratum is strongly dominated by common bazzania liverwort, with lesser amounts of pellucid plait moss, common broom moss, eastern haircap moss (*Polytrichum pallidisetum*), recurved brotherella moss (*Brotherella recurvans*), and white pincushion moss (*Leucobryum glaucum*).

Sites range from mesic to submesic and are characterized by the climatic conditions common to all red spruce types in the central Appalachians, including cold winter microclimates, low mean annual temperature, short growing seasons, frequent fog, and high annual precipitation, high relative humidity, and low evapotranspiration rates. Soils are typical of the red spruce zone, characterized as acidic, infertile, frigid silt or sandy loams, with thick surficial duff accumulations. Soils typically have high organic matter, high cation-exchange capacity, high exchangeable nitrogen, and generally lacking in micronutrients. Pennsylvanian sandstones of the Pottsville Group underlie this community at almost all sites, although outliers occur on older sandstone ridges to the east. See Chap. 3 for more details on soils supporting red spruce.

4.2.1.2 Red Spruce/Southern Mountain Cranberry Forest

Association CEGL007131. *Picea rubens – (Abies fraseri)/Vaccinium erythrocarpum/Dryopteris campyloptera/Hylocomium splendens* forest (Fig. 4.2). NatureServe Conservation Status: G2; NC: S2, TN: SNR, VA: S1, WV: S1.

This community occurs primarily in the southern Appalachians, but disjunct occurrences are present in the central Appalachians. There it is restricted to the very highest elevations and coolest climate niche within the red spruce zone in West Virginia, occurring primarily on ridgetops at elevations above 1,350 m (4,400 ft) with one at lower elevation occurrence (1,140 m [3,740 ft]) in a cold stream bottom. Extensive logging and fires significantly reduced the extent of this community type during the early part of the twentieth century. Well-developed, undisturbed examples of this community are extremely rare in the central Appalachians.

This community is characterized by a dense canopy of red spruce and much lower cover of yellow birch and red maple. Although Fraser fir is included parenthetically in the global name, and is abundant in Southern Appalachian examples, this species is not native to the central Appalachians and does not occur in stands of this association in West Virginia. Black cherry and American mountain-ash are occasionally present in the canopy. The subcanopy is much less dense, with yellow birch, red spruce, and red maple as co-dominants, and occasional low cover by American mountain-ash or American beech. The understory consists of sparse to dense southern mountain cranberry, on a luxuriant carpet of liverworts and mosses. Dominant species, with high constancy and cover, include red spruce, yellow birch, southern mountain

Fig. 4.2 Example of the Red Spruce/Southern Mountain Cranberry Forest association. Photo by Elizabeth Byers

cranberry, common bazzania liverwort, and pellucid plait moss. Diagnostic species include southern mountain cranberry, mountain holly, mountain woodfern, Clinton lily (*Clintonia borealis*), and painted trillium.

The shrub strata are dominated by southern mountain cranberry, which generally grows as a short shrub but can reach heights of more than 2 m (6.6 ft) where growing conditions are favorable. The tall shrub layer may also include mountain holly, striped maple, and regenerating canopy saplings. Occasionally, minniebush, great laurel, early azalea (*Rhododendron prinophyllum*), and alderleaf viburnum may be present in the tall shrub stratum. The short shrub stratum has a species distribution very similar to the tall shrub layer, including occasional trace amounts of late lowbush blueberry.

The herbaceous layer is sparse and includes evergreen woodfern, mountain woodfern, Canada mayflower, white woodsorrel, Clinton lily, and painted trillium. Less common herbaceous species include eastern hay-scented fern, treelike clubmoss (*Lycopodium dendroideum*), Appalachian rockcap fern (*Polypodium appalachianum*), running clubmoss (*Lycopodium clavatum*), tree clubmoss, whorled aster, and cinnamon fern (*Osmunda cinnamomea* var. *cinnamomea*).

Bryophytes and lichens make up a considerable percent of the vegetative coverage in this community, occurring on the surface of the soil, trees, and fallen logs.

Bryophyte cover is high in this forest type, with common bazzania liverwort dominant, followed by pellucid plait moss, common broom moss, and recurved brotherella moss. Other common bryophytes include white pincushion moss, eastern haircap moss, and beaked bow moss (*Dicranodontium denudatum*). The many-forked cladonia lichen (*Cladonia furcata*) is common on the ground, and fruticose or foliose lichens are common on tree trunks.

Soils are typical of the red spruce zone, with high organic matter and nitrogen, low pH, and generally low in micronutrients. Pennsylvanian sandstones of the Pottsville Group, often forming a resistant caprock on the ridges, underlie this community at most sites. The expression of this community farther south is described in Sect. 4.3.1.3.

4.2.1.3 Red Spruce – (Eastern Hemlock)/Great Laurel Forest

Association CEGL006152. *Picea rubens* – (*Tsuga canadensis*)/*Rhododendron maximum* forest (Fig. 4.3). NatureServe Conservation Status: G2G3; NC: S1, TN: SNR, WV: S2.

Fig. 4.3 Example of the Red Spruce – (Eastern Hemlock)/Great Laurel Forest association. Photo by James Vanderhorst

This association occurs in both the central and southern Appalachians. In the central Appalachians of West Virginia, it is environmentally restricted within a somewhat limited geographic range. Its former extent was probably limited to moist, sheltered areas in the lower elevation range of red spruce, and has been reduced by logging and subsequent fires (Allard and Leonard 1952; Clarkson 1964; Pielke 1981; Stephenson and Clovis 1983).

This species-poor forest type occurs primarily at low and middle elevations within the central Appalachian red spruce zone. It generally grows on moist, protected landforms and can include slope forests, boulder fields, ravines, and occasional ridges. This community has the lowest species richness of the central Appalachian red spruce forest types. It has the highest cover of tall shrubs, almost exclusively great laurel, and the lowest cover of herbaceous species.

The canopy layer is dominated by red spruce with lower cover by eastern hemlock, red maple, and yellow birch. Occasionally, sweet birch and downy serviceberry are also present in the canopy. The subcanopy is dominated by yellow birch and red spruce, with lower cover of eastern hemlock and red maple. Other subcanopy trees may include American mountain-ash, downy serviceberry, sweet birch, and Fraser magnolia.

The tall shrub layer is dense and strongly dominated by great laurel. Low cover of mountain holly and various tree seedlings may intermix, and occasionally a few shrubs of mountain laurel occur. The short shrub layer is much less dense but has the same dominance and species distribution as the tall shrubs. Occasional low cover of southern mountain cranberry, minniebush, and late lowbush blueberry may occur in the short shrub stratum.

The herbaceous layer is extremely sparse, with evergreen woodfern, mountain woodfern, whorled aster, white woodsorrel, and Appalachian rockcap fern. The bryophyte layer is strongly dominated by common bazzania liverwort, with much lower cover of pellucid plait moss, white pincushion moss, recurved brotherella moss, and common broom moss.

Soils are typical of the red spruce zone, with high organic matter and nitrogen, low pH, and generally low micronutrients. Pennsylvanian sandstones of the Pottsville Group underlie this community at most sites, with rare extensions onto the adjacent Mississippian Mauch Chunk shale. The expression of this community farther south is described in Sect. 4.3.1.8.

4.2.1.4 Red Spruce – Hemlock – Beech Forest

Association CEGL006029. *Picea rubens – Tsuga canadensis – Fagus grandifolia/ Dryopteris intermedia* forest (Fig. 4.4). NatureServe Conservation Status: G3; WV: S3. This association is also described in the Spodic Shale Upland Conifer Forest Ecological Site Description discussed in Chap. 3.

This community occurs in the Allegheny Mountain region of West Virginia, at elevations above 850 m (2,800 ft). Small additional stands may occur in the Allegheny

Fig. 4.4 Example of the Red Spruce – Hemlock – Beech Forest association. Photo by James Vanderhorst

Mountain region of Maryland and possibly in Pennsylvania. This forest type occupies relatively warmer, lower elevations within the red spruce zone in the central Appalachians. At the lower elevation margin of this community type, the forest grades into adjacent northern hardwoods forest, with a steadily decreasing spruce component. The Red Spruce – Hemlock – Beech forest association is characterized by a mixed canopy of red spruce, eastern hemlock, and deciduous trees, with a dense subcanopy, generally sparse shrub layer, and relatively diverse, fern-dominated herb layer. A less common variant of this association extends into the middle and upper elevations of the spruce zone along slightly richer substrates underlain by shale and limestone. Floristically, the richer variant has less hemlock but is otherwise quite similar to the lower elevation community. The similarities are probably due to a slight amelioration of environmental stresses at lower elevations and/or on richer substrates, which allows other species to compete successfully with red spruce.

Diagnostic species include American beech, eastern hemlock, black cherry, sugar maple (*Acer saccharum* var. *saccharum*), sweet birch, Fraser magnolia, striped maple, common greenbrier, evergreen woodfern, Canada mayflower, white wood-sorrel, and partridge-berry. Species richness is relatively high for the red spruce system, with the highest values on shale and limestone substrates.

The canopy is dominated by red spruce with eastern hemlock, yellow birch, and red maple. Less common are American beech, black cherry, and sweet birch, and occasional canopy species include sugar maple, Fraser magnolia, tuliptree (*Liriodendron tulipifera*), and cucumber magnolia (*Magnolia acuminata*). The subcanopy is relatively lush and exhibits co-dominance by several species, including red spruce, yellow birch, eastern hemlock, striped maple, red maple, American beech, and sugar maple. Less common subcanopy species include cucumber magnolia, sweet birch, Fraser magnolia, black cherry, hawthorn (*Crataegus* spp.), Allegheny serviceberry, white ash (*Fraxinus americana*), and American basswood (*Tilia americana*).

The shrub layers are relatively sparse in this community compared to other upland spruce communities. The tall shrub layer is dominated by regenerating tree saplings, especially red spruce, American beech, eastern hemlock, and striped maple. Shrubs that are occasionally present include mountain holly, great laurel, mountain laurel, and alderleaf viburnum. The short shrub layer is sparse and consists largely of regenerating tree species, with occasional presence of southern mountain cranberry, mountain holly, common greenbrier, mountain laurel, and minniebush.

This community has the richest herbaceous layer of any of the upland spruce communities in West Virginia and is strongly dominated by evergreen woodfern. Other common species include Canada mayflower, white woodsorrel, partridgeberry, eastern hay-scented fern, painted trillium, Indian cucumber-root (*Medeola virginiana*), whorled aster, mountain woodfern, nightcaps (*Anemone quinquefolia*), treelike clubmoss, jack-in-the-pulpit (*Arisaema triphyllum*), white-edge sedge, flattened oatgrass, sweet-scented bedstraw (*Galium triflorum*), shining clubmoss (*Huperzia lucidula*), running clubmoss, tree clubmoss, roundleaf orchid (*Platanthera orbiculata*), Appalachian rockcap fern, and heartleaf foamflower (*Tiarella cordifolia*). The non-vascular stratum has the lowest bryophyte cover among upland red spruce communities in West Virginia. Common bazzania liverwort and pellucid plait moss are the dominant species, followed by common broom moss, delicate fern moss (*Thuidium delicatulum*), recurved brotherella moss, and beaked bow moss.

Soils are typical of the red spruce zone, with high organic matter and nitrogen, low pH, and generally low micronutrients. This community has high concentrations of the macronutrient sulfur and the micronutrient iron. A variety of geologic formations underlie this community type, including Pennsylvanian Pottsville sandstone, Mississippian Mauch Chunk shale and Greenbrier limestone, Devonian Hampshire shale, and Silurian Tuscarora sandstone.

4.2.1.5 Red Spruce/Heath Rocky Woodland

Association CEGL006254. *Picea rubens/Kalmia latifolia – Menziesia pilosa* woodland (Fig. 4.5). NatureServe Conservation Status: G2; WV: S1.

This community is a West Virginia endemic occupying about 5 km^2 (1.9 mi^2) and restricted to a 50 km (31 mi) linear band along the Allegheny Front with a small outlier on Panther Knob in the Ridge and Valley region. This association has a limited habitat on exposed acidic bedrock and talus, within the restricted cool, moist climate of the red spruce zone. It is more threatened than other spruce communities not only because of its limited acreage but because of unique threats such as vacation home

Fig. 4.5 Example of the Red Spruce/Heath Rocky Woodland association. Photo by Elizabeth Byers

development on these highly scenic rocky ridgetops, industrial wind development along the Allegheny Front, and strip mining.

This red spruce rocky woodland occupies high elevations in the drier northeastern part of the red spruce range in West Virginia and is characterized by a stunted, open canopy of red spruce, with abundant heath shrubs and lichens. Large seasonal temperature variations characterize this habitat. In some areas, this community has been impacted by fire (for blueberry production) in the period following European settlement, and Native Americans may also have practiced burning in this community. Fire is likely to convert this community to the more common Blueberry Heath Barren (CEGL003958). Following the extensive logging and slash fires of 1885–1920, this community expanded to cover previously forested areas along the Allegheny Front. Soils are slowly accumulating again in the absence of widespread fires, with both in-situ organic matter deposition and windblown soil deposits as likely mechanisms. In more sheltered areas this community is following a slow successional pathway back to red spruce forest. However, the core of this red spruce community is maintained naturally by acidic bedrock outcrops and their shallow, infertile, and sometimes waterlogged soils.

Dominant species, with high constancy and cover, include red spruce and mountain laurel. Diagnostic species include mountain laurel, pitch pine, Allegheny serviceberry, catberry, late lowbush blueberry, minniebush, black huckleberry, black chokeberry (*Aronia melanocarpa*), Appalachian gooseberry (*Ribes rotundifolium*), fire cherry (*Prunus pensylvanica* var. *pensylvanica*), eastern teaberry (*Gaultheria procumbens*), bracken fern (*Pteridium aquilinum*), Appalachian rockcap fern, grey reindeer lichen (*Cladonia rangiferina*), and lesser rocktripe lichen (*Umbilicaria muehlenbergii*). Species richness in this type is the highest within the upland red spruce forest and woodland types with most of the diversity in the shrub strata.

This upland spruce type has a low, stunted, open woodland canopy strongly dominated by red spruce, with much lower cover of pitch pine, red maple, yellow birch, Allegheny serviceberry, American mountain-ash, and eastern hemlock. The tall shrub layer is dominated by mountain laurel with catberry, great laurel, and regenerating tree saplings. The short shrub layer is diverse and often includes late lowbush blueberry, minniebush, black huckleberry, catberry, black chokeberry, great laurel, and southern mountain cranberry. Occasional shrubs include Appalachian gooseberry, fire cherry, eastern teaberry, skunk currant, velvetleaf blueberry, mountain holly, striped maple, mountain maple, American witch-hazel, early azalea, and northern wild raisin.

The herbaceous layer is sparse and usually dominated by Canada mayflower, bracken fern, and Appalachian rockcap fern. Additional herbaceous species may include crinkled hairgrass (*Deschampsia flexuosa* var. *flexuosa*), wild sarsaparilla (*Aralia nudicaulis*), trailing arbutus (*Epigaea repens*), and brownish sedge (*Carex brunnescens*). The non-vascular stratum has significant diversity, much of which occurs as abundant crustose lichens on the rocky substrate. Nonvascular cover often includes grey reindeer lichen, lesser rocktripe lichen, pellucid plait moss, white pincushion moss, red-stemmed feather moss (*Pleurozium schreberi*) and common toadskin lichen (*Lasallia papulosa*), in addition to crustose lichens.

Soil is sparse in this rocky woodland, generally occurring in cracks and pore spaces between rocks. Some of the soil may be residual, sheltered from major fires in the last century. Soil chemistry characteristics include the highest values for organic matter within the red spruce zone. The macronutrients potassium and magnesium and the micronutrient manganese are high compared to other upland spruce communities, although they are low compared to other forest soils in West Virginia. Sulphur and iron are very low. Pennsylvanian sandstones of the Pottsville Group underlie this community.

4.2.2 Wetland Red Spruce Communities

Wetland red spruce communities in the central Appalachians are embedded in matrix upland forest systems, except where the land is kept open by human actions. Often, they occur adjacent to or in mosaics with other high elevation wetland types (Byers et al. 2007). Red spruce wetlands occur primarily in flat-lying headwater basins, ranging in size from <1 ha (2.5 ac) to about 3,000 ha (7,400 ac) of contiguous wetland in Canaan Valley. The flat-lying basins where many wetlands occur function as frost pockets, catching and pooling cold air that drains from the surrounding uplands. The wetlands are drained by low-gradient, meandering, small headwater streams. Drainage is impounded in these high, flat-lying basins by natural dams of resistant sandstone. Rainfall is plentiful, averaging about 1,300 mm/year (51 in/year). Communities in this system may have substrates of shallow to deep peat or, less commonly, mineral soil. Soils are acidic to circumneutral (i.e., nearly neutral, pH of 6.5 to 7.5).

Forested swamps occupy the less disturbed margins or slightly higher areas in open wetlands. Ombrotrophic bogs are rare but occur in undisturbed portions of a few of the larger wetlands. The more central, flood- or beaver-influenced portions contain shrub swamps, sedge fens, wet meadows, and open marshes. Undisturbed examples exist (e.g., Cranberry Glades), where old-growth swamp buffers the central peatlands, which have been dated to 10,000 years old (Darlington 1943). In pre-settlement time, some wetland mosaics in this system had significant forested components (e.g., Canaan Valley and Cranesville Swamp in West Virginia, Finzel Swamp and Hammel Glades in Maryland), while others (e.g., Cranberry Glades, Big Run Bog) were largely open peatlands with forested swamp only on the margins.

4.2.2.1 Red Spruce – Yellow Birch/Mannagrass Swamp

Association CEGL006556. *Picea rubens – Acer rubrum/Ilex verticillata* swamp forest (Fig. 4.6). NatureServe Conservation Status: G3; PA: S3, VA: S1, WV: S1.

This mixed woodland or forested swamp occurs on saturated and temporarily to semi-permanently flooded soils of the Allegheny Mountains region of West Virginia, at elevations between 770 and 1,220 m (2,500 and 4,000 ft). It also occurs in extreme west-central Virginia (Highland County), and outliers are reported from the Pocono Plateau and Ridge and Valley provinces of Pennsylvania (Fike 1999). In West Virginia, it is a small-patch type that occupies flat to very gently sloping land (0–2° slopes) in floodplains of streams and on the margins between upland forest and open wetland. When this community occurs in a floodplain setting, it often is characterized by fluvial morphological features such as backwater sloughs, levees, and meander scrolls. Microtopography is characterized by irregular moss-covered hummocks formed over buttressed tree roots, woody stem clusters, tip-up mounds, nurse logs, and decaying wood. Soils are poorly drained muck, peat, or silt/clay loam, with an average pH of 4.2.

The community is characterized by an open to closed canopy of red spruce, eastern hemlock, and yellow birch, with occasional additions of red maple, black ash (*Fraxinus nigra*), balsam fir, eastern white pine, or blackgum (*Nyssa sylvatica*). The subcanopy is similar in composition to the canopy. The tall-shrub layer is characterized by speckled alder, great laurel, the regenerating canopy species, and occasionally common winterberry. The short-shrub layer is sparse with species composition similar to the tall-shrub layer. The herbaceous layer is diverse and variable, including slender mannagrass (*Glyceria melicaria*), orange jewelweed (*Impatiens capensis*), cinnamon fern, American golden-saxifrage (*Chrysosplenium americanum*), arrowleaf tearthumb, rice cutgrass, white turtlehead (*Chelone glabra*),

Fig. 4.6 Example of the Red Spruce – Yellow Birch/Mannagrass Swamp community. Photo by Elizabeth Byers

skunk-cabbage, Canada mayflower, marsh-marigold (*Caltha palustris* var. *palustris*), sensitive fern (*Onoclea sensibilis*), and bristly-stalked sedge (*Carex leptalea* ssp. *leptalea*). Bryophytes are abundant and diverse, dominated by peatmosses (*Sphagnum palustre, Sphagnum fallax, Sphagnum squarrosum, Sphagnum divinum,* and *Sphagnum girgensohnii*) filling the mucky hollows, Appalachian leafy moss (*Rhizlomnium appalachianum*) in the seepy areas, and pellucid plait moss and common broom moss blanketing the woody hummocks.

4.2.2.2 Red Spruce – Hemlock/Great Laurel Swamp Forest

Association CEGL006277. *Picea rubens* – (*Tsuga canadensis*)/*Rhododendron maximum* swamp forest (Fig. 4.7). NatureServe Conservation Status: G2?; MD: SNR, NC: S1, NY: S1, PA: S1, TN: SH, VA: S1, WV: S2.

This red spruce wetland forest occurs in small patches in the central and southern Appalachian Mountains north to the high Allegheny Plateau of New York and Pennsylvania. In the central Appalachians, this acidic conifer swamp occurs on saturated, temporarily flooded, and semi-permanently flooded soils in headwater basins of the Allegheny Mountains region of West Virginia and Maryland, at elevations from 770 to 1,150 m (2,500–3,800 ft). It occupies flat to very gently sloping land (0–1° slopes)

Fig. 4.7 Example of the Red Spruce – Hemlock/Great Laurel Swamp Forest community. Photo by Brian Streets

along small streams, often in mixed wetland mosaics. It is a small-patch type which forms islands in open shrublands or peatlands, and occurs in backswamp locations, separated from the adjacent stream by a levee. It is also found along the margins of beaver-influenced wetlands. Microtopography is characterized by interfingering of wetter and drier areas, with irregular mossy hummocks formed over tree roots, tip-up mounds, nurse logs, and decaying wood. Tree roots are often buttressed and may form thick "root rafts" on top of mucky soils. Hollows are typically filled with standing water or muck. Anthropogenic disturbance includes historical logging and fires around 1900 and occasionally more recent logging. Soils are poorly drained muck, peat, or organic-rich silt loam, with an average pH of 3.8.

This evergreen swamp is characterized by an open to closed canopy of inundation-stressed trees and a dense great laurel shrub layer over a sparse herbaceous layer and abundant bryophytes. The canopy is dominated by red spruce and eastern hemlock with lower cover of red maple and yellow birch, and occasional presence of blackgum, American larch (*Larix laricina*), or eastern white pine. The subcanopy is dominated by eastern hemlock with yellow birch, red maple, and red spruce. Species that may be present with low cover in the subcanopy include white ash, sweet birch, blackgum, cucumber magnolia, Allegheny serviceberry, northern spicebush (*Lindera benzoin*), tuliptree, and American mountain-ash. The tall shrub layer is dominated by great laurel. Other commonly occurring species in the tall-shrub layer include common winterberry, eastern hemlock, red spruce, and American mountain-ash. Low cover of catberry and mountain laurel may be present. The short shrub layer species composition is similar to that of the tall-shrub stratum, with the occasional addition of northern wild raisin. The herbaceous ground layer is sparse and variable and often includes three-seeded sedge, cinnamon fern, red maple, skunk-cabbage, eastern hayscented fern, slender mannagrass, and partridge-berry. Nonvascular plants are dominated by peatmosses (*Sphagnum palustre, Sphagnum fallax, Sphagnum girgensohnii, Sphagnum recurvum, Sphagnum divinum* and *Sphagnum papillosum*), common bazzania liverwort, and pellucid plait moss.

4.2.2.3 Balsam Fir/Oatgrass Swamp

Association CEGL006592. *Abies balsamea – Picea rubens/Danthonia compressa – Lycopodium* spp./*Sphagnum* spp. swamp forest (Fig. 4.8). NatureServe Conservation Status: G2; WV: S2.

This acidic conifer woodland or forested swamp occurs on moist to saturated soils in headwater basins in the Allegheny Mountains region of West Virginia, at elevations from 960–1,130 m (3,150–3,700 ft). It is a small-patch community fed by slow seepage and rainfall. It occupies gently sloping land (1–10° slopes) along small headwater streams and in mixed wetland mosaics, often at the base of upland

Fig. 4.8 Example of the Balsam Fir/Oatgrass Swamp community. Photo by James Vanderhorst

slopes. Microtopography is characterized by irregular hummocks formed over tree roots, tip-up mounds, nurse logs, and decaying wood. Soils are moderately to poorly drained loamy soils with mottling in the upper 20 cm (8 in) and occasional gleyed horizons.

The canopy is open to closed and dominated by balsam fir and red spruce, with occasional presence of black cherry. Older balsam fir trees are stressed by the balsam woolly adelgid (*Adelges piceae*). The subcanopy is dominated by balsam fir with smaller amounts of red spruce, red maple, and yellow birch. The tall-shrub layer is also dominated by balsam fir and red spruce. The short-shrub layer contains regenerating canopy species and locally abundant velvetleaf blueberry and bushy St. John's-wort. The herbaceous ground layer is fairly diverse, typically including 25–50 species. The most abundant species are flattened oatgrass, tree clubmoss, and bristly dewberry. Other herbaceous species with high constancy include the regenerating canopy species, running clubmoss, long sedge (*Carex folliculata*), eastern hay-scented fern, bracken fern, and whorled aster. Nonvascular plants are dominated by peatmoss species and haircap moss species; white pincushion moss is also common. The community is characterized by a number of diagnostic species that highlight the slightly drier habitat of this swamp type, including white-edge sedge, hawthorn, running clubmoss, fan clubmoss (*Lycopodium digitatum*), partridge-berry, haircap mosses, black cherry, and bracken fern.

4.2.2.4 Balsam Fir/Winterberry Swamp

Association CEGL006591. *Abies balsamea – Picea rubens/Ilex verticillata/Sphagnum* spp. swamp forest (Fig. 4.9). NatureServe Conservation Status: G2; WV: S1.

This acidic conifer woodland swamp occurs on temporarily to semi-permanently flooded soils in frost-pocket headwater basins in the Allegheny Mountains region of West Virginia, at elevations from 980–1,120 m (3,200–3,600 ft). It is a small-patch community fed by seepage, occasional overflow from low-gradient headwater streams, and rainfall. It occupies flat to very gently sloping land (0–3° slopes) along small headwater streams, often in mixed wetland mosaics. The microtopography is characterized by interfingering of wetter and drier areas, with irregular hummocks formed over tree roots, tip-up mounds, nurse logs, and decaying wood. Beaver activity periodically causes inundation stress that kills trees in this community, and there is evidence of migration of balsam fir populations, possibly in response to fluctuating hydrology. Older balsam fir trees are also stressed by the balsam woolly adelgid. Soils are poorly to very poorly drained clayey or mucky soils with mottling in the upper 20 cm (8 in) and occasional gleyed horizons.

The canopy is open and dominated by stunted, inundation-stressed balsam fir and red spruce, with occasional presence of yellow birch and red maple. The subcanopy is dominated by balsam fir and eastern hemlock, with smaller amounts of red spruce and red maple. Serviceberry species may be present with very low cover in the subcanopy. The shrub layer is dominated by common winterberry, the regenerating

Fig. 4.9 Example of the Balsam Fir/Winterberry Swamp community. Photo by Leah Ceperley

canopy species, and speckled alder. Low cover of great laurel, bushy St. John's-wort, and velvetleaf blueberry may be present. The herbaceous ground layer is dense and diverse, typically including 30–40 species. The most abundant species are nodding sedge, long sedge, and bristly dewberry. Other herbaceous species with high constancy include regenerating woody species and arrowleaf tearthumb, cinnamon fern, stiff marsh bedstraw, crested shieldfern (*Dryopteris cristata*), three-seeded sedge, orange jewelweed, marsh blue violet (*Viola cucullata*), Fraser's marsh-St. John's-wort (*Triadenum fraseri*), soft rush, awl-fruit sedge, fowl mannagrass (*Glyceria striata*), dotted smartweed (*Polygonum punctatum*), American mannagrass (*Glyceria grandis* var. *grandis*), white woodsorrel, evergreen woodfern, and Canada mayflower. Nonvascular plants are dominated by peatmosses. Indicator species include hoary sedge (*Carex canescens*), long sedge, nodding sedge, American mannagrass, and Fraser's marsh-St. John's-wort.

4.2.2.5 Balsam Fir – Black Ash Swamp

Association CEGL006003. *Fraxinus nigra – Abies balsamea/Rhamnus alnifolia* swamp forest (Fig. 4.10). NatureServe Conservation Status: G1; PA: S1, WV: S1.

This community is a lush, circumneutral, seepage-fed, mixed woodland or forested swamp of the Allegheny Mountains in West Virginia. It is a late-successional, small-patch community limited to frost-pocket wetlands on the Mississippian Greenbrier

Fig. 4.10 Example of the Balsam Fir – Black Ash Swamp community. Photo by Elizabeth Byers

limestone, at elevations from 960 to 1,000 m (3,150–3,200 ft). The community occurs in flat headwater basins and backswamps along small streams. Microtopography is characterized by irregular hummocks formed over buttressed tree roots, tip-up mounds, nurse logs, and decaying wood. Soils are poorly drained muck or organic-rich silt loam over mottled or gleyed silty clay.

This rich swamp provides habitat for several rare shade-tolerant calciphile wetland species. The canopy is open to closed and dominated by stunted, inundation-stressed balsam fir, black ash, and eastern hemlock. Black ash is rapidly being lost from this community due to the emerald ash borer (*Agrilus planipennis*). Eastern hemlock and balsam fir are stressed by the hemlock woolly adelgid (*Adelges tsugae*) and balsam woolly adelgid, respectively. The subcanopy is dominated by balsam fir with varying amounts of yellow birch, eastern hemlock, red spruce, and black ash. The tall-shrub layer is dominated by speckled alder with locally abundant common winterberry and sometimes vigorous red spruce regeneration in this stratum. The short-shrub layer is dominated by alderleaf buckthorn (*Rhamnus alnifolia*) or rarely by silky dogwood (*Cornus amomum*). The herbaceous ground layer is extensive and diverse, typically including >50 species. The most abundant species are brome-like sedge (*Carex bromoides* ssp. *bromoides*) and fowl mannagrass. Other common species include Jack-in-the-pulpit, marsh-marigold, nodding sedge, crested shield-fern, orange jewelweed, soft rush, northern bugleweed, Canada mayflower, cinnamon fern, golden groundsel (*Packera aurea*), arrowleaf tearthumb, bristly dewberry, and roughleaf goldenrod. Grove meadow grass (*Poa alsodes*) is locally abundant. Dominant bryophytes are peatmosses, pellucid plait moss, delicate fern moss, and common bazzania liverwort. Diagnostic species are brome-like sedge, glade spurge (*Euphorbia purpurea*), black ash, purple avens (*Geum rivale*), bog Jacob's-ladder (*Polemonium vanbruntiae*), and alderleaf buckthorn.

4.2.2.6 Red Spruce/Heath Peat Woodland

Association CEGL006588. *Picea rubens/Rhododendron maximum – Kalmia latifolia/Eriophorum virginicum/Sphagnum* spp. swamp forest (Fig. 4.11). NatureServe Conservation Status: G2G3; PA: S2S3, WV: S2.

This acidic conifer woodland occurs on saturated soils in headwater basins of the Allegheny Mountains region of West Virginia, at elevations from 860 to 1,300 m (2,800–4,200 ft), and extends into eastern Pennsylvania. In West Virginia, it is a small-patch type that occupies flat to very gently sloping land (0–1° slopes) along the margins of open peatlands and in seepage-fed portions of wetland mosaics. Seepage from adjacent upland forest and the high water table in adjacent open wetlands keep the community wet enough to kill trees during wet years, leaving numerous snags. Microtopography is characterized by a mix of rounded peat hummocks and irregular moss-covered hummocks formed over tree roots, woody stem clusters, and decaying

Fig. 4.11 Example of the Red Spruce/Heath Peat Woodland community. Photo by Elizabeth Byers

wood. Bedrock is typically sandstone or shale. Soils are moderately to very poorly drained peat, underlain by clay-rich deposits, with an average pH of 3.6.

This community is characterized by an open canopy of stunted, inundation-stressed trees with a diverse shrub and herb layer growing on hummock-forming bryophytes. The canopy is dominated by eastern hemlock and red spruce, occasionally including low cover of red maple or yellow birch. Canopy height is <15 m (50 ft) and sometimes as low as 5 m (16 ft), approaching the threshold between woodland and shrubland physiognomy. The tall-shrub layer includes the canopy species along with great laurel, mountain laurel, catberry, and common winterberry. Other species that occasionally occur with low cover in the tall-shrub layer include northern wild raisin, mountain holly, red chokeberry (*Aronia arbutifolia*), red maple, American mountain ash, Allegheny serviceberry, and American witch-hazel. The short-shrub layer is similar in composition to the tall-shrub layer. The herbaceous layer typically includes bristly dewberry, cinnamon fern, bog goldenrod, and the regenerating canopy species. Species with lower cover often include three-seeded sedge, roundleaf sundew (*Drosera rotundifolia* var. *rotundifolia*), eastern hay-scented fern, nodding sedge, slender mannagrass, tawny cotton-grass, whorled aster, and soft rush. A lush bryophyte layer is dominated by peatmosses (*Sphagnum fallax, Sphagnum recurvum, Sphagnum divinum, Sphagnum affine,* and *Sphagnum capillifolium*) and common haircap moss (*Polytrichum commune*). Most of the diversity in this community type is in the shrub and herb layers.

4.2.2.7 Red Spruce/Three-Seeded Sedge Peat Woodland

Association CEGL006590. *Picea rubens/Carex trisperma/Sphagnum* spp. – *Polytrichum* spp. swamp forest (Fig. 4.12). NatureServe Conservation Status: G2; WV: S2.

This acidic conifer woodland swamp occurs on saturated and temporarily flooded organic soils in headwater basins of the Allegheny Mountains region of West Virginia, at elevations from 1,000 to 1,430 m (3,200–4,600 ft). It is a small-patch community that occupies flat to very gently sloping land (0–2° slopes) along the margins of open peatlands, forming narrow spits, fingers, or islands. It also occurs in peaty depressions within high plateau spruce forests. Microtopography is characterized by a mix of rounded peat hummocks and irregular moss-covered hummocks formed over tree roots, woody stem clusters, tip-up mounds, nurse logs, and decaying wood. Bedrock is typically sandstone or occasionally shale and may be encountered at <20 cm (8 in) depth. Soils are poorly drained muck, peat, or organic-rich silt/clay loam. Depth of organic soil varies greatly from 5 to 120 cm (2–47 in), with an average pH of 3.5.

The community is characterized by an open canopy of red spruce and a sparse shrub layer growing on irregular hummocks, with swales and hollows occupied by three-seeded sedge and peat-forming bryophytes. Additional canopy species that occur occasionally include eastern white pine, red maple, Allegheny serviceberry, pitch pine, American mountain ash, and yellow birch. The tall-shrub layer is also

Fig. 4.12 Example of a Red Spruce/Three-seeded Sedge Peat Woodland community. Photo by Elizabeth Byers

dominated by red spruce with occasional low cover by northern wild raisin, catberry, and black chokeberry. The short-shrub layer includes red spruce, velvetleaf blueberry, and late lowbush blueberry. The herbaceous layer is dominated by three-seeded sedge and may include low cover by bristly dewberry, cinnamon fern, creeping snowberry (*Gaultheria hispidula*), small cranberry (*Vaccinium oxycoccos*), and white-edge sedge. Bryophytes have high ground cover dominated by peatmosses with lesser amounts of haircap mosses, pellucid plait moss, and white pincushion moss.

4.2.2.8 Red Spruce/Southern Mountain Cranberry Swamp

Association CEGL006593. *Picea rubens/Vaccinium erythrocarpum/Sphagnum* spp. – *Bazzania trilobata* swamp forest (Fig. 4.13). NatureServe Conservation Status: G2; WV: S2.

This acidic conifer woodland or forested swamp occurs on saturated and temporarily flooded soils in headwater basins of the Allegheny Mountains region of West Virginia, at elevations from 1,140 to 1,400 m (3,700–4,600 ft) above sea level. It is a small-patch community maintained by slow seepage, low-energy overflow inundation, and rainfall. It occupies gently sloping land (0–6° slopes) on the margins between upland spruce forest and open beaver-influenced headwater

Fig. 4.13 Example of a Red Spruce/Southern Mountain Cranberry Swamp community. Photo by Brian Streets

wetlands, and alluvial bottoms along high-elevation meandering streams. Standing snags are common, the result of inundation stress during wet years and beaver-influenced water table fluctuations. Microtopography is characterized by irregular hummocks formed over tree roots, tip-up mounds, decaying wood, and around woody stem clusters. Soils are somewhat poorly drained peat, muck, or organic-rich mottled silt loam, generally underlain by clay.

The canopy is open to closed and strongly dominated by red spruce. The subcanopy is also dominated by red spruce, with lower cover by red maple and yellow birch. The tall-shrub layer is similar in composition to the subcanopy, with the occasional low cover by mountain holly, great laurel, or American mountain ash. The short-shrub layer is dominated by the diagnostic species southern mountain cranberry, with locally abundant mountain laurel. The herbaceous ground layer is variable and sparse; evergreen woodfern or New York fern (*Thelypteris noveboracensis*) may be locally abundant, with low cover of cinnamon fern, Canada mayflower, Clinton lily, and three-seeded sedge. Bryophytes have high ground cover with common bazzania liverwort blanketing the abundant downfall and peatmosses (*Sphagnum girgensohnii, Sphagnum fallax, Sphagnum palustre,* and *Sphagnum rubellum*) filling the mucky hollows; haircap mosses and pellucid plait moss are also common.

4.2.2.9 Pitch Pine – Red Spruce/Heath Peat Woodland

Association CEGL006587. *Pinus rigida – Picea rubens/Viburnum nudum* var. *cassinoides/Sphagnum* spp. swamp woodland (Fig. 4.14). NatureServe Conservation Status: G1G2; NY: SNR, WV: S1.

This acidic dwarf woodland swamp occurs on saturated and temporarily flooded soils in the Allegheny Mountains region of West Virginia, at elevations from 981 to 1,220 m (3,200–4,000 ft). This community occurs in narrow bands (10–200 m [33–650 ft] wide) immediately west of the Allegheny Front, between the upland forest of the summit ridge and open peatlands, with one stand to the west in the headwaters of Lindy Run. It is a small-patch type that occupies flat-lying land (<1° slope). Microtopography is characterized by irregular moss-covered hummocks formed over tree roots, woody stem clusters, tip-up mounds, and decaying wood. Soils are poorly drained peat. The underlying acidic sandstone bedrock (Pennsylvanian Allegheny and Pottsville Formations) is generally encountered at <70 cm (27 in) depth. Depth of organic soil is 20–70 cm (8–27 in), and soil pH averages 3.5.

Vegetation is characterized by an open canopy with an understory of ericaceous shrubs over a mat of peatmosses. The dwarfed canopy is dominated by red spruce and pitch pine, with low cover of red maple and eastern hemlock. The tall-shrub layer contains abundant catberry and mountain laurel and lower cover of great laurel and

Fig. 4.14 Example of the Pitch Pine – Red Spruce/Heath Peat Woodland community. Photo by Elizabeth Byers

the regenerating canopy species. The short-shrub layer is similar in composition to the tall-shrub layer, with the addition of abundant velvetleaf blueberry and low but consistent cover of black huckleberry, black chokeberry, late lowbush blueberry, northern wild raisin, and the dwarf-shrub small cranberry. The herbaceous layer is sparse, with bristly dewberry, eastern teaberry, goldthread, robin-run-away, tawny cotton-grass, trailing arbutus, tree clubmoss, cinnamon fern, white beakrush (*Rhynchospora alba*), three-seeded sedge, and roundleaf sundew. Bryophytes have high cover and are dominated by peatmosses (*Sphagnum rubellum, Sphagnum divinum, Sphagnum fallax,* and *Sphagnum papillosum*), common haircap moss, and eastern haircap moss. Indicators include goldthread, robin-run-away, roundleaf sundew, tawny cotton-grass, and catberry.

4.3 Southern Appalachians

4.3.1 Upland Red Spruce and Spruce-Fir Communities

In the southern Appalachians, communities with red spruce range from Mt. Rogers in southernmost Virginia, southwestward through North Carolina and Tennessee as far as the Great Balsam Mountains and Great Smoky Mountains. Upland communities reach their range limit a bit south of Richland Balsam and a few miles west of Kuwohi Mountain (a.k.a. Clingmans Dome), though comparable elevations continue along the Great Smoky Mountains crest. In between, they are present on Plott Balsam Mountain, Black Mountains, Unaka Mountain, Roan Mountain, and Grandfather Mountain, as a few small patches. The southern Appalachians contain higher elevations than the central Appalachians (highest peak—Spruce Knob at 1,482 m), extending upward to the highest peaks east of the Rocky Mountains and Black Hills. Even accounting for the lower latitude, the climate of the highest peaks is more extreme. A diverse array of topographic settings and substrates, as well as a broad range of altitude, leads to high community diversity. Eight distinct forest communities are recognized, all of them globally imperiled. One of the most important differences in vegetation between the central and southern Appalachians is the presence, often as a dominant or co-dominant tree, of Fraser fir and the absence of balsam fir. Fir in general is more extensive across the Southern Appalachians or was before the introduced balsam woolly adelgid devastated its populations. Fraser fir dominated the forest cover in the extreme climate of the highest peaks, where red spruce is much less abundant. Throughout the broad middle elevations of the spruce zone, Fraser fir was co-dominant or at least abundant, and yellow birch, the only common hardwood, is of limited importance. The communities are thus classified into elevational groups, with some dominated by Fraser fir at the highest elevations, those with mixed spruce and fir at middle elevations, and those with mixed red spruce and hardwoods or hemlock at the lower elevations.

The flora of the southern Appalachian spruce-fir forests is similar to that of the central Appalachians in having substantial northern affinities, as well as containing southern species. Many species are shared, but there is a larger number of southern species and a somewhat smaller number of northern species. Thus, alderleaf viburnum, white woodsorrel, Clinton lily, and painted trillium are common, but goldthread and lowbush blueberry are not components. Meanwhile, southern mountain cranberry, minniebush, Blue Ridge St. John's-wort, and other species shared with the central Appalachians are joined by more strictly southern endemics such as Roan rattlesnakeroot (*Ageratina roanensis*), Blue Ridge white heart-leaved aster (*Eurybia chlorolepis*), and, in some mountain ranges, by even narrower endemic species such as Rugel's ragwort (*Rugelia nudicaulis*).

A variety of successional states exist in the southern Appalachian spruce-fir forests. Several substantial areas were protected from logging by rugged terrain or came into conservation ownership before they were logged. The largest area is in the Great Smoky Mountains National Park, but old growth forests also remain in the Black Mountains, on Grandfather Mountain, Richland Balsam, Mt. Rogers, and in small pockets elsewhere (see Chap. 1). Most of the smaller protected areas are either on the highest peaks, where the dominant Fraser fir was less valued than red spruce as timber, or at the lower elevations where patchy distributions of red spruce made building logging railroads unprofitable. Nevertheless, a much larger acreage was devastated by early twentieth century logging and slash fires. Much of this acreage still does not support forest 100 or more years after the disturbance. There were some areas where logging occurred, but subsequent fires did not occur. In those areas, Fraser fir tended to dominate the regeneration and spruce has been slow to reclaim co-dominance. Because of the importance of Fraser fir in most communities, the impact of the balsam woolly adelgid was devastating for even the unlogged remnants. More details can be found in Chap. 5.

4.3.1.1 Fraser Fir Forest (Herb Subtype)

Association CEGL006049. *Abies fraseri/Viburnum lantanoides/Dryopteris campyloptera – Oxalis montana/Hylocomium splendens* forest (Fig. 4.15). NatureServe Conservation Status: G1; NC: S1, TN: S1, VA: S1.

This community is restricted to the highest peaks of North Carolina, Tennessee, and southernmost Virginia. It is generally above about 1,880 m (6,200 ft) in elevation, though it is a bit lower on the more northerly Mt. Rogers and on Grandfather Mountain, known for its extreme weather. Only a half dozen peaks support this community, although Fraser fir forests are abundant on these peaks. All occurrences of this community have been devastated by the balsam woolly adelgid, which killed all mature Fraser firs from the 1960 to 1980s. Prior to that, this community existed as dense forests periodically broken by wind throw gaps. Red spruce and yellow birch were minor but highly constant canopy components, while American mountain-ash dominated limited patches. Striped maple and mountain maple were the primary understory trees other than red spruce and Fraser fir. Alderleaf viburnum, southern

Fig. 4.15 Example of the Fraser Fir Forest (Herb Subtype) community. Photo by Gary Fleming

mountain cranberry, mountain highbush blueberry (*Vaccinium simulatum*), smooth highbush blueberry (*Vaccinium corymbosum*), and red elderberry (*Sambucus racemosa* var. *racemosa*) often formed an open shrub layer. Lush cover of mosses, ferns, herbs, and fir seedlings carpeted the ground. Because the smooth bark of mature Fraser fir provides especially good habitat for epiphytic mosses and liverworts, the tree trunks too were often lush and green. Mountain woodfern, evergreen woodfern, southern lady fern (*Athyrium asplenioides*), hay-scented fern, whorled aster, white wood sorrel, and Appalachian white snakeroot are often abundant. Stairstep moss, haircap mosses, and smoothcap mosses (*Atrichum* spp.) are often the dominant bryophytes.

Because the balsam woolly adelgid kills older trees but not seedlings and small saplings, advance regeneration in some of the devastated forests has developed into dense stands of young fir. Other areas became dominated by blackberries (*Rubus canadensis, Rubus alleghaniensis*). Some young fir stands are now maturing, but balsam woolly adelgids occasionally return and kill the maturing trees in some stands.

4.3.1.2 Fraser Fir Forest (Rhododendron Subtype)

Association CEGL006308. *Abies fraseri*/(*Rhododendron catawbiense, Rhododendron carolinianum*) forest. NatureServe Conservation Status: G1; NC: S1, TN: S1.

This community is not only restricted to the highest peaks of North Carolina and Tennessee, it is restricted to only a small portion of the area on those peaks, occurring on sharply convex slopes and near the edges of rock outcrops. As in the Herb Subtype, Fraser fir dominates the canopy, and red spruce, yellow birch, and American mountain ash occur as minor or patchy components. The canopy is often small and stunted. A dense shrub layer occurs in the forest, generally dominated by *Rhododendron* spp. Other shrubs may include minniebush or southern bush honeysuckle (*Diervilla sessilifolia*). Herbs are sparse and sometimes completely absent, except on small rock outcrops within the community. However, bryophytes may be abundant in some places, on the ground and on tree trunks. As with other Fraser fir forests, the balsam woolly adelgid has devastated all examples of this community. With less understory presence of Fraser fir, fewer examples have regenerated, and some examples have come to resemble heath bald vegetation.

4.3.1.3 Red Spruce – Fraser Fir Forest (Herb Subtype) or Red Spruce – Southern Mountain Cranberry Forest (Deciduous Shrub Understory Type)

Association: CEGL007131. *Picea rubens – (Abies fraseri)/Vaccinium erythrocarpum/Oxalis montana – Dryopteris campyloptera/Hylocomium splendens* forest (Fig. 4.16). Nature Serve Conservation Status: G2; NC: S2, TN: S1, VA: S1, WV: S1.

This community is the most extensive community in the middle elevations of the spruce-fir zone in the southern Appalachians, roughly from 1,680 to 1,880 m (5,500–6,200 ft). It is most extensive in North Carolina but extends into Tennessee and southernmost Virginia. The highest elevation spruce forests in West Virginia are in the same IVC community, hence it is described above for the central Appalachians.

In the southern Appalachians, prior to balsam woolly adelgid disturbance, it had a dense canopy consisting of both red spruce and Fraser fir. At present, most stands are somewhat open, but regeneration of fir and red spruce from existing seedlings has closed the canopy in some areas. Yellow birch is a highly constant component but limited in cover. Striped maple and mountain maple are the primary understory trees other than red spruce and Fraser fir. Alderleaf viburnum, southern mountain cranberry, mountain highbush blueberry, smooth highbush blueberry, and red elderberry are frequent shrubs, but are not usually dense. Where the canopy is broken, blackberries sometimes become dense. Beneath intact canopies, lush ground cover of mosses, ferns, herbs, and red spruce and Fraser fir seedlings is common. Mountain woodfern, evergreen woodfern, southern lady fern, hay scented fern, whorled aster, white wood sorrel, Appalachian white snakeroot and, in the Smokies, Rugel's ragwort, are frequent and abundant vascular plants. Stairstep moss, haircap mosses, and smoothcap mosses are often the dominant bryophytes. *Bazzania* liverworts, which are often dominant in the central Appalachians, are not a major component.

Fig. 4.16 Example of the Red Spruce – Fraser Fir Forest (Herb Subtype) or Red Spruce – Southern Mountain Cranberry Forest (Deciduous Shrub Understory Type) community. Photo by Gary Fleming

4.3.1.4 Red Spruce – Fraser Fir Forest (Rhododendron Subtype) or Red Spruce – Fraser Fir Forest (Evergreen Shrub Understory Type)

Association CEGL007130. *Picea rubens* – (*Abies fraseri*)/(*Rhododendron catawbiense, Rhododendron maximum*) forest (Fig. 4.17). NatureServe Conservation Status: G1; NC: S1, TN: S1, VA: S1.

Fig. 4.17 Example of the Red Spruce – Fraser Fir Forest (Rhododendron Subtype) or Red Spruce – Fraser Fir Forest (Evergreen Shrub Understory Type) community. Photo by Gary Fleming

Like the Herb Subtype, this community occurs in the middle elevations of the spruce zone, roughly from 1,680 to 1,880 m (5,500–6,200 ft). It is restricted to a few areas on sharply convex slopes and near edges of rock outcrops. The canopy may be a mix of red spruce and Fraser fir or may be more strongly dominated by red spruce. The canopy may be dense or somewhat open, and the trees may be somewhat stunted. Allegheny serviceberry may occur in the understory, along with red spruce, Fraser fir, or yellow birch. The shrub layer is dense and usually dominated by Catawba rhododendron or great laurel. Shrubs common in the other spruce communities may also be present, as well as evergreen fetterbush (*Pieris floribunda*), deerberry (*Vaccinium stamineum*), and other species. Herbs are generally sparse. Mountain woodfern, brownish sedge, or various species shared with the Herb Subtype may be present in small numbers.

4.3.1.5 Red Spruce – Fraser Fir Forest (Boulderfield Subtype)

Association CEGL007128. *Picea rubens/Ribes glandulosum* forest. NatureServe Conservation Status: G1; NC: S1.

This community occurs on talus or relict periglacial boulderfields and is an extension of the yellow birch-dominated boulderfield forests that occurs in similar settings at slightly lower elevations. It is one of the rarest of spruce communities, with only a few examples definitively known, exclusively in North Carolina. The only extensive stand is on Grandfather Mountain.

The boulderfields consist of large loose rock fragments that entirely cover the ground surface. Boulders generally rest on other boulders, with extensive void space between them. Soil consists of organic matter in pockets between rocks and on their surfaces. Despite the limited rooting environment, this community generally has a well-developed closed or somewhat open canopy. The canopy consists almost exclusively of red spruce and yellow birch. Both species are capable of establishing in shallow organic deposits on raised surfaces, then wrapping their roots around the object to seek deeper levels. The understory is dominated by mountain maple, with American mountain ash the only other frequent species. Alderleaf viburnum is the most constant and abundant species in the sparse shrub layer, but skunk currant, characteristic of boulderfield environments, may also be present. The herb layer consists primarily of species that are able to live on bare rock. There is often extensive moss cover. Appalachian rockcap fern is extensive in most places. Other herbs typical of spruce-fir forests are present in limited favorable soil pockets, including mountain woodfern, evergreen woodfern, white woodsorrel, and shining clubmoss.

4.3.1.6 Red Spruce – Fraser Fir Forest (Birch Transition Herb Subtype)

Association CEGL006256. *Picea rubens* – (*Betula alleghaniensis*, *Aesculus flava*)/ *Viburnum lantanoides*/*Solidago glomerata* forest. NatureServe Conservation Status: G1; NC: S1, TN: SNR.

This community occurs in the lower elevations of the red spruce zone, where red spruce tends to be co-dominant with yellow birch and other northern hardwood species. It typically occurs around 1,370–1,680 m (4,500–5,500 ft) elevation and is recognized only in North Carolina and Tennessee. Though fundamentally an elevation-driven transitional community between spruce-fir and northern hardwood forests, this community generally occurs with a patchy distribution, interfingering with both higher and lower communities where it occurs.

The forest is usually dense where not recently disturbed, consisting predominantly of red spruce and yellow birch; Fraser fir may be present in small numbers. Yellow buckeye (*Aesculus flava*), red maple, sugar maple, beech, red oak, or sweet birch may be present, and any of these species may occasionally be abundant. The understory often includes striped maple and mountain maple as well as the various canopy species. Shrubs may be sparse to moderate in density. Southern mountain cranberry, alderleaf viburnum, and mountain holly are the most frequent. The herb layer may be dense or sparse. Most of the frequent species are shared with higher elevation spruce-fir forests, such as white wood sorrel, evergreen woodfern, shining clubmoss, whorled aster, and Clinton lily, but species more typical of northern hardwoods communities, such as Canada mayflower, cucumber root, Blue Ridge white heart-leaved aster, and New York fern are also often present.

This community is one of the more abundant red spruce communities of the southern Appalachians, though its overall coverage is not well known. In the dendritically dissected topography of the Blue Ridge, lower elevation is associated with increased acreage, but this is offset by the discontinuous occurrence of the community. In some places, the increased proximity to land clearing and settlement at the lower elevations have led to greater loss of this community. But in more remote mountain ranges, logging railroads often were built at higher elevation and patches were left uncut on rugged or steep lands below the logged areas. Though the patchy and discontinuous distribution is sometimes interpreted as a result of logging history, this does not generally appear to be the case. It is also present in the unlogged parts of the Great Smoky Mountains and other mountain ranges. This complex natural distribution complicates decisions about appropriate sites for red spruce restoration activities.

4.3.1.7 Red Spruce – Fraser Fir Forest (Birch Transition Shrub Subtype)

Association CEGL004983. *Picea rubens* – (*Betula alleghaniensis, Aesculus flava*)/ *Rhododendron* (*maximum, catawbiense*) forest (Fig. 4.18). NatureServe Conservation Status: G1?; NC: S1, TN: SNR, VA: S1.

This community occurs in the lower elevations of the red spruce zone, where red spruce tends to co-dominate with yellow birch and other northern hardwood species. It typically occurs around 1,370–1,680 m (4,500–5,500 ft) elevation on more convex ridges and around rock outcrops, and is known in North Carolina, Tennessee, and southernmost Virginia. It is a lower elevation analogue to the Red Spruce – Fraser Fir Forest (Shrub Subtype) association. It is much less extensive on the landscape than the Red Spruce – Fraser Fir Forest Birch Transition Herb Subtype association, but its overall abundance is not well known.

This community is a dense to open forest, with red spruce and yellow birch generally dominant. Red oak, red maple, beech, and eastern hemlock may also be present. The understory may include serviceberry, striped maple, and mountain maple. The shrub layer is dense and usually dominated by great laurel but may have abundant Catawba rhododendron or mountain laurel. Deciduous shrubs such as alderleaf viburnum, southern mountain cranberry, or mountain holly may be present in smaller amounts. The herb layer is sparse, containing mainly species typical of other spruce-fir forests, such as mountain woodfern, evergreen woodfern, and whorled aster. Appalachian rockcap fern may be abundant where rock cover is high.

Fig. 4.18 Example of the Red Spruce – Fraser Fir Forest (Birch Transition Shrub Subtype) community. Photo by Gary Fleming

4.3.1.8 Red Spruce – Fraser Fir Forest (Low Rhododendron Subtype) or Red Spruce Forest (Protected Slope Type)

Association CEGL006152. *Picea rubens* – (*Tsuga canadensis*)/*Rhododendron maximum* forest. NatureServe Conservation Status: G2?; NC: S1, TN: SNR, WV: S1.

This community occurs at lower elevations in moist, but not wet, sites with sheltered topography, such as heads of ravines. It is recognized in both the central and southern Appalachians and was described above as Red Spruce – (Eastern Hemlock)/ Great Laurel Forest. In the southern Appalachians, it generally occurs at elevations that do not support continuous spruce-fir forests (as low as 1,200 m [3,900 ft]) in sites that have cold air drainage or a cooler microclimate. It is generally surrounded by northern hardwoods, sometimes by oak forests, but may extend uphill to mix with other spruce-fir forests. Most known examples are in small patches and, while some may have been reduced or destroyed by past logging, it was always naturally limited in extent.

The southern Appalachian examples have dense to open canopies that are dominated by red spruce in combination with eastern hemlock or yellow birch. Most are marked by a dense shrub layer dominated by great laurel. Mountain laurel may co-dominate, and lower elevation mesophytic shrubs such as dog hobble (*Leucothoe fontanesiana*) may possibly occur. Herbs are sparse to nearly absent. Bryophytes are not as prominent as they are in higher elevation forests, nor as prominent as they are in the central Appalachians. Given the sheltered topography, wind throw, lightning, and other natural disturbances are more limited than in many communities, and trees can grow quite large.

4.3.2 Wetland Red Spruce Communities

Though the southern Appalachians have a great diversity of wetland communities, only one has enough red spruce to be considered a spruce community. Examples of several additional open bog/poor fen communities may contain scattered stunted stems of red spruce, but the species is too limited in cover and frequency to include here. These wetlands are notable, however, because they support red spruce at elevations below the lowest upland populations, and often at some distance from them. They presumably represent Pleistocene relict populations, often containing other disjunct species of northern affinities, now rare in the southern Appalachians. Though these wetlands are likely sites of cold air drainage and accumulation, the survival of red spruce in them, including through the warmer, drier period between the Pleistocene and the present, suggests a broader physiological tolerance in red spruce than is generally recognized.

4.3.2.1 Swamp Forest – Bog Complex (Spruce Subtype) or Red Spruce – (Eastern Hemlock)/Great Laurel Swamp Forest

Association CEGL006277. *Picea rubens* – (*Tsuga canadensis*)/*Rhododendron maximum* saturated forest (Fig. 4.19). NatureServe Conservation Status: G2?; MD: SNR, NC: S1, NY: S1, PA: S1, TN: SH, VA: S1, WV: S2.

This community, as defined in the IVC, is the widest-ranging spruce community, occurring in the central and southern Appalachians, and further north in Pennsylvania and New York. While it may be more broadly defined than other communities, its limited species richness makes it appear similar in widely separated locations. Its expression in the central Appalachians is described above.

In the southern Appalachians, this community occurs in small to medium-sized flat valley bottoms at moderate elevations from 1,640 m (5,300 ft) down to as low as 1,100 m (3,600 ft). Such settings are rare in a region where most streams at these elevations have high gradients and narrow valleys. These settings appear perched, generally dropping into a more typical steep gorge or ravine downstream. Amid the complex geology of the region, it is generally unclear if they are caused by more resistant rock in the valley, by resistant rock at the outlet preventing downcutting, or if they represent relicts of valleys formed at a higher base level.

In contrast to the central Appalachians, the southern Appalachian examples tend to be heterogeneous. Microtopography caused by stream deposition or hummock

Fig. 4.19 Example of the Swamp Forest – Bog Complex (Spruce Subtype) or Red Spruce – (Eastern Hemlock)/Great Laurel Swamp Forest community. Photo by Gary Fleming

and hollow topography often amounts to 1 m (3.3 ft) or more of fine scale relief, and wetness varies from permanently saturated and periodically shallowly flooded in low areas to moist but somewhat drained in the higher areas. Areas of groundwater seepage are often also visible. Soils are predominantly mineral, but they have numerous local pockets of shallow to deep muck in the wetter areas.

The name Swamp Forest – Bog Complex reflects the heterogeneity in vegetation. The community is predominantly forested, with a canopy dominated by red spruce or, rarely, a mix of red spruce with eastern hemlock. In one unique example Carolina hemlock (*Tsuga caroliniana*) is co-dominant in the canopy. Red maple, yellow birch, sweet birch, and other species may occur in the canopy, in addition to serviceberry frequently present in the midstory. The forest matrix generally has a dense shrub layer dominated by great laurel. Mountain laurel, wetland shrubs such as northern wild raisin, black elderberry (Sambucus nigra ssp. canadensis), and black chokeberry, and high elevation shrubs such as alderleaf viburnum and red elderberry are generally present. Disjunct shrubs of northern affinities, Canada yew (*Taxus canadensis*), and long-stalked holly (*Ilex collina*), are present in some examples. Interspersed with this forested matrix are small openings, too small to recognize as separate communities, in wetter microsites. Most are only a few meters across, but they form a small break in the canopy. Peat moss often covers these openings. Widespread wetland species such as cinnamon fern, royal fern, slender mannagrass, and northern roughleaf goldenrod (*Solidago patula*) are frequent in them. Less frequent species include several wetland orchids and several species associated with larger bog communities.

4.4 Discussion

4.4.1 Rare Plants of the Red Spruce Ecosystem

Five globally rare plants have important populations within central Appalachian red spruce communities. White monkshood is found in small forest seeps and Blue Ridge St. John's-wort is found in sunnier seep areas. Glade spurge and bog Jacob's-ladder grow together in shaded limestone-influenced spruce-fir swamps. Long-stalk holly prefers high-gradient cool stream edges. Two additional globally rare plants, bog bluegrass (*Poa paludigena*) and Tennessee pondweed (*Potamogeton tennesseensis*) have been recorded, but their known populations are small and red spruce communities are not their essential habitat. The globally rare rock skullcap (*Scutellaria saxatilis*) is found in rocky forests in a variety of forest types, including red spruce.

A number of state-rare species within the central Appalachian red spruce wetland communities have northern affiliations, including some that are disjunct, such as balsam fir, American larch, and bog rosemary. Additional state-rare species with primarily northern distributions include black ash, Bartram shadbush (*Amelanchier bartramiana*), creeping snowberry, small cranberry, goldthread, robin-run-away, bog buckbean (*Menyanthes trifoliata*), black-girdle bulrush (*Scirpus atrocinctus*), lake-bank sedge (*Carex lacustris*), slender sedge (*Carex lasiocarpa*), and few-flower sedge (*Carex pauciflora*).

In West Virginia, 145 state-rare plant species have been documented in red spruce wetlands, including 60 critically imperiled species, 56 imperiled species, and 29 vulnerable species. This represents a remarkable 31% of West Virginia's rare flora; nearly one-third of the rare plants in the state may be found on 0.1% of its land surface (Byers et al. 2007).

The southern Appalachians have several globally rare species. Most of these are southern Appalachian endemics, such as bent avens (*Geum geniculatum*), Smoky Mountain mannagrass (*Glyceria nubigena*), small mountain bittercress (*Cardamine clematitis*), Clingman's hedge-nettle (*Stachys clingmanii*), Grandfather Mountain leptodontium moss (*Leptodontium vituscululoides* var. *sulfureum),* the liverwort *Plagiochila exigua*, and other bryophytes. A few plants, such as white monkshood and Blue Ridge Saint John's-wort are examples of species shared with the central Appalachians.

A large number of additional state-rare species also occur. These include species that are common far to the north, but which occur rarely as disjunct populations or which extend southward as small populations along the Appalachians. Examples include blunt-lobe grapefern (*Botrychium oneidense*), bluejoint reedgrass (*Calamagrostis canadensis*), northern beech fern (*Phegopteris connectilis*), necklace sedge (*Carex projecta*), slender wood reedgrass (*Cinna latifolia*), clasping twisted-stalk (*Streptopus amplexifolius*), rosy twisted-stalk (*Streptopus lanceolatus*), and bryophytes such as peak moss (*Brachydontium trichoides*) and the liverwort *Bazzania nudicaulis*.

The relationship of rare plant species to spruce communities varies. Some of the species noted here occur only in spruce communities while others also occur in open bogs and seeps, in rock outcrop communities, or in northern hardwoods forests, and some may be more concentrated in those other communities. Of particular note is that some of the rare species associated with red spruce wetlands in West Virginia occur in open wetlands in North Carolina but are not found in red spruce forests there. Some species that are relatively common in red spruce wetlands in West Virginia, such as three-seeded sedge, are state-rare in North Carolina.

4.4.2 Associated Plant Communities

While red spruce forests are often the predominant plant communities in the elevational zones and on the landforms where they occur, there are sometimes associated communities occurring in small patches among them. These communities often harbor a disproportionate number of rare species. Some of these species are shared with adjacent spruce communities while others are not. Because these communities are often small, the surrounding forest can have important effects on their composition.

In the central Appalachians, some rocky or barren areas within the red spruce ecosystem have resulted from anthropogenic fires, but the disjunct occurrence of circumboreal lichens (*Arctoparmelia centrifuga*, *Clauzadeana maculata*, *Fuscidea praeruptorum*) and timber rattlesnakes (*Crotalus horridus*) attest to the natural status of the rock fields at some scale. These associated communities include High Elevation Sandstone Boulderfield, Mountain Laurel – Black Huckleberry Heath Barren, and Blueberry Heath Barren.

On the broad ridgetops and plateau areas of the central Appalachians, characteristic wetland communities are embedded within the red spruce forest. Many of these wetlands do not occur outside of the red spruce forest in this region. They are often characterized by soils that are too saturated to support tree species, and typically are bordered by red spruce swamp communities. They include the globally critically imperiled Bog-rosemary Peatland and Barbara's-buttons Ice-scour Prairie; globally imperiled communities including Cranberry – Beakrush Peatland, Nodding Sedge – Prickly Bog Sedge Seep, and Star Sedge Fen; and globally vulnerable communities of Cottongrass Fen, Golden Saxifrage Seep, American Bur-reed Marsh, Hairy-fruit Sedge Floodplain Prairie, Rough Sedge Seep, and Twisted Sedge Rivershore.

Other special wetland types that are embedded in red spruce forests include the state-rare Montane Tall Sedge Fen, Tamarack Swamp, Bluejoint Wet Meadow, Silvery Sedge Fen, Blueberry – Bracken Fern Shrub Swamp, Chokeberry – Wild Raisin Peatland, Red Maple – Blackgum/Peatmoss Swamp, and Quaking Aspen Swamp. Wetland communities that occur within the red spruce ecosystem but are also distributed more widely in the central Appalachians include Softstem Bulrush Marsh, Bushy St. Johnswort Shrub Swamp, Goldenrod Wet Meadow, Meadowsweet Shrub Swamp, Rice Cutgrass Marsh, Silky Willow Shrub Swamp, Speckled Alder

Alluvial Thicket, Threeway Sedge Fen, Tussock Sedge Wet Meadow, and Woolgrass Wet Meadow.

In the higher-relief southern Appalachians, rock outcrop communities are often associated with spruce-fir forests. High Elevation Rocky Summit (Typic Subtype) (Michaux's Saxifrage – Wretched Sedge – Poverty Oatgrass – Mountain Dwarf-dandelion Grassland), High Elevation Rocky Summit (High Peak Subtype) (Michaux's Saxifrage – Wretched Sedge – Whorled Wood Aster – Clustered Goldenrod Grassland), and High Elevation Rocky Summit (High Peak Lichen Subtype) (Common Toadskin Lichen – Carolina Rocktripe Lichen Nonvascular Vegetation) are sparsely vegetated or mixed-structure open communities on rock outcrops. These communities, especially the first, are notable for the number of rare plant species associated with them. Of particular importance is a suite of globally rare, narrow endemic, often federally listed species, including Roan Mountain bluet (*Houstonia montana*), Blue Ridge goldenrod (*Solidago spithamaea*), spreading avens (*Geum radiatum*), and Heller's blazing star (*Liatris helleri*). These combine with northern species at their southern range limit, such as highland rush (*Oreojuncus trifidus*) and deerhair bulrush (*Trichophorum cespitosum*). Less rare northern species, such as three-tooth cinquefoil (*Sibbaldiopsis tridentata*), also occur in these communities. Some of these species occur in Arctic tundra or in the alpine tundra of New England, and some of the southern endemic species are close relatives of species in these habitats. For example, spreading avens is closely related to the globally imperiled mountain avens (*Geum peckii*) of New Hampshire and Nova Scotia. It is widely believed that, during the Pleistocene, the climate was cold enough that the southern Appalachians had a timberline and that there was alpine tundra vegetation on the high peaks. That tundra vegetation disappeared from the southern Appalachians as spruce-fir forests migrated uphill, but some of its flora survives on the open rock outcrops.

In addition to rock outcrop communities, non-forest communities in the spruce-fir zone also include shrub-dominated Heath Bald communities of several types and the enigmatic Grassy Bald communities. The three Grassy Bald communities: Grassy Bald (Grass Subtype) (Flattened Oatgrass – Threetooth Cinquefoil Grassland, Grassy Bald (Sedge Subtype) (Pennsylvania Sedge Grassland), and Grassy Bald (Alder Subtype) (*Alnus viridis crispa*/*Carex pensylvanica* Shrubland), cover significant acreage in association with spruce-fir forests in just a couple of places. They harbor rare species, most of which are not in adjacent forests, including globally rare endemics such as Gray's lily (*Lilium grayi*) and long distance disjuncts such as green alder (*Alnus viridis crispa*). They likely represent remnants of Pleistocene alpine tundra. Several Heath Bald communities may also be associated with red spruce. These dense shrublands generally are not known for supporting rare species. An additional notable community associated with red spruce forests is known in the literature as Beech Gap – Northern Hardwood Forest (Beech Gap Subtype) (American Beech/Pennsylvania Sedge – Appalachian White Snakeroot Forest). They occur in ridgetop saddles (gaps) or exposed ridges in the elevational range of spruce-fir forests.

At their lower elevation boundary, spruce forests interfinger with various typical Northern Hardwood and Hemlock – Northern Hardwood Forest communities dominated by yellow birch, eastern hemlock, yellow buckeye, sugar maple, and beech, or on warmer slope aspects, forests dominated by red oak. These communities are much more extensive in the central and southern Appalachians than are spruce-fir forests, both downhill in the same mountain ranges and in ranges that do not have spruce-fir forests. It is notable that, where spruce-fir forests are present uphill, scattered individuals of red spruce often are found in these lower elevation communities. Most occur only as saplings, but occasional trees reach the canopy.

4.4.3 Management Implications

Plant communities are important to consider during three management stages. First, land acquisition decisions should be based in part on an inventory of existing plant communities, including their rarity and condition, as these are important conservation targets. Second, restoration activities should be guided by the plant communities that a particular habitat niche is likely to support. Third, recreation or other development activities should avoid disturbance to rare and threatened plant communities, especially to those that occur only in small patches.

As detailed in this chapter, red spruce communities range from those dominated by red spruce or Fraser fir in contiguous forest to small-patch wetlands. When considering restoration or other management actions at the project scale, it is important to recognize the particular environmental setting of the project and the specific plant communities that occur in that environment. Knowing the potential plant communities provides essential information on rare species, rare habitats, disturbance regimes, soil types, and reference conditions. A crucial implication of spruce community diversity is that reference condition and appropriate restoration goals are not the same for every site. The appropriate amount of spruce in the canopy varies, with the lowest elevation communities having substantial hardwood components and the highest southern Appalachian communities dominated by Fraser fir. Shrub cover and composition also varies, and high shrub cover does not necessarily indicate a need for management. In addition to avoiding harm to the many rare species that inhabit spruce forests, managers need to avoid harm to rare communities.

When considering restoration of small-patch wetland communities within the red spruce ecosystem, it is important to survey the site for legacy erosion and entrenchment of streams from extensive logging and fires a century ago. Today, many embedded wetlands are drained by entrenched or incised streams that are disconnected from their floodplains. Today's wetland plant communities may be drier and more vulnerable than the original well-connected wetlands. In these cases, restoration of hydrology is key. The target plant community of the successfully restored area may be significantly wetter than that of pre-restoration wetlands.

Current threats and stresses include climate change, air pollution, ongoing effects and new infestations of balsam woolly adelgid, and, especially on private lands, conversion of land use to residential or second homes, coal mining, and wind energy development. Non-native invasive plants are less problematic than in forest ecosystems with richer soils or milder climates, but they can still be a serious problem, especially along road corridors and in wetlands or forest openings. An increase in management activity would exacerbate this threat, as heavily used trails and especially roads often act as corridors for invasion, heavy equipment is a major vector of spread, and canopy opening can trigger invasion. Excessive deer herbivory, while not a direct threat to red spruce seedlings, negatively impacts the ecosystem by reducing or removing palatable shrubs and herbs, including rare species. Deer trails are active vectors for non-native invasive species. There are a number of tree pathogens and invasive insects in addition to balsam woolly adelgid that impact components of spruce forests, including hemlock woolly adelgid, emerald ash borer, beech bark disease, and beech leaf disease.

Plant communities directly reflect specific environmental gradients and ecological processes and are a key measure of beta diversity. While rare species may be considered the fine filter of biodiversity, plant communities are the equally important coarse filter ensuring conservation of habitat niches and the many species they support. Knowledge of the plant communities associated with a piece of land is fundamental to management for biodiversity conservation.

References

Allard HA, Leonard EC (1952) The Canaan and Stony River Valleys of West Virginia, their former magnificent spruce forests, their vegetation and floristics today. Castanea 17:1–60

Byers EA, Vanderhorst JP, Streets BP (2007) Classification and conservation assessment of high elevation wetland communities in the Allegheny Mountains of West Virginia. West Virginia Division of Natural Resources Report, Elkins, West Virginia

Byers EA, Vanderhorst JP, Streets BP (2010) Classification and conservation assessment of upland red spruce communities in West Virginia. West Virginia Division of Natural Resources Report, Elkins, West Virginia

Clarkson RB (1964) Tumult on the Mountains: lumbering in West Virginia 1770–1920. McClain Printing Company, Parsons, West Virginia

Darlington HC (1943) Vegetation and substrate of Cranberry Glades, West Virginia. Bot Gaz 104:371–393

Faber-Langendoen D, Baldwin K, Peet RK et al (2018) The EcoVeg approach in the Americas: U.S., Canadian and international vegetation classifications. Phytocoenologia 48:215–237

Federal Geographic Data Committee (2008) National vegetation classification standard, ver 2. Federal Geographic Data Committee Vegetation Subcommittee Document FGDC-STD-005-2008, Reston, Virginia

Fike J (1999) Terrestrial & palustrine plant communities of Pennsylvania. Pennsylvania Department of Conservation and Natural Resources, The Nature Conservancy, and Western Pennsylvania Conservancy Report, Harrisburg, Pennsylvania

Fleming GP, Moorhead WH III (1996) Ecological land units of the Laurel Fork Area, Highland County, Virginia. Virginia Department of Conservation and Recreation Division of Natural Heritage Technical Report 96–08, Richmond, Virginia

Grossman DH, Faber-Langendoen D, Weakley AS et al (1998) International classification of ecological communities: terrestrial vegetation of the United States. Volume I. The national vegetation classification system: development, status, and applications. The Nature Conservancy, Arlington, Virginia

Jennings M, Faber-Langendoen D, Peet R et al (2004) Guidelines for describing associations and alliances of the U.S. national vegetation classification, ver 4.0. The Ecological Society of America Vegetation Classification Panel Report

Jennings M, Loucks O, Glenn-Lewin D et al (2002) Standards for the associations and alliances of the U.S. national vegetation classification, ver 1.0. The Ecological Society of America Vegetation Classification Panel Report

Jennings M, Loucks O, Glenn-Lewin D et al (2003) Guidelines for describing associations and alliances of the U.S. national vegetation classification, ver 2.0. The Ecological Society of America Vegetation Classification Panel Report

NatureServe (2022) NatureServe network conservation status ranks and biodiversity location data. NatureServe, Arlington, Virginia. https://explorer.natureserve.org/. Accessed 30 Nov 2022

Pielke RA (1981) The distribution of spruce in west-central Virginia before lumbering. Castanea 46:201–216

Schafale MP (2024) Classification of the natural communities of North Carolina, fourth approximation. North Carolina Department of Environment and Natural Resources Natural Heritage Program, Raleigh, NC

Stephenson SL, Clovis JF (1983) Spruce forests of the Allegheny Mountains in central West Virginia. Castanea 48:1–12

Stevens CE (1969) A native red spruce stand in Rockingham County, vol 3. Jeffersonia

Chapter 5
Ecosystem Dynamics

Melissa A. Thomas-Van Gundy, Beverly Collins, Peter S. White, and Saskia L. van de Gevel

5.1 Introduction

Red spruce (*Picea rubens*)-dominated forest and woodland dynamics emerge from interactions among the life histories of the constituent species, biogeography, and impacts from historical anthropogenic and natural disturbances. In this chapter, we build on descriptions of species biology, community biogeography, and regional forest history to describe components of spruce-dominated ecosystem dynamics, including disturbance regimes and altered dynamics due to loss of key species from logging, invasive insects, pollution, or climate change. Our use of the term dynamics implies the influence of time on these processes, and at a minimum a disturbance regime should include descriptions of frequency and scale or patch size of disturbance events. In some areas or timeframes, pre-European settlement for instance, we may only have qualitative descriptions of disturbance regime components. Quantitative measures or qualitative descriptions may be used to help design restoration treatments or determine restoration success.

Higher elevations and steeper slopes of the central and southern Appalachians were some of the last to be impacted by wide-scale, exploitative logging (see Chap. 1). However, use of narrow-gauge railroads to reach even remote sites resulted in near complete removal of all economically viable trees in the region (Lewis 1998; Fig. 5.1).

M. A. Thomas-Van Gundy (✉)
USDA Forest Service, Northern Research Station, Parsons, WV, USA
e-mail: melissa.thomasvangundy@usda.gov

B. Collins
Western Carolina University, Department of Biology, Cullowhee, NC, USA

P. S. White
University of North Carolina at Chapel Hill, Department of Biology, Chapel Hill, NC, USA

S. L. van de Gevel
Virginia Tech, College of Natural Resources and Environment, Blacksburg, VA, USA

D. J. Brown et al. (eds.), *Ecology and Restoration of Red Spruce Ecosystems in the Central and Southern Appalachians*, https://doi.org/10.1007/978-3-032-19616-3_5

In addition to direct effects of exploitative logging, secondary events had significant effects on red spruce ecosystems. Common across the various anthropogenic disturbances is change from pre-Euro-American settlement frequency, scale, and magnitude of disturbance. Notable examples include changes in wildfire intensity, deer numbers, and domestic livestock grazing. Due to large amounts of logging slash and use of coal-fueled logging equipment, fires that occurred during and just after exploitative harvest were outside normal fire regime intensity and return interval. Deer numbers plummeted after exploitative logging, but rebounded with re-introductions and re-growth of the hardwood forest. Livestock grazing also occurred in many communities on constructed pastures and by free-range animals (Pyle and Schafale 1988).

Within the broad classification of the red spruce ecosystem, upland forest and woodland communities range from nearly pure red spruce to mixtures of red spruce and hardwood species. Generally, these forests are found on ridgetops, side slopes, and soils influenced by terrestrial processes. In the central Appalachians, these forests range across middle and upper elevations (1,070–1,400 m [3,500–4,500 ft]) and the highest elevations (above 1,370 m [4,500 ft]) on gentle slopes and on exposed ridges and summits (Fig. 5.2 and 5.3). Red spruce becomes mixed with northern hardwoods at lower elevations, but above 850 m (2,800 ft), where the climate is generally warmer. The southern Appalachian Mountains reach their highest elevations and greatest extent in eastern Tennessee, western North Carolina, and southwestern

Fig. 5.1 Construction of a railroad grade near Cranberry Glades, Pocahontas County, West Virginia, around 1911. Photo by USDA Forest Service

Virginia (Fig. 5.3). Red spruce forest can be found above about 1,370 m (4,500 ft) elevation, but only becomes widely dominant above 1,525–1,680 m (5,000–5,500 ft) elevation (Cogbill and White 1991). Ten named mountain areas in the southern Appalachians surpass the 1,680 m (5,500 ft) elevation threshold for continuous spruce forest, contain all of the southern Appalachian region's spruce-fir and red spruce forests, and are also referred to as *sky islands* (Table 5.1).

Red spruce forests also occur in small patches within the matrix of hardwood-dominated forests (central Appalachians) and *sky islands* (southern Appalachians). These wetland patches are largely found in landscape positions with low slope. Hydrologic regimes range from stream and river flooding to wetlands fed by groundwater seepage and rainfall. Protection of the hydrologic regime is important in sustaining the dynamics and ecosystem service of these wetland communities. For more detailed descriptions of community types for upland and wetland communities, see Chap. 4.

For this chapter, we reviewed and synthesized the literature on the dynamics of central and southern upland Appalachian red spruce and spruce-fir forests under five topics. We begin with a section on gap dynamics in old-growth forests that describes the natural dynamics of these forests and contrasting life strategies of the major tree

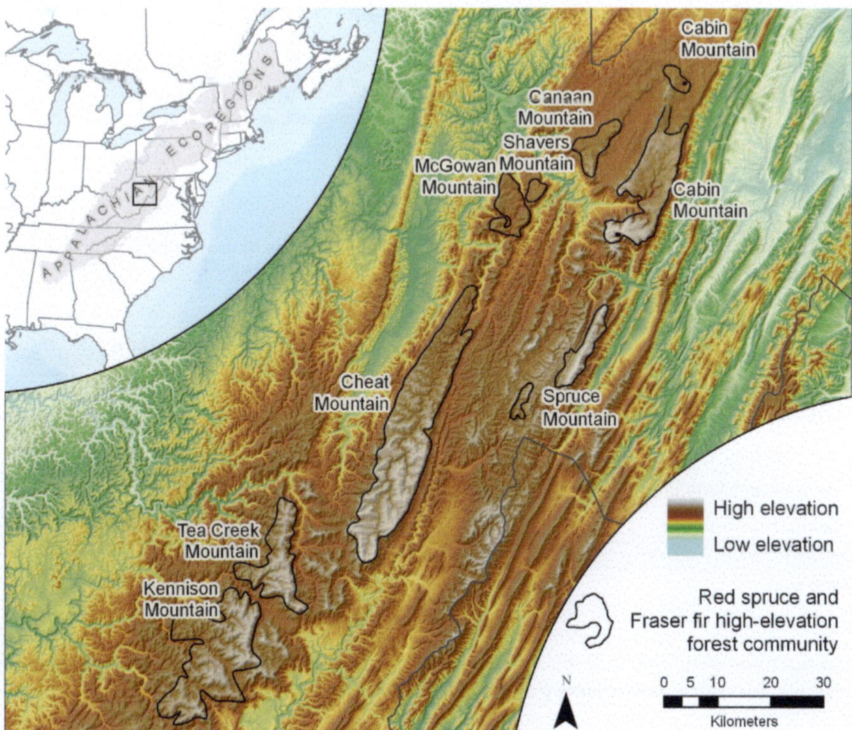

Fig. 5.2 Red spruce (*Picea rubens*) forests of the central Appalachians

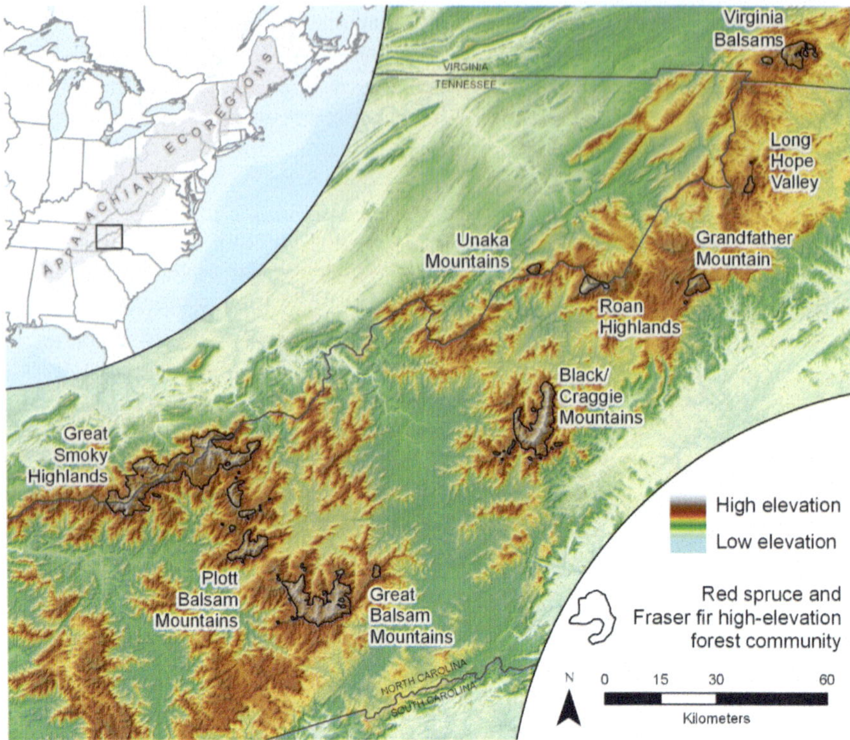

Fig. 5.3 Red spruce (*Picea rubens*) and Fraser fir (*Abies fraseri*) *sky islands* of the southern Appalachians

Table 5.1 Topographic characteristics of the ten mountain regions in the southern Appalachians that have spruce-fir or red spruce (*Picea rubens*) forest

Region	Area (km^2)	Number	Max. elev. (m)
Great Smoky Mountains	40.0	10	2,024
Great Balsam Mountains	29.4	20	1,954
Black Mountains and Black Craggies	26.8	6	2,037
Roan Highlands	8.8	3	1,916
Plott Balsam Mountains	5.7	1	1,918
Grandfather Mountain	1.1	1	1,818
Virginia Balsams (Mt. Rogers, Mt. Pisgah, and Whitetop)	0.48	5	1,743
Total	121.1	46	

Modified from White (1984). Area = area above 1,680 m (5,500 ft); Number = number of peaks in the site; Max. elev. = maximum elevation reached at the site

species: red spruce, Fraser fir (*Abies fraseri*), yellow birch (*Betula alleghaniensis*), and pin cherry (*Prunus pensylvanica*). We then discuss four topics that address human impacts on these dynamics: broad-scale logging and remnant old-growth forests; the balsam woolly adelgid (*Adelges piceae*); pollutant deposition and exposure; and climate change. We also discuss potential interactions among these disturbances, a topic that has been addressed directly in some studies we reviewed (e.g., Soulé et al. 2012; Stehn et al. 2013). Here, we treat these topics in terms of their potential impacts on community dynamics. In Chap. 2, several of the same topics are reviewed from the perspective of how they act as physiological stressors on red spruce individuals.

5.2 Disturbance Regimes

The impacts of disturbance regimes over time define ecosystem services and influence structure and composition at the landscape scale (see Sidebar 5.1). Disturbances interact with current vegetation, topography, climate and other factors, resulting in a range of landscape patterns. A disturbance regime, unlike a disturbance event, is defined by frequency, return interval, rotation period, size, intensity, severity, and residuals (Turner 2010). Additionally, a disturbance regime can be described by the agent or source, patterns, seasonality, duration, interactions, and variability of disturbances (Keane 2013).

Dynamics of spruce forests are linked to the biology of red spruce and co-occurring species. In general, red spruce has a shallow root system and is extremely shade tolerant compared to northern hardwood species. It is slow-growing and long-lived, but can respond to release; seeds need moist substrate for best germination (Blum 1990, see Chap. 2). These characteristics make red spruce susceptible to wind-throw, but enable it to regenerate from seeds or advance regeneration following disturbances and management actions that open the canopy or remove litter. However, saplings have greater net photosynthesis, leaf conductance, and xylem water potential on cloudy days compared to clear days, suggesting that red spruce may be less able than other species to respond to large disturbances that create high light, high vapor pressure deficits, and low soil moisture conditions (Berry and Smith 2013). This adaptation to low light levels and moist conditions means red spruce can be out-competed by faster-growing species, especially following broad-scale disturbances. In general, dynamics of red spruce-dominated communities differ with the characteristics of associated species and the types and scales of disturbances. As these differ subtly between the central and southern Appalachian Mountains, we discuss the regions separately below.

Sidebar 5.1: Disturbance Regimes 101 Early tenets of ecology viewed succession as a sequential linear process whereby a plant community develops into a stable climax state controlled by climate (Clements 1936). This highly deterministic idea (i.e., that there is an end point to succession) has largely been replaced by a more dynamic view of plant communities as described by H. C. Cowels, H. A. Gleason, and others (see historical overview by Kingsland 1991). Gleason (1926) contended that a plant association can be mapped, and that the association "…represents the result of an environmental sorting of a population…" that is, the species making up the community are all able to persist under the conditions at that location over time. One key difference between Clements and Gleason is the acknowledgement of disturbance and how it affects plant community development over time. In many ecosystems, communities are likely to experience multiple disturbances and may never reach a stable end point defined by Clements as climax. In both views, spatial and temporal scales are important but also often unstated.

For land managers working on time scales ranging from years to decades and spatial scales of tens to thousands of hectares, the principles of disturbance ecology are important to consider. Disturbance should be viewed as a series of events and not individual occurrences, and in aggregate they are fundamental to spatial and temporal heterogeneity across a landscape (Turner 2010). In disturbance ecology, disturbance regimes are defined by many factors including agent, frequency, return interval, size, intensity, severity, seasonality, and pattern (Table 5.2). Restoration actions designed to emulate natural disturbances benefit from an assessment of the historical range of variability in a landscape through determination of disturbance regimes for the area being considered for management (Engstrom et al. 1999; Greenberg and Collins 2016). For example, the information in Tables 5.3 and 5.4 give land managers an idea of the historic disturbance dynamics of red spruce-dominated forests.

Another important idea to consider is the concept of the shifting mosaic steady state of a landscape (Bormann and Likens 1979; Turner 2010). This concept integrates both time and space, describing various vegetation states of a landscape where individual stands are at different successional stages and changing over time, however, the proportion of the landscape in those successional stages remains constant (Turner 2010). In restoration, this idea can help define goals and frame monitoring efforts at the landscape scale. Monitoring of many restoration actions will focus on a sample of treated or planted trees, but landscape scale goals can also add to the determination of success or progress. For example, the Monongahela NF Forest Plan includes desired conditions for the spruce and spruce-hardwood forest restoration management prescription for tracking progress across age or successional classes. The desired conditions are stated as percentages of the area by successional stage, with 3–8% of the area in early successional habitat and 60–80% in late successional forest (USDA Forest Service 2006). As these forests trend toward multi-aged conditions, the spatial locations of those age classes will create a shifting mosaic of conditions.

Table 5.2 Definitions of terms related to disturbance regimes

Term	Definition
Agent	Factor or origin causing the disturbance (Keane 2013)
Frequency	How often an event occurs (Keane 2013); probability of an event, or the mean or median number of events per unit of time (Turner 2010)
Return interval	Mean or median time between events, the range or variability may also be important (Turner 2010)
Size	Area disturbed in an event, can be expressed in area or percentage of area (Turner 2010)
Intensity	Magnitude of the disturbance (Keane 2013); physical energy of or released by the event not the ecological impact (Turner 2010)
Severity	Effect or impact of the event on the community or a given component of the community (Turner 2010; Keane 2013)
Seasonality	Time of year the event occurs (Keane 2013)
Pattern	Spatial arraignment of disturbance effects (Keane 2013)

5.2.1 *Canopy Gap Dynamics*

Red spruce-dominated ecosystems of the central and southern Appalachians are driven largely by disturbances that result mainly in small canopy gaps after the death or injury of individual or groups of trees (Table 5.3). Disturbances that can lead to the downing of snags or living trees include wind-throws, ice storms (Nicholas and Zedaker 1989), and the debris avalanches and down-watershed flood scour that accompany intense rainfall events (Harmon et al. 1984; White et al. 1985). The humidity and dampness of high elevation forests mean that natural fires are rare. Post logging fires in the early 1900s, however, were significant and are described more fully below. Native insects and fungi might cause individual tree deaths, but do not cause stand-initiating disturbance to the forest canopy (Bruck et al. 1989; Blum 1990). However, there is a record of two large-scale damaging insect events in West Virginia during the broad-scale logging of the original forest. One outbreak period, from 1882 to 1886, was attributed to a spruce bark beetle (*Polygraphus rufipennis*) and a second, from 1890 to 1803, was attributed to a pine bark beetle (Hopkins 1899). Given that the damaging beetles were found during timber harvest, it is not clear whether these beetles had broad-scale impacts; in 1899, living spruce were healthy and free from insect damage (Hopkins 1899). The non-native balsam woolly adelgid, however, has significantly affected spruce-fir forests of southern Appalachian forests and is discussed more fully below.

5.2.1.1 Southern Appalachians

Although shade tolerant, both red spruce and Fraser fir also depend on canopy gaps for long-term survival and growth into the canopy. White et al. (1985) documented

Table 5.3 Estimates of major natural disturbances in the old-growth spruce-fir forests of Mt. Collins in the Great Smoky Mountains as reported in White et al. (1985). Note that the area affected only includes vegetation types and topographic positions appropriate for each disturbance type, and that year of the last event is as of 1985

Disturbance	Area affected (%)	Return interval (yr)	Recovery time (yr)
Fire	10^{-5}	$>10^7$	10^2
Debris avalanches			
Headwalls	1	10^2–10^3	10^2–10^3
Scour channels	1	10^2	10^2
Wind			
Windfall > 200 m^2	0.1	$>10^3$	10^2
Fir patches	10	50–75	20
Gaps < 200 m^2	5–20	10^2	50

the dynamics of an old-growth spruce-fir forest in the Great Smoky Mountains. This study is notable in that it used stands that had not yet suffered mortality from the balsam woolly adelgid and focused on the middle part of the gradient where red spruce, Fraser fir, and yellow birch are co-dominant (1,740–1,830 m [5,700–6,000 ft]). Small canopy gaps (≤0.02 ha [0.05 ac]) dominated the landscape and disturbance regime, with gap (complete or nearly complete reductions in canopy) creation rates of just under 1% of the canopy per year (see Table 5.3). Several methods were used to measure gaps and calculate opening size, and depending on the method used, gaps formed in the most recent decade covered 6–17% of the study area (Table 5.4). This translates to a return interval, at the scale of canopy trees, of between 111 and 178 years. As suggested by these results, red spruce and Fraser fir trees that have reached the canopy in southern Appalachian forests have been found to have periods of growth suppression and release that correlate with gap size (Reams et al. 1993; Wu et al. 1999). For example, periods of red spruce suppression averaged 19 years (range 4–79 years), while the duration of growth release periods averaged 29 years (Wu et al. 1999), which is consistent with a small gap disturbance regime and shade tolerance of this species.

Interestingly, for the small gaps studied by White et al. (1985), the species capturing a gap was not predictable from the size or age of that gap although the three main species had different competitive strategies. Fraser fir had the highest density of seedlings and saplings in the understory, but had a lifespan, and consequently a canopy residence time, only about half as long as red spruce (140 versus over 250 years). Red spruce was well distributed in forest understories but had lower density of seedlings and saplings. Korstian (1937) attributed this to the fact that red spruce seed crops come at longer intervals than those of Fraser fir (also see Nicholas et al. 1992; Johnson and Smith 2005). As might be expected from the difference in lifespan, however, red spruce held onto its place in the canopy (canopy residence time) about twice as long as Fraser fir. Yellow birch, moderately tolerant of shade, has small but widely dispersed seeds (Houle and Payette 1990) and seedlings had

Table 5.4 Summary of selected gap size characteristics and mortality rates in red spruce-dominated forests of the central and southern Appalachians from published literature

Location	Community type	Mean gap size (range) in m^2	Annual mortality	Proportion of area in gaps	Other information	Reference
West Virginia	Red spruce-northern hardwoods	53.4 (6.2–276.4)	1.4%	4.7%	Second-growth stands; gaps generally elliptical	Rentch et al. (2010)
West Virginia	Red spruce-northern hardwoods	50.6 (5–459)		7.6%	Second-growth; about 79% of gaps smaller than 75 m^2; average gap age 19 years (SE 6.8)	Lutz (2018)
Tennessee and North Carolina	Red spruce-Fraser fir		1%		Smaller trees experience higher mortality	Busing and Wu (1990)
Great Smoky Mountains NP/ Tennessee and North Carolina	Red spruce-Fraser fir	60–174		6–17% in gaps 1–10 years of age	Return interval of 111–178 years	White et al. (1985)
Great Smoky Mountains NP/ Tennessee	Red spruce-Fraser fir		2% of red spruce canopy trees		Compared mortality before and after adelgid-caused fir mortality	Busing and Pauley (1994)

very low survivorship in the shade but had the highest growth rates in gaps. In short, yellow birch compensates for a smaller starting size (little advance regeneration) with faster growth rates, whereas red spruce and Fraser fir start from larger sizes (advance regeneration), but have lower growth rates. Fraser fir captures most gaps, but the other two species can hold onto canopy positions twice as long.

In larger windfall gaps (>0.02 ha [0.05 ac]), on open soil after debris avalanche, and on flood scour along streams, pin cherry becomes important in spruce forest dynamics. This species, too, has distinctive life history characteristics, the most important of which is its ability to form a persistent soil seedbank of dormant seeds that can survive 100 or more years before germination (Tierney and Fahey 1998). In large disturbance patches, faster litter decomposition rates, due to higher forest floor temperatures from increased solar exposure, and less competition from larger trees result in higher soil solution nitrate, which triggers pin cherry germination (Marks 1974; Nyland et al. 2007). As a result of its rapid colonization, pin cherry serves to conserve nitrogen on watersheds (Marks 1974; Fahey et al. 1998; Nyland et al. 2007). Pin cherry growth rates are the highest and life span the shortest (<80 years) of the four species profiled here. The large disturbances that lead to pin cherry colonization are rare, however, and succession more often leads to dominance by red spruce, Fraser fir, and yellow birch.

Additional spruce-fir forest dynamic patterns are found on high summits, ridges, and upper slopes. On the highest summits in the southern Appalachians, where Fraser fir forms nearly pure stands (White et al. 1985), upper slopes are wind-exposed, and blowdowns affect larger patches of trees. Although seedling survival can be low in direct sunlight (Johnson and Smith 2005), Fraser fir forms a high density of seedlings and saplings so that, after wind disturbance, the ensuing stand of maturing trees is of similar heights and interlocking crowns, making the forest vulnerable to larger windthrow patches. Aerial photographs taken before balsam woolly adelgid invasion show these larger patches of windthrow in pure Fraser fir forest (Peter White, personal observation). Evergreen rhododendron (*Rhododendron* spp.) populations can also influence spruce-fir dynamics on upper slopes and convex ridges. These shrubs cast deep shade and produce thick, slow to decay, leaf litter, thereby acidifying soils (Nilsen et al. 1999; Wurzburger and Hendrick 2007). At least at small scales, this reduces tree seedling establishment and leads to stable shrub communities, called heath balds, on the most convex ridges and upper slopes (White et al. 2001). Several areas with severe logging impacts are dominated by *Rhododendron* thickets with some trees today, and it is unclear whether past logging altered the density of the heath shrubs. However, no evidence indicates that past logging altered the shrub pattern at the landscape scale (White et al. 2001; Woodbridge and Dovciak 2022).

5.2.1.2 Central Appalachians

In the central Appalachians, Fraser fir does not occur and balsam fir (*Abies balsamea*) is a very small component of spruce-dominated forests. Very few red spruce-dominated old-growth forests remain in the central Appalachian region for study

Fig. 5.4 Contemporary forest structure and composition at Gaudineer Scenic Area, Monongahela NF. Approximately 20 ha (50 ac) of this scenic area is virgin red spruce-northern hardwood forest. Many of the largest red spruce (*Picea rubens*) trees have died from natural causes. The complexity of the forest structure, size of trees on the forest floor, and standing snags are examples of old-growth conditions possible in this forest type. Photo by USDA Forest Service

of disturbance patterns or species composition (Fig. 5.4). A 1920 report based on timber inventory data collected for the West Virginia Pulp and Paper Company for its holdings in Pocahontas, Randolph, and Webster counties gives us a glimpse of the original red spruce forest's character (Sterling 1920). These lands, the headwaters of the Cheat and Elk rivers, included about 52,000 ha (130,000 ac) of cut over, burned, and unharvested forests. The primary purpose of the survey was to quantify current reproduction on the cut over lands to calculate future value and growth of the second-growth forest. The virgin unharvested areas were described as small parcels with harvests occurring rapidly. Contracted foresters divided the area into three forest types based on dominant species (softwood, softwood-hardwood, and hardwood) and noted that existing reproduction under mature overstory generally matched the overstory in species composition. This suggests the original forest had reached understory reinitiation stage of development (Oliver and Larson 1996) and was a stable and self-sustaining community. Based on second-growth spruce-dominated forests and old-growth forests in other regions, and this glimpse of the unharvested areas of West Virginia, we can generalize that the pre-European settlement red spruce forests were dominated by a disturbance regime of small canopy gaps. As found in other forest types, it is likely that trees would have been taller and larger in old-growth forests and canopy gaps were likely larger on average compared to canopy gaps in current second-growth forests (Clebsch and Busing 1989; Ziegler 2000).

Canopy gap formation in second growth red spruce-northern hardwood forests of West Virginia is estimated to be about 1.4% of stand area per year based on ages of existing gaps (Rentch et al. 2010). Like in other red spruce-dominated forests, most gaps (65%) were created by the death of one or two trees; however, in the West Virginia study gaps were the result of the death of American beech (*Fagus grandifolia*) by beech bark disease (Rentch et al. 2010). Gap creation rates only tell part of the stand dynamics story even in forests dominated by a long lived and shade tolerant species such as red spruce. Gap closure rates differ by species composition of the stand. Where present, hardwoods in the canopy can quickly close a canopy gap by lateral expansion of branches. Consequently, existing red spruce seedlings and saplings will likely require more than one release event to reach the upper canopy status (Rentch et al. 2007). Gaps formed by the gradual decline and death of diseased trees, such as American beech following beech bark disease or eastern hemlock (*Tsuga canadensis*) following hemlock woolly adelgid (*Adelges tsug*ae), differ in character from gaps formed by sudden windthrow due to gradual changes in light levels. Additionally, repeated disturbances are more likely to expand existing gaps than create new gaps as trees on the edges of existing gaps are susceptible to uprooting by wind (Worrall et al. 2005).

The culmination of a disturbance regime dominated by small gap dynamics results in an uneven-aged forest structure with a reverse J-shaped distribution when stems per hectare are plotted by size class. To explore the structure of stands likely to be included in restoration efforts, a second growth (approximately 80 years old) red spruce-dominated forest in West Virginia was compared to an old-growth reference stand. Although the distribution of stem diameters showed second-growth structure and maximum DBH for red spruce trees was smaller than the reference stand, the study stand did have some characteristics of an old-growth forest, including similar basal area, snag density, and downed coarse woody debris (Schuler et al. 2002).

Although wind throw is the main creator of canopy gaps, ice storm damage is also very common in central hardwood and central Appalachian forests (Lafon 2016) and southern Appalachians (Nicholas and Zedaker 1989), and results in a disturbance regime of intermediate severity (partial canopy removal or reduction) but potentially high frequency. The Appalachian Mountains are uniquely situated for continental air masses to interact with high elevation and complex terrain to produce weather events with heavy ice accumulations (Changnon and Karl 2003). Based on the analysis of climate and weather data, the Appalachian Mountains can experience up to four or five days with freezing rain annually (Changnon 2003). Combining tree ring records with documented accounts of ice storms in newspaper accounts has shown a return interval of 25 years for damaging ice events in southwest Virginia between 1914 and 1998 (Lafon and Speer 2002). Further, dendrochronology exposes a combination of ice storm effects on trees, with heavily damaged trees showing reduced growth and other stems showing increased radial growth (Lafon and Speer 2002). Even though the variation between sites and individual trees will occur, red spruce is considered resistant to ice or glaze damage (Lemon 1961) and in general this disturbance regime sustains shade tolerant and slow growing late successional forest species.

5.2.1.3 Influence of Rhododendron Species

Some upland communities dominated by red spruce include a significant heath family (*Ericaceae*) shrub component (Fig. 5.5). These communities are found at middle and lower elevations in red spruce-dominated and red spruce-northern hardwood forests, generally in sheltered locations such as concave sideslopes and cove landforms. While the canopy disturbance regime is still dominated by wind throw in these upland communities, the role of rhododendron species (predominantly *Rhododendron maximum*) is not trivial. Early descriptions of spruce communities noted that great laurel and red spruce combined to retain moisture and maintain cooler air temperatures (Allard and Leonard 1952). Studies in the southern Appalachians documented the influence of rhododendron species in lowering seedling density and light levels, and found limited recruitment of trees into gaps created where rhododendron dominated the shrub layer (Beckage et al. 2000). Rhododendron's ability to stabilize microsite shade, moisture, and temperature suggests that rhododendron could be considered a foundation species that adds stability in forests impacted by pests and climate change (Dudley et al. 2020).

Great laurel is widespread and abundant in the mountains of both the central and southern Appalachians, meeting the first criterion for a foundation species. Rhododendron species also can replace red spruce on toeslopes after fire, making red spruce reestablishment there more difficult due to negative impacts of dense rhododendron on soil and light, which hinders regeneration. Leaves and roots of great laurel are slow to decay and produce soils with lower nutrient levels, but higher organic matter, than those in sites without rhododendron (Beckage et al. 2000). Soil nitrogen, for example, is impacted by rhododendron through creation of compounds that inhibit nitrification (Wurzburger and Hendrick 2007). Although air under rhododendron thickets can be higher in moisture, soil moisture is lower (Clinton and Vose 1996), pH is lower, and aluminum content is higher (Horton et al. 2009) compared to deciduous overstories without rhododendron shrub layers. These conditions might also

Fig. 5.5 Example of a great laurel (*Rhododendron maximum*) understory in a red spruce (*Picea rubens*) forest along the Yellow Creek Trail in the Otter Creek Wilderness, Monongahela National Forest, Randolph County, West Virginia. Photo by Kelly Bridges

be expected in spruce forests, as red spruce also has recalcitrant leaf litter and the ability to acidify soils.

Once rhododendron is established it is difficult to remove either by natural disturbances or by management practices such as fire, cutting, or herbicide treatments (Harrell and Zedaker 2010). Consequently, rhododendron shrub layers afford resilience and resistance to changes in other ecosystem components even with loss or reduction of tree canopy (Beckage et al. 2000), although they also lower plant diversity and affect species composition through reduced seed germination and seedling recruitment (Stehn et al. 2011). Further, rhododendron's shallow rooting habit increases the potential for landslides when soils are saturated (Hwang et al. 2015), and severe droughts can lead to high fuel loading and flammability in dense rhododendron patches (Stottlemyer et al. 2009). However, current understanding of rhododendron effects on ecosystems is largely based on thickets in hardwood forests.

5.2.2 Industrial Logging Era (1890–1940)

Harvesting of the historical red spruce forest began in the late 1800s and became particularly intense from about 1920–1940 as roads and railroads penetrated even the most remote high elevations (Fig. 5.6; Pyle and Schafale 1988; Lewis 1998), resulting in clearcutting of whole watersheds. Accidental fires were frequent, but logged sites were also burned to remove slash, which, given the region's high rainfall, led to soil erosion. Many logged areas were burned twice (Korstian 1937). As a result, red spruce regeneration failed widely in the southern Appalachians due to loss of the seed bank and seed trees (Korstian 1937). In the Great Smoky Mountains, the lower limit of spruce-fir forests was displaced upward by approximately 200 m (650 ft) and the overall extent of this forest type was reduced by about one-third (Hayes et al. 2007) after broad-scale harvesting. Over a century after this historical logging, some logged, burned, and eroded slopes in the Smokies have not yet developed continuous tree cover (Lindsay and Bratton 1979; Peter White, personal observation).

Fig. 5.6 Sherwood Forest burn on the Pisgah NF; photo taken in April 1938. Bracken fern (*Pteridium aquilinum*) and blackberry (*Rubus* spp.) dominated the site after the fire and pin cherry (*Prunus pensylvanica*) is sparse or absent at this elevation (approximately 1,680 m [5,500 ft]). See Minckler (1939) for additional information. Photo by USDA Forest Service

The failure of post-logging regeneration has been related to several traits of red spruce: high fire sensitivity because of shallow roots, relatively thin bark early in life, slow growth rates, and resinous exudates (Korstian 1937). Further, red spruce requires damp substrates for seedling survival, and logging can reduce moisture in the forest floor (Brooks and Kyker-Snowman 2008). Mechanical disturbance from logging practices of the period damaged existing regeneration of red spruce and other shade tolerant species, and red spruce subsequently was outcompeted by hardwoods that sprouted prolifically from surviving root systems (unlike red spruce which do not sprout from roots or stumps) or those with light, wind-borne seed.

The few direct assessments of succession after logging have found both broadscale patterns and site-specific differences. Compared to those that were unlogged, logged sites in the southern Appalachians can have lower abundance of red spruce (Smith and Nicholas 1999; Soulé et al. 2012) and younger-aged spruce (Smith and Nicholas 1999), but higher tree diameter growth rates and recruitment into the canopy across all species (Smith and Nicholas 1999), as well as release of Fraser fir from the understory (Soulé et al. 2012). However, on Mt. Rogers in southwestern Virginia, maximum elevation is relatively low (1,747 m [5,700 ft]) and the elevation gradient from red spruce to Fraser fir dominance, which is typical in other areas of the southern Appalachians, does not occur. Here, logged stands, which ranged in age from 60 to 92 years, were found to be dominated by Fraser fir (importance value [i.e., the sum of relative density, relative frequency, and relative dominance] = 72), followed by red spruce (importance value = 21), mountain ash (*Sorbus americana*; importance value = 5.4), and yellow birch (importance value = 1; Stephenson and Adams 1984). Compared to a survey in the 1950s, mountain ash and yellow birch had increased in importance. Despite the observed site-level differences, Smith and Nicholas (1999) concluded that it likely would be many decades before the "forest communities resembled those in old growth areas."

From studies on second growth forests, we can construct general models of post-logging succession in which effects of logging vary with magnitude of logging and pre-logging composition (and thus elevation). These general dynamics for the main upland spruce communities are depicted in the conceptual state and transition models in Fig. 5.7. In the broad-scale logging era, red spruce failed to reproduce, or at least failed to reproduce as quickly as other species (Korstian 1937) where harvests were most intense. Where harvests occurred with partial removal of canopy trees and minimal soil and understory damage, post-logging forest dynamics depended on composition. If the understory was dominated by Fraser fir, as is typical at low and mid-elevations within the range of red spruce forests, then Fraser fir increased. If hardwoods were present in the overstory and understory, as is typical at lower elevations within the range of red spruce forests, hardwood sprouting and seedling establishment likely dominated the resulting young forest. In contrast, if large areas were cut, and particularly if the soil was damaged by fire and soil erosion, colonization by early successional species like pin cherry and blackberry (*Rubus* spp.) predominated. At higher elevations and in forests in which Fraser fir was a dominant or co-dominant and with advance Fraser fir regeneration, balsam woolly adelgid

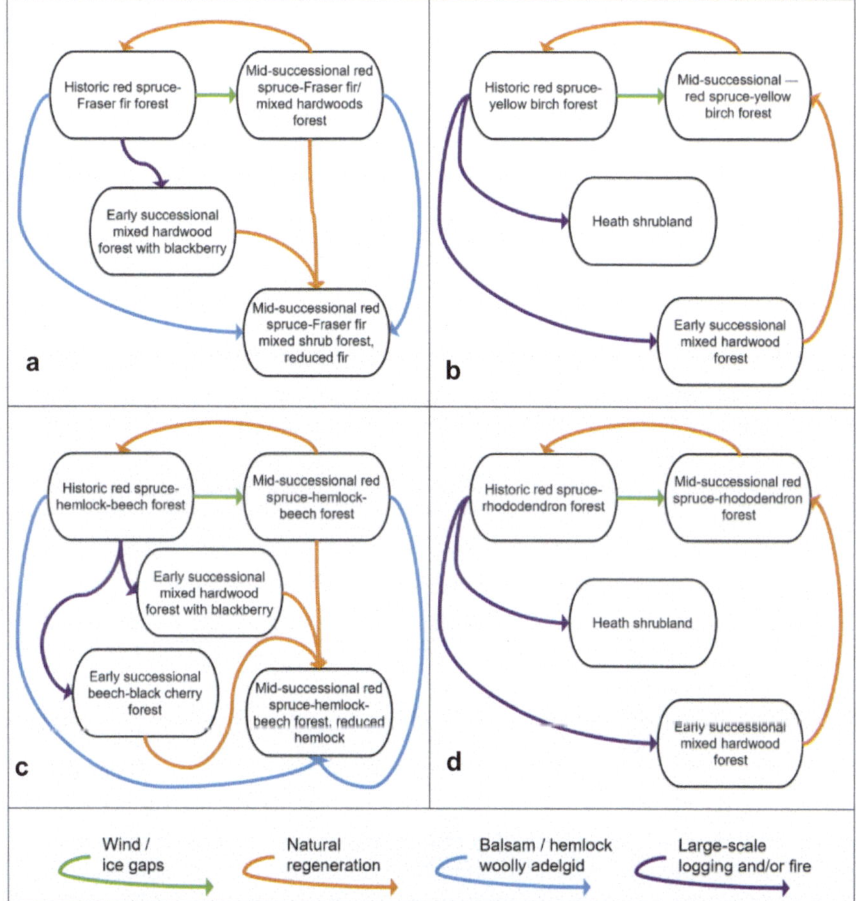

Fig. 5.7 Conceptual model of historical red spruce (*Picea rubens*)-dominated forest states and transitions through biotic and abiotic disturbances. Historical red spruce-Fraser fir (*Abies fraseri*) forest (A), historical red spruce-yellow birch (*Betula alleghaniensis*) forest (B), red spruce-hemlock (*Tsuga*)-beech (*Fagus*) forest (C), and historical red spruce-rhododendron (*Rhododendron*) forest (D)

creates a second episode of disturbance, leading to a recovering stand of young Fraser fir (further discussed in Sect. 5.2.3).

5.2.3 *Balsam Woolly Adelgid—Southern Appalachians*

The balsam woolly adelgid, a Eurasian invasive species, was first detected in the Black Mountains of North Carolina in 1957, apparently after it had been transported to the region on infested nursery stock (Eagar 1984; McManamay et al. 2011). Over the next three decades, balsam woolly adelgids spread throughout the range of Fraser fir, causing near complete mortality of canopy Fraser fir trees (Allen and Kupfer 2000, 2001). While the balsam woolly adelgid has relatively minor effects on fir seedlings and saplings, mature trees typically succumb within seven years of infestation because this sap-sucking insect feeds in bark crevices that develop over time, setting off a wound reaction in fir stems that greatly damages the function of the sapwood (Eagar 1984). The adelgid disperses on wind currents and eddies produced by ridgelines which caused balsam woolly adelgid populations to settle first in the lower elevation part of the Fraser fir range, followed by movement upslope (Allen and Kupfer 2000, 2001). From its initial establishment, the geographic pattern of balsam woolly adelgid-caused mortality followed from its pattern of dispersal. For example, Allen and Kupfer (2000, 2001) used remote sensing imagery to document the pattern of Fraser fir mortality from 1988 to 1998 in the Great Smoky Mountains, finding nearly complete canopy tree mortality moving from east to west and from the lower elevation part of the Fraser fir range to higher elevations. As part of a larger assessment, aerial photography and ground-based sampling in the mid-1980s documented 26,610 ha (65,752 ac) of spruce-fir forest with 24% classed as severe mortality, 6% as heavy, and 70% as light mortality (Dull et al. 1988).

Smaller stems do not become vulnerable to the balsam woolly adelgid until they are large enough to develop bark crevices; thus, the initial waves of Fraser fir mortality left behind a high density of younger stems. These are now contributing to recovery of the Fraser fir population (Boner 1979; Smith and Nicholas 1998, 1999; Jenkins 2003; Moore et al. 2008; Bowers et al. 2010; Lusk et al. 2010; McManamay et al. 2011; Soulé et al. 2012; Stehn et al. 2013; Kaylor et al. 2017). For example, in spruce-fir forests of the Great Smoky Mountains, biomass was stable overall (around 260 megagrams [Mg]/ha [105 Mg/ac]), but Fraser fir biomass increased from 3.3 to 12.7 Mg/ha (1.3–5.1 Mg/ac) between 1993 and 2003 (Moore et al. 2008). However, Fraser fir recovery may be weaker in the lower part of its elevation range (McManamay et al. 2011).

It is uncertain whether patches of healthy Fraser fir will survive long enough to provide seeds for future generations and open the possibility of long-term persistence (Dale et al. 1991; Smith and Nicholas 2000; Moore et al. 2008). Eagar (1984) noted that the time to develop bark fissures is shorter than the time required for cone production, suggesting a likely decline of Fraser fir populations in the future despite perceived recovery. On the other hand, it is possible that the balsam woolly adelgid and Fraser fir will continue to coexist through a patch dynamic in which the balsam woolly adelgid becomes locally scarce (since it does not thrive on younger stems), allowing Fraser fir stands time to reproduce because there is a delayed recolonization by balsam woolly adelgids, resulting in increasing patchiness of Fraser fir populations. Additionally, clones of Fraser fir showed potential resistance to the balsam woolly adelgid through production of sesquiterpenes, which may be useful indicators of genetically resistant genotypes (Thomas et al. 2022).

Although the balsam woolly adelgid does not attack red spruce directly, the sudden loss of Fraser fir canopy has been shown to shift the cause of red spruce mortality

(Busing and Clebsch 1988; Busing et al. 1988; Busing and Wu 1990; Pauley and Clebsch 1990; Busing and Pauley 1994; Wu et al. 1999; Busing 2004). In spruce-fir forests of Great Smoky Mountains National Park, increases in red spruce tree mortality rates over time were attributed to the sudden death of Fraser fir exposing spruce to more wind damage, especially on exposed topographic positions (Busing 2004). Interestingly, dendrochronological analysis of those dead red spruce at Mt. Collins in the Great Smoky Mountains showed that mortality was not associated with reduced radial growth. This finding contradicts the hypothesis that acid rain was causing the red spruce mortality because, if acid rain was responsible, reduced radial growth would accompany increased mortality. The sharp increase of Fraser fir and red spruce mortality produced a large increase (93–132%) in red spruce sapling density (Busing et al. 1988). Nonetheless, live basal area in these stands declined by 50%.

5.2.4 Pollutant Deposition and Exposure

Pollutant deposition and exposure have been the subject of much research in the central and southern Appalachians (Eagar and Adams 1992; Rosenberg and Butcher 2010; see also Chap. 3). Pollutant deposition and exposure in the Appalachians increased from the 1960s to the 1980s and then declined from the 1990s onward, as the Clean Air Act (1970, as amended 1977 and 1990) led to reduction of emissions from industrial point sources in regions to the west and northwest of the Appalachians (Moore et al. 2002; Lawrence and Bailey 2021). Along with rainfall and cloud immersion, pollutant deposition and exposure increase with elevation. The elevation gradient has therefore been used to determine how spruce forests respond to various levels of exposure (e.g., Berry et al. 2014). Spruce-fir forests in the southern Appalachians are immersed in clouds 25–40% of the time and one third to one-half of all water at upper elevation sites derives from cloud immersion (Aneja et al. 1992; Reinhardt and Smith 2008). Cloud-intercepted water is more acidic (pH = 2.5–4.5) than rainwater (pH = 3.5–5.5; Aneja et al. 1992).

As reviewed in Chap. 3, acid exposure resulted in leached cation nutrients from red spruce needles, made soil cations vulnerable to leaching, reduced cation retention in soils, and led to a reduction in calcium to aluminum ratios to potentially toxic aluminum levels (e.g., McLaughlin et al. 1991). These effects were superimposed on soils that are naturally acidic and poorly buffered, and a number of studies have shown declining nutrient cations and low calcium to aluminum ratios that are correlated with toxicity in soils (Robarge et al. 1989; Joslin and Wolfe 1992; Bintz and Butcher 2007; Rosenberg and Butcher 2010; Stehn et al. 2013). There are five documented physiological effects of this acidification on red spruce (McLaughlin et al. 1993, 1995; McLaughlin and Percy 1999): (1) a decrease in cold hardiness and, at least in the northern Appalachians, subsequent winter damage (Thornton et al. 1994); (2) an increase in respiration and a decrease in photosynthesis (McLaughlin et al. 1990, 1991, 1993); (3) membrane disruption and foliar injury, including needle speckling (DeHayes et al. 1999); (4) lessened response to water stress because of effects on leaf stomata (Borer et al. 2005); and (5) potential aluminum toxicity to growth of fine roots (Schier 1985). McLaughlin et al. (1995) tied many of these changes to reduced calcium availability, especially membrane-associated calcium. Work by Stehn et al. (2010, 2011, 2013) suggests heterogeneous effects of acidification that affect regeneration, including local hotspots of high nitrogen and increased competition, hotspots of aluminum levels toxic to sensitive species, and acidification that favors members of the heath family (e.g., *Rhododendron*) whose acidic and slow decaying leaf litter reinforces the acidification effect.

A similar loss of cations and decrease in the calcium to aluminum ratio in red spruce tissues as soils become acidified was found at Whitetop Mountain, Virginia (Joslin et al. 1992). In contrast, another study that sampled sites in Virginia, West Virginia, and Tennessee found that foliar and soil cations were only weakly correlated, aluminum levels were "well below toxic levels," and cation concentrations were not indicative of deficiencies (Bryant et al. 1997). However, the authors noted that they had no high elevation sites, suggesting that perhaps the greatest acidification results occur only at the highest elevations (>1,800 m [6,000 ft]). As described below, red spruce and spruce-fir forests are generally considered to be "healthy," with the most dramatic changes occurring during recovery from balsam woolly adelgid infestation and recovery from earlier exploitative logging. However, pollutant deposition and exposure, though at reduced levels since the 1990s, continues. Further, impacts from climate change are just beginning to be studied.

Although elevated mortality of red spruce was widespread in the northern Appalachians in the 1960s to the 1980s, Johnson et al. (1992), summarizing a decade of research on southern Appalachian forests, concluded that widespread mortality had not occurred in this region (e.g., Busing et al. 1988; McLaughlin et al. 1990, 1998; Cook and Zedaker 1992; Johnson et al. 1992; Busing and Pauley 1994; Smith and Nicholas 1999). Further, evidence for red spruce growth declines linked to pollution is mixed with some studies finding growth declines, particularly at the highest elevation study sites (McLaughlin et al. 1990, 1994, 1998; McLaughlin and Percy 1999; White et al. 2014), and some studies finding no growth declines (Busing and Wu 1990; Cook and Zedaker 1992; Busing 2004; Soulé 2011). Indeed, research since 2000 suggests that there is no ongoing decline in red spruce growth (Bowers et al. 2010; Lusk et al. 2010; Mathias and Thomas 2018). Nowacki et al. (2010), stated "no problems found for red spruce recruitment under current climate" and suggested that red spruce populations were increasing, perhaps recovering, albeit slowly, in some of the areas that had been lost during the wave of clearcut logging and logging slash fires in the late 1800s to early 1900s. In addition, a positive trend in radial growth rates of red spruce was found on Grandfather Mountain in the southern Appalachians (Soulé 2011). From these studies, it does not appear that pollutant deposition and exposure have, by themselves, caused a change in the dynamics of forests dominated or co-dominated by red spruce. However, research has also suggested that there is a more chronic, longer-term issue: ongoing ecosystem acidification. In the long-term, this could slow element recycling rates and has the potential to affect community dynamics in the future, particularly if tolerances of some species are exceeded and species differ in their responses (Nodvin et al. 1995; Pardo et al. 2018). Acidification stress will likely interact with climate change. Further, stream ecosystems have experienced greater acidification effects and have had a slower response to recent reductions in pollutant emissions because differing rates of soil retention of sulfur and nitrogen pollutants create a lag in release and cycling of these nutrients (Pardo and Duarte 2007; Fakhraei et al. 2016), meaning forest restoration may fail to result in immediate changes in stream chemistry. Given these longer-term uncertainties, monitoring of key species and environmental factors will be needed.

5.2.5 Climate Change

Red spruce and spruce-fir forests are the highest elevation forested ecosystems in the central and southern Appalachians and are dependent on the cool and wet conditions that prevail at these elevations. As a result, they form a series of *sky islands* where mountains reach sufficient elevation, from Shenandoah National Park in Virginia and south to the high mountains of North Carolina and Tennessee. In the central Appalachians of West Virginia red spruce-dominated forests are found over larger areas and are less isolated, although still found at higher elevations. A straightforward expectation is that climate warming will cause these islands to become smaller as the ideal environmental conditions move to higher elevations (Potter et al. 2010). Red spruce and spruce-fir forests will likely disappear from mountains that are not tall enough to provide the environmental conditions necessary for these forests. For example, Walter et al. (2017) predicted that the area of habitat supportive of red spruce would decline from the present day to the year 2100, but "the magnitude of this decline depended on the level of carbon emissions, and there was considerable variability between climate models."Red spruce has been ranked as the most vulnerable of 40 tree species to climate change, followed by two other frequent associates of red spruce, yellow birch, and balsam fir (Rogers et al. 2016). Thus, red spruce and spruce-fir forests have frequently been called the most threatened forest ecosystem in the eastern U.S. (e.g., McLaughlin et al. 1998). Please also see Chap. 7 for more in-depth discussions of the topics covered here and for more information.

Current models for red spruce documented in the Forest Ecosystem Atlas (https://www.fs.usda.gov/nrs/atlas/tree/97) conclude that habitat suitability is expected to decline range-wide, and that red spruce is considered to have low adaptability to change and poor ability to adapt to climate change (Peters et al. 2020). However, red spruce climate adaptability can be explored at different scales. In the central Appalachian Mountains, climate forecasting results suggested that red spruce in northern latitudes on south aspects or central latitudes on north aspects will likely be the most resilient to future climate change due to increases in monthly minimum temperatures and extended growing seasons (Yetter et al. 2021). Site-specific dendroclimatic results and future growth projections can assist with identifying locations that are most suitable for future red spruce restoration activities.

Climate change has the potential to reduce the amount of area suitable for supporting red spruce, and thus effects of climate change on the distribution of suitable habitat is an important management consideration. However, research in northeastern spruce-fir forests supports the need to consider the historical range of spruce in predicting future distributions. Along these lines, Andrews et al. (2021) found that models that included historical data predicted persistence of spruce-fir, while models that did not include historical data predicted extirpation of spruce-fir forests. Further, mountain spruce-fir ecotones in New England have been found to be changing in complex ways, with some expanding downslope, rather than upslope (Foster and D'Amato 2015, 2018), likely due to ongoing response to historical disturbances. Declining range is not, however, the entire story, as researchers have noted that red spruce only partially occupies some suitable habitats because of past logging impacts (Walter et al 2017). There are opportunities to expand spruce and fir forests through restoration of sites where this forest failed to regenerate after the intense logging of the early 1900s (Morin and Widmann 2010; Nowacki et al. 2010; Yetter et al. 2021).

Although species distribution models have produced useful perspectives, more complex models that incorporate important factors like pollutant deposition, disturbance rates, demographic processes like seed production and establishment, competitive interactions (for example, with evergreen shrubs), and ecosystem processes such as nutrient cycling are needed. For example, Koo et al. (2011, 2014, 2015) built a more complex model to include these kinds of factors by adding demographic modeling and individual tree growth to a habitat model. Under climate change projections and three pollution scenarios (10% increase, no change, and 10% decrease) they found that lower elevation populations of red spruce are most at risk (Koo et al. 2015). Clouds and precipitation are important to these forests but harder to predict than temperature change. At the very least, ecological complexity and the uncertainty of future climate changes and emission levels make a strong case for further research and a long-term program of monitoring red spruce populations and environmental parameters, even as restoration efforts occur.

Because red spruce and spruce-fir forests are dependent on high moisture supply, the effect of climate change on precipitation and cloud immersion is critical (Berry et al. 2014). Red spruce radial growth can be positive with warmer temperatures as long as moisture is not limiting (Soulé 2011; Ribbons 2014), although warmer temperatures may increase competition with warm-adapted species moving into red spruce territory. Berry and Smith (2012, 2013) found that xylem water potentials decreased with lower cloud immersion and, as a result, photosynthesis, carbon gain, leaf conductance, and growth of seedlings and saplings decreased. Red spruce has also been found to grow faster when the previous winter and current growing season were warmer (Soulé 2011). There was a positive correlation with CO_2 and a negative correlation with sulfur and nitrogen oxide exposure (Soulé 2011). White et al. (2014) showed that the correlation between red spruce radial growth and temperature changed after logging, where the correlation was positive before the 1930s and then became negative, possibly because of an effect of logging on soil properties, leading to drier conditions during warm temperatures. As noted above, McLaughlin et al.

(1994) hypothesized that red spruce growth became insensitive to climate variation under acidification effects.

Some researchers have asserted that the average cloud ceiling sets the lower boundary of spruce-fir forests and, in a Vermont study, that climate change is causing that ceiling to move to higher elevations (Richardson et al. 2003). In the Great Smoky Mountains, low elevation sites tracked regional weather patterns more closely than high elevation sites, suggesting the importance of cloud immersion and high rainfall in buffering temperature change (Fridley 2009). Low elevations also warmed during regional warm periods, but high elevation sites experienced less warming. Red spruce and spruce-fir forests at lower elevations may be more vulnerable to the rate and magnitude of temperature change than those at higher elevations. Patterns of cloud immersion and precipitation will likely change as a function of climate change. However, this reasoning also depends on whether regionally warm periods are a good analogy for future warming—that is, the nighttime-daytime contrast in temperatures and seasonal differences in warming may be more important in future climates than they are in contemporary warm periods.

In complex mountain landscapes, the degree of climate change will not be constant across topographic gradients. Rather, the degree of change is itself a function of those topographic variables. Some sites may be relatively stable and resilient over time, while others change more dramatically. This question has been examined directly in the central and southern Appalachians. Temperature across the region at a 4 km (2.5 mi) scale showed clear differences in the degree of temperature variability in the years 1950–2005; mountaintops had more stable temperatures over time (Nelson 2017). Temperature sensitivity is a factor for choosing restoration sites using fine-scale environmental analysis and for continued monitoring of environmental change at small scales to analyze the effect of changing climates on active restoration.

Disjunct low elevation red spruce populations in wetlands may be particularly vulnerable to future climates that are drier and warmer. The Alarka Creek headwaters basin is among the lowest elevation and southernmost of these wetlands in the southern Appalachians (Schafale and Weakley 1990; White and Cogbill 1992; Collins et al. 2010). This site has a mixed canopy of red spruce, American serviceberry (*Amelanchier arborea*), sweet birch (*Betula lenta*), and other hardwoods, with red spruce more abundant than hardwoods in the larger size classes (Collins et al. 2010). The understory has a patchy, but dense, shrub layer of great laurel (*Rhododendron maximum*) and mountain laurel (*Kalmia latifolia*) on gully tops (Collins et al. 2010). Vegetation and tree ring analyses found that red spruce seedlings and saplings ranged from 0.5 to s4 m (1.6–13.1 ft) tall; overstory tree ages ranged up to 155 years with a linear population structure, suggesting steady spruce recruitment over the last hundred years (Collins et al. 2010). However, drops in diameter growth in the 1930s and late 1990s correlated with widespread drought and a regional pine beetle outbreak, respectively (Collins et al. 2010). Thus, while red spruce in sheltered basins with cold air drainage, such as the Alarka basin, may persist at lower elevations, they may be sensitive to periods of extreme drought or pest outbreaks.

5.3 Management Implications

Efforts are underway to restore and manage red spruce and spruce-fir forests using non-commercial and commercial methods (see Chaps. 8 and 9). Active restoration efforts rely, in part, on restoring historical disturbance regimes, including fine-scale canopy gaps and soil disturbance associated with tree blowdowns. Non-commercial actions include canopy release of red spruce stems to increase their height growth and accelerate their advancement to the overstory. These targeted treatments mimic the multiple release events resulting from ice storms and wind throw that make up the dominant disturbance regime of these forests.

In the central Appalachians, two examples of simulated stand-level forest management show potential success in increasing stand-level red spruce basal area and individual tree diameter growth through thinning. Using actual stand data, Schuler et al. (2002) compared two thinning treatments to no management over 50 years. The time to develop old-growth structural characteristics could be shortened by thinning smaller-sized trees and removing at most 50% of the existing basal area (Schuler et al. 2002). In a larger study involving more study areas, simulated treatments where 50% of the basal area was removed to release understory red spruce resulted in a near doubling of red spruce basal area (Rentch et al. 2007). The treatment in this simulation was a one-time release event or liberation cut where all trees that overtopped targeted red spruce were removed. A similar release of targeted red spruce from overtopping hardwoods was also tried in a field experiment in West Virginia (Rentch et al. 2016). Three levels of release were implemented on target red spruce trees of >2.54 cm (1 in) DBH, overtopping trees were killed through stem injection of herbicide, and height and DBH of target trees tracked for six years. The treatment that provided the highest level of release resulted in the largest change in height and DBH growth, although increases were of small magnitude (Rentch et al. 2016).

Traditional silviculture guides, developed without considering restoration, suggest single tree selection and group selection for commercial management (Blum et al. 1983). Even-aged practices are mentioned but a caution is given that planting may be necessary due to the failure of red spruce to compete (Blum et al. 1983). Red spruce has been successfully regenerated and managed using a variety of thinning or partial cutting methods in northern areas (e.g., Pothier and Prévost 2008; Olsen et al. 2014). Studies utilizing operational-scale treatments such as commercial timber harvesting to promote seedling development are lacking for the central Appalachian red spruce stands. Using commercial treatments to accelerate the development of advance spruce regeneration in hardwood-dominated stands can substantially shorten the time needed to increase the coverage of red spruce forest communities across the landscape. However, specific guidelines for promoting the development of advance reproduction and subsequent ascension into the canopy are lacking for central and southern Appalachian forests.

Unlike the red spruce forests of central and southern Appalachia, which are relatively rare and largely second growth, current red spruce-dominated forests in New England have undergone multiple harvests, resulting in simplified age structures and altered species composition (Seymour 2005). To mimic natural stand dynamics in these forests with silvicultural practices, a hybrid system that combines shelterwood with reserves and group selection has been proposed. In this system, similar to the German *Femelschlag*, canopy gaps are created and over several cutting cycles are expanded in size (Seymour 2005). Ten-year results of this expanding-gap method of group selection in Maine show that larger gaps (>0.1 ha [0.25 ac]) favored mid-successional species and gaps smaller than that threshold favored shade tolerant species (Arseneault et al. 2011).

Given the unknowns, any restoration action for spruce forests will need to be adaptive to changing conditions and will benefit from intensive and detailed monitoring. Regardless of the restoration action chosen, non-commercial release of individual red spruce trees to commercial scale expanding gap shelterwood prescriptions should consider the range of attributes, such as gap size or return intervals, and not manage strictly for the average. More details of current restoration efforts are found in Chap. 8.

5.4 Knowledge Gaps and Research Needs

While silvicultural prescriptions for commercial management of red spruce-dominated forests are in use in forests in New England and Canada, demonstrated success for commercial scale management is lacking for the central and southern Appalachians. Management goals will differ when active management is used for restoration rather than production of pulp or lumber. Research efforts are underway at the Kumbrabow State Forest in West Virginia to test a version of expanding gap shelterwood (*Femelschlag*) harvest for restoration of red spruce at a commercial scale.

While environmental regulations have resulted in reduced deposition of sulfur and nitrogen, there may be lingering effects of pollutants on red spruce forests, and questions remain about the recovery of these systems from decades of acidic deposition. There are also gaps in our knowledge of the impacts other pollutants have had or are continuing to have on these forests. These effects could also amplify predicted impacts of climate change. Red spruce-dominated forests were refugia for past climate change events but impacts from current anthropogenic climate change might override this capacity.

References

Allard HA, Leonard EC (1952) The Canaan and the Stony River valleys of West Virginia, their former magnificent spruce forests, their vegetation and floristics today. Castanea 17:1–60
Allen TR, Kupfer JA (2000) Application of spherical statistics to change vector analysis of Landsat data: southern Appalachian spruce–fir forests. Remote Sens Environ 74:482–493
Allen TR, Kupfer JA (2001) Spectral response and spatial pattern of Fraser fir mortality and regeneration, Great Smoky Mountains, USA. Plant Ecol 156:59–74

Andrews C, Foster JR, Weiskittel A et al (2021) Integrating historical observations alters projections of eastern North American spruce–fir habitat under climate change. Ecosphere 13:e4016

Aneja VP, Robarge WP, Claiborn CS et al (1992) Chemical climatology of high elevation spruce-fir forests in the southern Appalachian Mountains. Environ Pollut 75:89–96

Arseneault, JE Saunders MR, Seymour RS et al (2011) First decadal response to treatment in a disturbance-based silviculture experiment in Maine. Forest Ecol Manag 262:404–412

Beckage B, Clark JS, Clinton BD et al (2000) A long-term study of tree seedling recruitment in southern Appalachian forests: the effects of canopy gaps and shrub understories. Can J Forest Res 30:1617–1631

Berry ZC, Smith WK (2012) Cloud pattern and water relations in *Picea rubens* and *Abies fraseri*, southern Appalachian Mountains, USA. Agric for Meteorol 162–163:27–34

Berry ZC, Smith WK (2013) Ecophysiological importance of cloud immersion in a relic spruce–fir forest at elevational limits, southern Appalachian Mountains, USA. Oecologia 173:637–648

Berry ZC, Hughes NM, Smith WK (2014) Cloud immersion: an important water source for spruce and fir saplings in the southern Appalachian Mountains. Oecologia 174:319–326

Bintz WW, Butcher DJ (2007) Characterization of the health of southern Appalachian red spruce (*Piceae rubens*) through determination of calcium, magnesium, and aluminum concentrations in foliage and soil. Microchem J 87:170–174

Blum BM (1990) Red spruce *Picea rubens* Sarg. In: Burns RM, Honkala BH (eds) Silvics of North America. Volume 1. Conifers. USDA Forest Service Handbook No. 654, Washington, DC, pp 250–259

Blum BM, Benzie JW, Merski E (1983) Eastern spruce–fir. In: Burns RM (ed) Silvicultural systems for the major forest types of the United States. USDA Forest Service Agricultural Handbook No. 445, Washington, DC, pp 128–130

Boner RR (1979) Effects of Fraser fir death on population dynamics in Southern Appalachian boreal ecosystems. University of Tennessee, Knoxville, Tennessee, Thesis

Borer CH, Schaberg PG, DeHayes DH (2005) Acidic mist reduces foliar membrane-associated calcium and impairs stomatal responsiveness in red spruce. Tree Physiol 25:673–680

Bormann FH, Likens GE (1979) Catastrophic disturbance and the steady state in northern hardwood forests. Am Sci 67:660–669

Bowers TA, Allen T, Bruck RI (2010) Evidence of montane spruce-fir forest recovery on the high peaks and ridges of the Black Mountains, North Carolina: recent trends, 1986–2003. In: Rentch JS, Schuler TM (eds) Proceedings from the conference on the ecology and management of high-elevation forests in the central and southern Appalachian Mountains. USDA Forest Service Northern Research Station General Technical Report NRS-P-64, Newtown Square, Pennsylvania, p 203

Brooks RT, Kyker-Snowman TD (2008) Forest floor temperature and relative humidity following timber harvesting in southern New England, USA. For Ecol Manag 254:65–73

Bruck RI, Robarge WP, McDaniel A (1989) Forest decline in the boreal montane ecosystems of the southern Appalachian Mountains. Water Air Soil Pollut 48:161–180

Bryant KN, Fowlkes AJ, Mustafa SF et al (1997) Determination of aluminum, calcium, and magnesium in Fraser fir, balsam fir, and red spruce foliage and soil from the southern and middle Appalachians. Microchem J 56:382–392

Busing RT (2004) Red spruce dynamics in an old southern Appalachian forest. J Torrey Bot Soc 131:337–342

Busing RT, Clebsch EEC (1988) Fraser fir mortality and the dynamics of a Great Smoky Mountains fir-spruce stand. Castanea 53:177–182

Busing RT, Pauley EF (1994) Mortality trends in a southern Appalachian red spruce population. Forest Ecol Manag 64:41–45

Busing RT, Wu X (1990) Size-specific mortality, growth, and structure of a Great Smoky Mountains red spruce population. Can J Forest Res 20:206–210

Busing RT, Clebsch EEC, Eagar CC et al (1988) Two decades of change in a Great Smoky Mountains spruce-fir forest. Bull Torrey Bot Club 115:25–31

Changnon SA (2003) Characteristics of ice storms in the United States. J Appl Meteorol 42:630–639
Changnon SA, Karl TR (2003) Temporal and spatial variations of freezing rain in the contiguous United States: 1948–2000. J Appl Meteorol 42:1302–1315
Clebsch EEC, Busing RT (1989) Secondary succession, gap dynamics, and community structure in a southern Appalachian cove forest. Ecology 70:728–735
Clements FE (1936) Nature and structure of the climax. J Ecol 24:252–284
Clinton BD, Vose JM (1996) Effects of *Rhododendron maximum* L. on *Acer rubrum* L. seedling establishment. Castanea 61:38–45
Cogbill CV, White PS (1991) The latitude-elevation relationship for spruce-fir forest and treeline along the Appalachian Mountain chain. Vegetatio 94:153–175
Collins BS, Schuler TM, Ford WM et al (2010) Stand dynamics of relict red spruce in the Alarka Creek headwaters, North Carolina. In: Rentch JS, Schuler TM (eds) Proceedings from the conference on the ecology and management of high-elevation forests in the central and southern Appalachian Mountains. USDA Forest Service Northern Research Station General Technical Report NRS-P-64, Newtown Square, Pennsylvania, pp 22–27
Cook ER, Zedaker SM (1992) The dendroecology of red spruce decline. In: Eagar C, Adams MB (eds) Ecology and decline of red spruce in the Eastern United States. Springer, New York, pp 192–231
Dale VH, Gardner RH, DeAngelis DL et al (1991) Elevation-mediated effects of balsam woolly adelgid on southern Appalachian spruce–fir forests. Can J Forest Res 21:1639–1648
DeHayes DH, Shaberg PG, Hawley GJ et al (1999) Acid rain impacts on calcium nutrition and forest health: alteration of membrane-associated calcium leads to membrane destabilization and foliar injury in red spruce. Bioscience 49:789–800
Dudley MP, Freeman M, Wenger S et al (2020) Rethinking foundation species in a changing world: the case for *Rhododendron maximum* as an emerging foundation species in shifting ecosystems of the southern Appalachians. For Ecol Manag 472:118240
Dull CW, Ward JD, Brown HD et al (1988) Evaluation of spruce and fir mortality in the southern Appalachian Mountains. USDA Forest Service Southern Region Protection Report R8-PR 13, Atlanta, Georgia
Eagar C (1984) Review of the biology and ecology of the balsam woolly aphid in southern Appalachian spruce–fir forests. In: White PS (ed) The southern Appalachian spruce–fir ecosystem: its biology and threats. National Park Service Southeast Region Research/Resources Management Report SER-71, Atlanta, Georgia, pp 36–50
Eagar C, Adams MB (eds) (1992) The ecology and decline of red spruce in the Eastern United States. Springer, New York
Engstrom RT, Gilbert S, Hunter ML Jr et al (1999) Practical applications of disturbance ecology to natural resource management. In: Szaro RC, Johnson NC, Sexton WT, Malk AJ (eds) Ecological stewardship: a common reference for ecosystem management, vol 2. Elsevier Science Ltd., Oxford, United Kingdon, pp 313–330
Fahey TJ, Battles JJ, Wilson GF (1998) Responses of early successional northern hardwood forests to changes in nutrient availability. Ecol Monogr 68:183–212
Fakhraei H, Driscoll CT, Renfro JR et al (2016) Critical loads and exceedances for nitrogen and sulfur atmospheric deposition in Great Smoky Mountains National Park United States. Ecosphere 7:e01466
Foster JR, D'Amato AW (2015) Montane forest ecotones moved downslope in northeastern USA in spite of warming between 1984 and 2011. Glob Change Biol 21:4497–4507
Foster JR, D'Amato AW (2018) Unexpected expansion of spruce and fir varies at ecotones across the eastern US. Paper presented at the American Geophysical Union, Fall Meeting
Fridley J (2009) Downscaling climate over complex terrain: high finescale (<1000 m) spatial variation of near-ground temperatures in a montane forested landscape (Great Smoky Mountains). J Appl Meteorol Climatol 48:1033–1049
Gleason HA (1926) The individualistic concept of the plant association. Bull Torrey Bot Club 53:7–26

Greenberg, CH, Collins, BS (2016) Natural disturbances and historic range of variation: type, frequency, severity, and post-disturbance structure in central hardwood forests USA. Springer, New York

Harmon ME, Bratton SP, White PS (1984) Disturbance and vegetation response in relation to environmental gradients in the Great Smoky Mountains. Vegetatio 55:129–139

Harrell C, Zedaker S (2010) Effects of prescribed burning, mechanical, and chemical treatments to curtail rhododendron dominance and reduce wildfire fuel loads. In: Stanturf JA (ed) Proceedings of the 14th biennial southern silvicultural research conference. USDA Forest Service Southern Research Station General Technical Report SRS–121, Asheville, North Carolina, p 545

Hayes M, Moody A, White PS et al (2007) The influence of logging and topography on the distribution of spruce-fir forests near their southern limits in Great Smoky Mountains National Park, USA. Plant Ecol 189:59–70

Hopkins AD (1899) Report on investigations to determine the cause of unhealthy conditions of the spruce and pine from 1880–1893, West Virginia Exp. Sta. Bulletin 56

Horton JL, Clinton BD, Walker JF et al (2009) Variation in soil and forest floor characteristics along gradients of ericaceous, evergreen shrub cover in the southern Appalachians. Castanea 74:340–352

Houle G, Payette S (1990) Seed dynamics of *Betula alleghaniensis* in a deciduous forest of north-eastern North America. J Ecol 78:677–690

Hwang T, Band LE, Hales TC et al (2015) Simulating vegetation controls on hurricane-induced shallow landslides with a distributed ecohydrological model. J Geophys Res Biogeosci 120:361–378

Jenkins MA (2003) Impact of the balsam woolly adelgid (*Adelges piceae* Ratz.) on an *Abies fraseri* (Pursh) Poir. dominated stand near the summit of Mount LeConte, Tennessee. Castanea 68:109–118

Johnson AH, McLaughlin SB, Adams MB et al (1992) Why are red spruce declining at high elevations? Synthesis and conclusions from epidemiological and mechanistic studies of red spruce decline. In: Eagar C, Adams MB (eds) Ecology and decline of red spruce in the Eastern United States. Springer, New York, pp 385–412

Johnson DM, Smith WK (2005) Refugial forests of the southern Appalachians: photosynthesis and survival in current-year *Abies fraseri* seedlings. Tree Physiol 25:1379–1387

Joslin JD, Wolfe MH (1992) Red spruce soil solution chemistry and root distribution across a cloud water deposition gradient. Can J Forest Res 22:893–904

Joslin JD, Kelly JM, Van Miegroet H (1992) Soil chemistry and nutrition of North American spruce-fir stands: evidence for recent change. J Environ Qual 21:12–30

Kaylor SD, Hughes MJ, Franklin JA (2017) Recovery trends and predictions of Fraser fir (*Abies fraseri*) dynamics in the southern Appalachian Mountains. Can J Forest Res 47:125–133

Keane R (2013) Disturbance regimes and the historical range of variation in terrestrial ecosystems. In: Encyclopedia of biodiversity, 2nd edn. Elsevier, pp 568–581

Kingsland SE (1991) Defining ecology as a science. In: Real LA, Brown JH (eds) Foundations of ecology: classic papers with commentaries. University of Chicago Press, Chicago, Illinois, pp 1–13

Koo K-A, Patten BC, Creed IF (2011) *Picea rubens* growth at high versus low elevations in the Great Smoky Mountains National Park: evaluation by systems modeling. Can J Forest Res 41:945–962

Koo KA, Madden M, Patten BC (2014) Projection of red spruce (*Picea rubens* Sargent) habitat suitability and distribution in the southern Appalachian Mountains, USA. Ecol Model 293:91–101

Koo KA, Patten BC, Madden M (2015) Predicting effects of climate change on habitat suitability of red spruce (*Picea rubens* Sarg.) in the southern Appalachian Mountains of the USA: understanding complex systems mechanisms through modeling. Forests 6:1208–1226

Korstian CF (1937) Perpetuation of spruce on cut-over and burned lands in the higher southern Appalachian Mountains. Ecol Monogr 7:125–167

Lafon CW (2016) Ice storms in central hardwood forests: the disturbance regime, spatial patterns, and vegetation influences. In: Greenberg CH, Collins BS (eds) Natural disturbances and historic range of variation. Springer, New York, pp 147–166

Lafon CW, Speer JH (2002) Using dendrochronology to identify major ice storm events in oak forests of southwestern Virginia. Climate Res 20:41–54

Lawrence GB, Bailey SW (2021) Recovery processes of acidic soils experiencing decreased acidic deposition. Soil Syst 5:36

Lemon PC (1961) Forest ecology of ice storms. Bull Torrey Bot Club 88:21–29

Lewis RL (1998) Transforming the Appalachian countryside. University of North Carolina Press, Chapel Hill, North Carolina

Lindsay MM, Bratton SP (1979) The vegetation of grassy balds and other high elevation disturbed areas in the Great Smoky Mountains National Park. Bull Torrey Bot Club 106:264–275

Lusk L, Mutel M, Walker E et al (2010) Forest change in high-elevation forests of Mt. Mitchell, North Carolina: re-census and analysis of data collected over 40 years. In: Rentch JS, Schuler TM (eds) Proceedings from the conference on the ecology and management of high-elevation forests in the central and southern Appalachian Mountains. USDA Forest Service Northern Research Station General Technical Report NRS-P-64, Newtown Square, Pennsylvania, pp 104–112

Lutz A (2018) Characterizing red spruce (*Picea rubens* Sarg.) advanced reproduction in a high elevation stand in West Virginia. Thesis, West Virginia University, Morgantown, West Virginia

Marks PL (1974) The role of pin cherry (*Prunus pensylvanica* L.) in the maintenance of stability in northern hardwood ecosystems. Ecol Monogr 44:73–88

Mathias JM, Thomas RB (2018) Disentangling the effects of acidic air pollution, atmospheric CO_2, and climate change on recent growth of red spruce trees in the central Appalachian Mountains. Glob Change Biol 24:3938–3953

McLaughlin S, Percy K (1999) Forest health in North America: some perspectives on actual and potential roles of climate and air pollution. Water Air Soil Pollut 116:151–197

McLaughlin SB, Andersen CP, Edwards NT et al (1990) Seasonal patterns of photosynthesis and respiration of red spruce saplings from two elevations in declining southern Appalachian stands. Can J Forest Res 20:485–495

McLaughlin SB, Andersen CP, Hanson PJ et al (1991) Increased dark respiration and calcium deficiency of red spruce in relation to acidic deposition at high-elevation southern Appalachian Mountain sites. Can J Forest Res 21:1234–1244

McLaughlin SB, Blasing TJ, Downing DJ (1994) Two hundred year variation of southern red spruce radial growth as estimated by spectral analysis: comment. Can J Forest Res 24:2299–2304

McLaughlin SB, Joslin JD, Robarge W et al (1998) The impacts of acidic deposition and global change on high elevation southern Appalachian spruce-fir forests. In: Mickler RA, Fox S (eds) The productivity and sustainability of southern forest ecosystems in a changing environment. Springer, New York, pp 255–277

McLaughlin SB, Tjoelker MG, Roy WK (1993) Acid deposition alters red spruce physiology: laboratory studies support field observations. Can J Forest Res 23:380–386

McLaughlin SB, Wullschleger S, Stone A et al (1995) Effects of acid deposition on calcium nutrition and health of southern Appalachian spruce fir forests. Paper presented at the International Union of Forest Resource Organizations, New Brunswick, Canada, 7 Sep 1994. https://www.osti.gov/biblio/32571/. Accessed 31 Jul 2024

McManamay RH, Resler LM, Campbell JB et al (2011) Assessing the impacts of balsam woolly adelgid (*Adelges piceae* Ratz.) and anthropogenic disturbance on the stand structure and mortality of Fraser fir [*Abies fraseri* (Pursh) Poir.] in the Black Mountains, North Carolina. Castanea 76:1–19

Minckler LS (1939) Spruce type problem analysis: analysis of problems in the reforestation of the spruce type of the southern Appalachians. USDA Forest Service Unpublished File Report

Moore JA, Bartlett JG, Boggs JL et al (2002) Abiotic factors. In: Wear DN, Greis JG (eds) Southern forest resource assessment. USDA Forest Service Southern Research Station General Technical Report SRS-53, Asheville, North Carolina, pp 429–452

Moore PT, Van Miegroet H, Nicholas NS (2008) Examination of forest recovery scenarios in a southern Appalachian *Picea-Abies* forest. Forestry 81:183–194

Morin RS, Widmann RH (2010) A comparison of the status of spruce in high-elevation forests on public and private land in the southern and central Appalachian Mountains. In: Rentch JS, Schuler TM (eds) Proceedings from the conference on the ecology and management of high-elevation forests in the central and southern Appalachian Mountains. USDA Forest Service Northern Research Station General Technical Report NRS-P-64, Newtown Square, Pennsylvania, pp 134–139

Nelson AH (2017) Locating stable sites: climate refugia in the southern Appalachians. Clemson University, Clemson, South Carolina, Thesis

Nicholas NS, Zedaker SM (1989) Ice damage in spruce–fir forests of the Black Mountains, North Carolina. Can J Forest Res 19:1487–1491

Nicholas NS, Zedaker SM, Eagar C et al (1992) Seedling recruitment and stand regeneration in spruce-fir forests of the Great Smoky Mountains. Bull Torrey Bot Club 119:289–299

Nilsen ET, Walker JF, Miller OK et al (1999) Inhibition of seedling survival under *Rhododendron maximum* (Ericaceae): could allelopathy be a cause? Am J Bot 86:1597–1605

Nodvin SC, Van Miegroet H, Lindberg SE et al (1995) Acidic deposition, ecosystem processes, and nitrogen saturation in a high elevation southern Appalachian watershed. Water Air Soil Pollut 85:1647–1652

Nowacki G, Carr R, Van Dyck M (2010) The current status of red spruce in the eastern United States: distribution, population trends, and environmental drivers. In: Rentch JS, Schuler TM (eds) Proceedings from the conference on the ecology and management of high-elevation forests in the central and southern Appalachian Mountains. USDA Forest Service Northern Research Station General Technical Report NRS-P-64, Newtown Square, Pennsylvania, pp 140–162

Nyland RD, Bashant AL, Heitzman EF et al (2007) Interference to hardwood regeneration in Northeastern North America: pin cherry and its effects. North J Appl Forestry 24:52–60

Oliver CD, Larson BC (1996) Forest stand dynamics, updated edition. Wiley, Incorporated, New York

Olsen MG, Meyer SR, Wagner RG et al (2014) Commercial thinning stimulates natural regeneration in spruce–fir stands. Can J Forest Res 44:173–181

Pardo LH, Duarte N (2007) Assessment of effects of acidic deposition on forested ecosystems in Great Smoky Mountains National Park using critical loads for sulfur and nitrogen. Report to the National Park Service, South Burlington, Vermont

Pardo LH, Duarte N, Van Miegroet H et al (2018) Critical loads of sulfur and nitrogen and modeled effects of deposition reduction for forested ecosystems of Great Smoky Mountains National Park. USDA Forest Service Northern Research Station General Technical Report NRS-180, Newtown Square, Pennsylvania

Pauley EF, Clebsch EEC (1990) Patterns of *Abies fraseri* regeneration in a Great Smoky Mountains spruce-fir forest. Bull Torrey Bot Club 117:375–381

Peters MP, Prasad AM, Matthews SN et al (2020) Climate change tree atlas, ver 4. USDA Forest Service Northern Research Station and Northern Institute of Applied Climate Science, Delaware, Ohio. https://www.nrs.fs.fed.us/atlas. Accessed 1 Aug 2024

Pothier D, Prévost M (2008) Regeneration development under shelterwoods in a lowland red spruce—balsam fir stand. Can J Forest Res 38:31–39

Potter KM, Hargrove WW, Koch FH (2010) Predicting climate change extirpation risk for central and southern Appalachian forest tree species. In: Rentch JS, Schuler TM (eds) Proceedings from the conference on the ecology and management of high-elevation forests in the central and southern Appalachian Mountains. USDA Forest Service Northern Research Station General Technical Report NRS-P-64, Newtown Square, Pennsylvania, pp 179–189

Pyle C, Schafale MP (1988) Land use history of three spruce-fir forest sites in southern Appalachia. J Forest Hist 32:4–21

Reams GA, Nichols NS, Zedaker SM (1993) Two hundred year variation of southern red spruce radial growth as estimated by spectral analysis. Can J Forest Res 23:291–301

Reinhardt K, Smith WK (2008) Leaf gas exchange of understory spruce–fir saplings in relict cloud forests, southern Appalachian Mountains, USA. Tree Physiol 28:113–122

Rentch JS, Schuler TM, Ford WM et al (2007) Red spruce stand dynamics, simulations, and restoration opportunities in the central Appalachians. Restor Ecol 15:440–452

Rentch JS, Schuler TM, Nowacki GJ et al (2010) Canopy gap dynamics of second-growth red spruce-northern hardwood stands in West Virginia. Forest Ecol Manag 260:1921–1929

Rentch JS, Ford WM, Schuler TS et al (2016) Release of suppressed red spruce using canopy gap creation—ecological restoration in the central Appalachians. Nat Areas J 36:29–37

Ribbons RR (2014) Disturbance and climatic effects on red spruce community dynamics at its southern continuous range margin. PeerJ 2:e293

Richardson AD, Denny EG, Siccama TG et al (2003) Evidence for a rising cloud ceiling in Eastern North America. J Clim 16:2093–2098

Robarge WP, Pye JM, Bruck RI (1989) Foliar elemental composition of spruce-fir in the southern Blue Ridge province. Plant Soil 114:19–34

Rogers BM, Jantz P, Goetz SJ et al (2016) Vulnerability of tree species to climate change in the Appalachian landscape conservation cooperative. In: Hansen AJ, Monahan WB, Olliff ST et al (eds) Climate change in wildlands. Island Press, Washington, DC, pp 212–233

Rosenberg MB, Butcher DJ (2010) Investigation of acid deposition effects on southern Appalachian red spruce (*Picea rubens*) by determination of calcium, magnesium, and aluminum in foliage and surrounding soil using ICP-OES. Instrum Sci Technol 38:341–358

Schafale MP, Weakley AS (1990) Classification of the natural communities of North Carolina, third approximation. North Carolina Department of Environment and Natural Resources Natural Heritage Program, Raleigh, North Carolina

Schier GA (1985) Response of red spruce and balsam fir seedlings to aluminum toxicity in nutrient solutions. Can J Forest Res 15:29–33

Schuler TM, Ford WM, Collins RJ (2002) Successional dynamics and restoration implications of a montane coniferous forest in the central Appalachians, USA. Nat Areas J 22:88–98

Seymour RS (2005) Integrating natural disturbance parameters into conventional silvicultural systems: experience from the Acadian forest of Northeastern North America. In: Peterson CE, Maguire DA (eds) Balancing ecosystem values: innovative experiments for sustainable forestry, proceedings of a conference. USDA Forest Service Pacific Northwest Research Station General Technical Report PNW-635, Portland, Oregon, pp 41–48

Smith GF, Nicholas NS (1998) Patterns of overstory composition in the fir and fir-spruce forests of the Great Smoky Mountains after balsam woolly adelgid infestation. Am Midl Nat 139:340–352

Smith GF, Nicholas NS (1999) Post-disturbance spruce-fir forest stand dynamics at seven disjunct sites. Castanea 64:175–186

Smith GF, Nicholas NS (2000) Size- and age-class distributions of Fraser fir following balsam woolly adelgid infestation. Can J Forest Res 30:948–957

Soulé PT (2011) Changing climate, atmospheric composition, and radial tree growth in a spruce-fir ecosystem on Grandfather Mountain, North Carolina. Nat Areas J 31:65–74

Soulé PT, White PB, van de Gevel SL (2012) Succession and disturbance in an endangered red spruce–Fraser fir forest in the southern Appalachian Mountains, North Carolina, USA. Endangered Species Research 18:17–25

Stehn SE, Jenkins MA, Webster CR et al (2013) Regeneration responses to exogenous disturbance gradients in southern Appalachian *Picea-Abies* forests. Forest Ecol Manag 289:98–105

Stehn SE, Webster CR, Jenkins MA et al (2010) Influence of acid deposition on regeneration dynamics along a disturbance intensity gradient. In: Rentch JS, Schuler TM (eds) Proceedings from the conference on the ecology and management of high-elevation forests in the central and southern Appalachian Mountains. USDA Forest Service Northern Research Station General Technical Report NRS-P-64, Newtown Square, Pennsylvania, p 241

Stehn SE, Webster CR, Jenkins MA et al (2011) High elevation ground-layer plant community composition across environmental gradients in spruce-fir forests. Ecol Res 26:1089–1101

Stephenson SL, Adams HS (1984) The spruce-fir forest on the summit of Mount Rogers in southwestern Virginia. Bull Torrey Bot Club 111:69–75

Sterling EA (1920) Report of reproduction study, WV pulp and paper company lands, Pocahontas, Randolph, and Webster counties, W. VA. James D. Lacey and Company, New York

Stottlemyer AD, Shelburne VB, Waldrop TA et al (2009) Fuel characterization in the southern Appalachian Mountains: an application of landscape ecosystem classification. Int J Wildland Fire 18:423–429

Thomas A, Tilotta DC, Frampton J et al (2022) Sesquiterpene induction by the balsam woolly adelgid (*Adelges piceae*) in putatively resistant Fraser fir (*Abies fraseri*). Forests 13:716

Thornton FC, Joslin JD, Pier PA et al (1994) Cloudwater and ozone effects upon high elevation red spruce: a summary of study results from Whitetop Mountain, Virginia. J Environ Qual 23:1158–1167

Tierney GL, Fahey TJ (1998) Soil seed bank dynamics of pin cherry in a northern hardwood forest, New Hampshire, U.S.A. Can J Forest Res 28:1471–1480

Turner MG (2010) Disturbance and landscape dynamics in a changing world. Ecology 91:2833–2849

USDA Forest Service (2006) Monongahela National Forest land and resource management plan. https://www.fs.usda.gov/Internet/FSE_DOCUMENTS/stelprdb5330420.pdf. Accessed 26 Jul 2024

Walter JA, Neblett JC, Atkins JW et al (2017) Regional- and watershed-scale analysis of red spruce habitat in the southeastern United States: implications for future restoration efforts. Plant Ecol 218:305–316

White PB, Soulé P, van de Gevel S (2014) Impacts of human disturbance on the temporal stability of climate-growth relationships in a red spruce forest, southern Appalachian Mountains, USA. Dendrochronologia 32:71–77

White PS (ed) (1984) The southern Appalachian spruce–fir ecosystem: its biology and threats. National Park Service Southeast Region Research/Resources Management Report SER-71, Atlanta, Georgia

White PS, Cogbill CV (1992) Spruce-fir forests of eastern North America. In: Eagar C, Adams MB (eds) Ecology and decline of red spruce in the Eastern United States. Springer, New York, pp 3–39

White PS, MacKenzie MD, Busing RT (1985) Natural disturbance and gap phase dynamics in southern Appalachian spruce–fir forests. Can J Forest Res 15:233–240

White PS, Wilds SP, Stratton DA (2001) The distribution of heath balds in the Great Smoky Mountains. J Veg Sci 12:453–466

Woodbridge M, Dovciak M (2022) Logging legacies in a plant diversity hotspot: altered distributions and abundance patterns of the shrub layer in the southern Appalachians. Forest Ecol Manag 516:120245

Worrall JJ, Lee TD, Harrington TC (2005) Forest dynamics and agents that initiate and expand canopy gaps in *Picea-Abies* forests of Crawford Notch, New Hampshire, USA. J Ecol 93:178–190

Wu X, McCormick JF, Busing RT (1999) Growth pattern of *Picea rubens* prior to canopy recruitment. Plant Ecol 140:245–253

Wurzburger N, Hendrick RL (2007) Rhododendron thickets alter N cycling and soil extracellular enzyme activities in southern Appalachian hardwood forests. Pedobiologia 50:563–576

Yetter E, Chhin S, Brown JP (2021) Dendroclimatic analysis of central Appalachian red spruce in West Virginia. Can J Forest Res 51:1607–1620

Ziegler SS (2000) A comparison of structural characteristics between old-growth and postfire second-growth hemlock–hardwood forests in Adirondack Park, New York, U.S.A. Glob Ecol Biogeogr 9:373–389

Chapter 6
Wildlife

Donald J. Brown, Hannah L. Clipp, Corinne A. Diggins, Craig Roghair, C. Andrew Dolloff, Chad M. Landress, Matthew McKinney, Jakob T. Goldner, and W. Mark Ford

6.1 Introduction

The central and southern Appalachian Mountain region is among the most biodiverse temperate zone regions in the world (Delcourt and Delcourt 1998; Olson and Dinerstein 2002; Pickering et al. 2002). For vertebrate wildlife, it is a region of particularly high diversity for fishes and amphibians (Etnier and Starnes 1993; Petranka 1998; Jenkins et al. 2015), and many northern breeding bird and small mammal species reach their southern extent in the region (Steele and Powell 1999; Merker and Chandler 2020). Red spruce (*Picea rubens*) and spruce-fir forests occupy a small proportion of the central and southern Appalachians and generally occur only at high elevations, which have comparatively harsh climates and typically small and ephemeral surface water systems (reviewed in Chap. 1). As a result of the restricted occurrence and limiting environmental conditions of high elevation red spruce and spruce-fir forests, wildlife diversity is comparatively low in these forest types. However, they represent optimal habitat for a variety of wildlife species, some

D. J. Brown (✉)
USDA Forest Service, Pacific Northwest Research Station, Amboy, WA, USA
e-mail: donald.brown2@usda.gov

H. L. Clipp
USDA Forest Service, Northern Research Station, Delaware, OH, USA

Present Address:
Appalachian Mountain Club, Jackson, NH, USA

C. A. Diggins
US Fish & Wildlife Service, Science Applications, Albuquerque, NM, USA

C. Roghair
USDA Forest Service, Southern Research Station, Blacksburg, VA, USA

C. A. Dolloff
USDA Forest Service, Southern Research Station (retired), Blacksburg, VA, USA

D. J. Brown et al. (eds.), *Ecology and Restoration of Red Spruce Ecosystems in the Central and Southern Appalachians*, https://doi.org/10.1007/978-3-032-19616-3_6

of which have highly restricted distributions, and thus these forest types are important contributors to regional wildlife diversity (Fig. 6.1).

In this chapter, we review wildlife diversity in red spruce and spruce-fir forests of the central and southern Appalachians, with a focus on species that are strongly associated with these forest types. Sections are organized by taxonomic groups, including fish, herpetofauna (amphibians, reptiles), birds, mammals, and lastly, invertebrates. As discussed in Chap. 1, the extent of red spruce and spruce-fir forests in the region has been dramatically reduced since European settlement. Consequently, local occurrence and abundance of wildlife species that are strongly associated with these forest types has likely been greatly reduced (Chaplin et al. 2000; Diggins 2023). Our knowledge synthesis provides managers and biologists with a single reference to assist

Fig. 6.1 Habitat for the federally threatened Cheat Mountain salamander (*Plethodon nettingi*) in a high elevation red spruce (*Picea rubens*) forest in West Virginia. Photo by Donald Brown

C. M. Landress
USDA Forest Service, Monongahela National Forest, Bartow, WV, USA

M. McKinney
West Liberty University, College of Sciences, West Liberty, WV, USA

J. T. Goldner
West Virginia Division of Natural Resources, Elkins, WV, USA

W. M. Ford
US Geological Survey, Virginia Cooperative Fish and Wildlife Research Unit, Blacksburg, VA, USA
e-mail: wmford@vt.edu

with identifying local wildlife species that would most benefit from restoration and management of these forest types.

6.2 Fish

Although the central and southern Appalachian Mountain region is known for its tremendous freshwater fish diversity (Etnier and Starnes 1993), fish assemblages that occur in small, high gradient first- and second-order headwater streams within red spruce and spruce-fir forests are fairly depauperate, particularly in the Blue Ridge region and the few red spruce forest occurrences in the Ridge and Valley region. Brook trout (*Salvelinus fontinalis*) are perhaps the most notable fish species in red spruce forests (see Sidebar 6.1) and occur throughout the Appalachian Mountains with their southern distributional terminus in northern Georgia where summer maximum water temperature does not exceed 21 °C (70 °F). Although a number of fish species co-exist with brook trout at lower elevations, the occurrence of fishes other than brook trout in red spruce stands above 1,200 m (3,900 ft) is limited; thus, many cold-water high elevation streams are commonly and appropriately known as *trout streams* because often the only fish inhabitants are brook trout (Fig. 6.2). A notable exception, the western blacknose dace (*Rhinichthys obtusus*) has been found at elevations of > 1,200 m (3,900 ft) in the Great Smoky Mountains National Park in North Carolina and Tennessee (Simbeck 1990).

Fig. 6.2 Brook trout (*Salvelinus fontinalis*) from a red spruce (*Picea rubens*) headwater stream in West Virginia. Photo by Chad Landress

Sidebar 6.1 Focal Species—Brook Trout (*Salvelinus fontinalis*)

Brook trout (*Salvelinus fontinalis*) in the central and southern Appalachians are glacial refugees at the southern extent of their range (Karas 1997). As cold-water specialists, brook trout are capable of occupying extreme headwaters at high elevations (Etnier and Starnes 1993; Jenkins and Burkhead 1994). The cold, clean waters flowing through and from expansive red spruce (*Picea rubens*) forests fed prime brook trout streams prior to the mid-1800s (Kennedy 1853; McNeill 1877; Zurbuch 2015). However, the exploitative logging and subsequent wildfires that reduced the extent of red spruce forest from the mid-1800s to 1930s (McNeill 1877; Yetter et al. 2021) degraded habitat through sedimentation, streambed scouring, and loss of forest shade, which increased stream temperatures and subsequently reduced the distribution and abundance of brook trout populations throughout the region (Marschall and Crowder 1996; Zurbuch 2015). Brook trout and red spruce share more contemporary environmental stressors as well, including increased headwater acidification from atmospheric deposition on already naturally acidic systems (Menendez 1976; Herlihy et al. 1993). These stressors can push brook trout to lower elevations, whereas warming waters from elevation drop and/or climate change chase them upslope into an increasingly narrow and restricted zone of remaining habitability (McDonnell et al. 2015, Zurbach 2015). In the central Appalachians, acidic runoff from surface coal mines is also a contemporary stressor that has degraded aquatic habitat quality and, in some instances, eliminated brook trout populations entirely (Petty et al. 2005, 2010). Throughout the central and southern Appalachians in red spruce forests, road culverts that limit or block passage represent a significant movement barrier to brook trout and other fish in many watersheds (Poplar-Jeffers et al. 2009). As a result, brook trout in the central and southern Appalachians today share the same fate as red spruce and the biotic communities therein: small, isolated, high elevation populations occupying a fraction of their historical range (Hudy et al. 2008) that are vulnerable to extirpation (Etnier and Starnes 1993; Kazyak et al. 2022) and consequently are in severe need of conservation and restoration (Habera and Moore 2005; Eastern Brook Trout Joint Venture 2018).

The shared histories, ecological intersections, and overlapping threats to brook trout and red spruce are striking and suggest that conservation and restoration efforts for each may be enhanced by more direct consideration of the other. Despite the obvious intersection between strategies (Clingerman 2008), we were unable to locate any examples where both brook trout and red spruce were explicitly considered together during conservation or restoration planning in the Blue Ridge portion of the southern Appalachians. In contrast, addition of limestone fines to ameliorate geologic, atmospheric, and mine-induced aquatic acidity to improve brook trout habitat conditions and populations has a long history of application in central Appalachian streams (Fig. 6.3;

Downey et al. 1994; Hudy et al. 2000), including in streams in red spruce forests (Zurbuch et al. 1996). Increases in stream pH from lime application have improved conditions for brook trout and other biota (Clayton et al. 1998), though full system recovery remains elusive with regard to water chemistry, disconnectivity of treated streams, macroinvertebrate recovery, physical barriers to fish movement, and thermal conditions in most red spruce-dominated watersheds (McClurg et al. 2007). Shingleton and Brown (2010) estimated that restoration of mine-impacted cold-water streams in West Virginia would provide > 300,000 additional angler-days with a considerable economic benefit. Reforestation of red spruce on former surface mines (Rhodes and Barton 2024) in West Virginia and red spruce forest enhancement and restoration throughout the central and southern Appalachians (Thomas-Van Gundy and Sturtevant 2014) are now being implicitly linked to improved brook trout habitat and populations (Ray and Pool 2018).

Low fish diversity should not be interpreted to mean that streams in high elevation red spruce and spruce-fir forests are unimportant to fish. Appalachian headwater streams are highly connected to their surroundings (Richardson 2019), and the condition of headwater forests impacts the quality and quantity of water delivered downstream to fish and other aquatic biota (MacDonald and Coe 2007). A headwater spruce-fir stand with its dense evergreen canopy provides cooling shade, and the thick, spongy forest floor of organic material and moss acts as a filter for organic and inorganic chemicals (Cormack 1949; Gregory et al. 1991; Ball et al. 2010; Zurbuch 2015). The cold, clean water flowing from these forests is important for a variety of fish species living downstream, including the economically important non-native wild trout fishery for rainbow trout (*Oncorhynchus mykiss*) and brown trout (*Salmo trutta*) in the region (Ahn et al. 2000).

In the southern Appalachians, high elevation spruce-fir forests typically only contain brook trout and potentially blacknose dace. At slightly lower elevations, sculpin species such as the mottled sculpin (*Cottus bairdi*) are known to co-occur with brook trout (Jenkins and Burkhead 1994). In the Allegheny Mountains and Plateau portion of the central Appalachians, where the downslope limit of red spruce is 400–500 m (1,300–1,650 ft) lower than in the southern Appalachians, greater fish species richness can be found within the larger second- and third-order systems flowing through red spruce forests near the interface with northern hardwood forests (e.g., Blackwater River, Cherry River, Gauley River, Upper Greenbrier, Upper Shavers Fork, and Williams River in West Virginia). Fish therein can include: least brook lamprey (*Lampetra aepyptera*), American eel (*Anguilla rostrata*), white sucker (*Catostomus commersonii*), torrent sucker (*Thoburnia rhothoeca*), northern hogsucker (*Hypentelium nigricans*), bigmouth chub (*Nocomis platyrhynchus*), creek chub (*Semotilus atromaculatus*), central stoneroller (*Campostoma anomalum*), longnose dace (*Rhinichthys cataractae*), blacknose dace,

Fig. 6.3 Application of crushed limestone into a headwater stream impacted by acid deposition. Photo by Eric Gladwell

rosyface shiner (*Notropis rubellus*), New River shiner (*Notropis scabriceps*), rosyside dace (*Clinostomus funduloides*), cutlip minnow (*Exoglossum maxillingua*), river chub (*Nocomis micropogon*), mountain redbelly dace (*Chrosomus oreas*), common shiner (*Luxilus cornutus*), striped shiner (*Luxilus chrysocephalus*), pearl dace (*Margariscus margarita*), sharpnose darter (*Percina oxyrhynchus*), candy darter

(*Etheostoma osburni*), fantail darter (*Etheostoma flabellare*), mottled sculpin, Blue Ridge sculpin (*Cottus caeruleomentum*), Potomac sculpin (*Cottus girardi*), and the non-native brown trout and rainbow trout (Stauffer et al. 1995; Clayton et al. 1998; Petty et al. 2005).

6.3 Herpetofauna

Species richness of amphibians and reptiles is high and moderate, respectively, in the central and southern Appalachians relative to other regions in the U.S. (Jenkins et al. 2015). In particular, the Appalachians represent a global biodiversity hotspot for salamanders (Rissler and Smith 2010; Kozak 2017) and are the potential origin of the lungless salamander family Plethodontidae (Beachy and Bruce 1992). Although only a small proportion of herpetofaunal species in the central and southern Appalachians are strongly associated with red spruce or spruce-fir forests, these forest types are considered optimal habitat for some high elevation-endemic salamander species of conservation concern, such as the federally threatened Cheat Mountain salamander (*Plethodon nettingi*; Pauley 2022). In addition, many generalist amphibian and reptile species can be found in these forest types.

To identify herpetofaunal species associated with red spruce and spruce-fir forests, we consulted the Partners in Amphibian and Reptile Conservation (PARC) habitat management guidelines for the northeastern and southeastern U.S. (Bailey et al. 2006; Mitchell et al. 2006). PARC represents a broad partnership of herpetofaunal species experts from across research and management sectors in the U.S. and develops products useful for management of these species, such as designation of priority conservation areas and conservation action plans. The PARC guidelines provide species-specific associations with habitat types in defined geographic regions, including red spruce and spruce-fir forests, and qualitatively rank them as optimal/primary, suitable, marginal, and non-habitat. We excluded species only found in the northern Appalachians based on distribution maps in Powell et al. (2016) and reviewed the scientific literature for all species for which spruce-fir was designated as optimal habitat. The PARC guidelines list spruce-fir forests as habitat for 40 amphibian and 13 reptile species in the central and southern Appalachians (Table 6.1). Based on our assessment of the PARC habitat designations, we consider spruce-fir forests to be habitat for 40 amphibian and 12 reptile species, including optimal habitat for 7 amphibian species (all salamanders), suitable habitat for 23 amphibian and 4 reptile species, and marginal habitat for 10 amphibian and 8 reptile species.

6.3.1 Amphibians—Salamanders

The PARC guidelines consider spruce-fir forests to be optimal, suitable, and marginal habitat for 11, 19, and 2 salamander species, respectively (Table 6.1). After consulting

Table 6.1 Amphibian and reptile species considered to be associated with red spruce (*Picea rubens*) or spruce-fir forests in the central and southern Appalachian regions. Spruce-fir forests were qualitatively ranked as optimal, suitable, or marginal habitat for each species by Partners in Amphibian and Reptile Conservation (PARC) habitat management guidelines (Bailey et al. 2006; Mitchell et al. 2006). We reviewed the scientific literature for all species where spruce-fir was designated as optimal habitat in one or both of the PARC guidelines and modified habitat designations for several of the species. Many additional amphibian and reptile species have been documented in spruce-fir forests, but are not generally considered to be associated with the forest type. Scientific and common names follow Nicholson (2025) except where noted

Scientific name	Common name	Appalachia distribution	Spruce-fir association
Amphibians—Salamanders			
Ambystoma maculatum	Spotted salamander	Central/Southern	Suitable
Ambystoma opacum	Marbled salamander	Central/Southern	Suitable
Desmognathus aeneus	Seepage salamander	Southern	Suitable
Desmognathus carolinensis	Carolina mountain dusky salamander	Southern	Suitable
Desmognathus folkertsi	Dwarf black-bellied salamander	Southern	Suitable
Desmognathus fuscus	Northern dusky salamander	Central/Southern	Suitable
Desmognathus imitator	Imitator salamander	Southern	Optimal
Desmognathus marmoratus	Northern shovel-nosed salamander	Central/Southern	Suitable[e]
Desmognathus monticola	Northern seal salamander	Central/Southern	Suitable
Desmognathus ochrophaeus	Allegheny Mountain dusky salamander	Central/Southern	Optimal
Desmognathus ocoee	Ocoee salamander	Southern	Suitable
Desmognathus orestes	Blue Ridge dusky salamander	Central [c]/Southern ePara>ePara>	Suitable
Desmognathus organi[a]	Northern pygmy salamander	Southern	Optimal
Desmognathus quadramaculatus[b]	Black-bellied salamander	Central/Southern	Suitable
Desmognathus santeetlah	Santeetlah dusky salamander	Southern	Suitable
Desmognathus welteri	Black Mountain salamander	Central/Southern	Suitable
Desmognathus wrighti	Pygmy salamander	Southern	Optimal[f]
Eurycea bislineata	Northern two-lined salamander	Central	Suitable

(continued)

Table 6.1 (continued)

Scientific name	Common name	Appalachia distribution	Spruce-fir association
Eurycea wilderae	Blue Ridge two-lined salamander	Southern	Suitable
Gyrinophilus porphyriticus	Spring salamander	Central/Southern	Suitable[g]
Hemidactylium scutatum	Four-toed salamander	Central/Southern	Suitable
Notophthalmus viridescens	Eastern newt	Central/Southern	Suitable[g]
Plethodon cinereus	Eastern red-backed salamander	Central/Southern	Suitable
Plethodon glutinosus	Northern slimy salamander	Central/Southern	Marginal
Plethodon jordani	Red-cheeked salamander	Southern	Suitable
Plethodon nettingi	Cheat Mountain salamander	Central	Optimal
Plethodon punctatus	Cow Knob Salamander	Central	Suitable
Plethodon teyahalee	Southern Appalachian salamander	Southern	Marginal[h]
Plethodon wehrlei	Wehrle's salamander	Central/Southern[d]	Optimal
Plethodon welleri	Weller's salamander	Southern	Optimal[f]
Plethodon yonahlossee	Yonahlossee salamander	Southern	Suitable
Pseudotriton ruber	Red salamander	Central/Southern	Marginal
Amphibians—Anurans			
Anaxyrus americanus	American toad	Central/Southern	Suitable[i]
Dryophytes chrysoscelis	Cope's gray treefrog	Central/Southern	Marginal[i]
Dryophytes versicolor	Gray treefrog	Central/Southern[e]	Marginal
Lithobates clamitans	North American green frog	Central/Southern	Marginal
Lithobates palustris	Pickerel frog	Central/Southern ePara>ePara>	Marginal
Lithobates sylvaticus	Wood frog	Central/Southern	Marginal
Pseudacris brachyphona	Mountain chorus frog	Central/Southern	Marginal
Pseudacris crucifer	Spring peeper	Central/Southern	Marginal

(continued)

Table 6.1 (continued)

Scientific name	Common name	Appalachia distribution	Spruce-fir association
Reptiles—Squamates			
Coluber constrictor	North American racer	Central/Southern	Marginal
Crotalus horridus	Timber rattlesnake	Central/Southern	Marginal
Diadophis punctatus	Ring-necked snake	Central/Southern	Marginal
Lampropeltis triangulum	Eastern milksnake	Central/Southern	Suitable[j]
Storeria occipitomaculata	Red-bellied snake	Central/Southern	Suitable[j]
Thamnophis saurita	Eastern ribbonsnake	Central/Southern	Marginal
Thamnophis sirtalis	Common gartersnake	Central/Southern	Suitable
Reptiles—Turtles			
Chelydra serpentina	North American snapping turtle	Central/Southern	Marginal
Chrysemys picta	Painted turtle	Central/Southern	Marginal
Clemmys guttata	Spotted turtle	Central	Marginal
Glyptemys insculpta	Wood turtle	Central	Suitable
Terrapene carolina	Eastern box turtle	Central/Southern	Marginal

[a]In Bailey et al. (2006), species had not yet been described and was considered to be *Desmognathus wrighti*
[b]Pyron and Beamer (2022) proposed that *Desmognathus quadramaculatus* be split into four distinct species, which was adopted by Nicholson (2025): Kanawha black-bellied salamander (*D. kanawha*), Pisgah black-bellied salamander (*D. mavrokoilius*), Smoky Mountains black-bellied salamander (*D. gvnigeusgwotli*), and southern black-bellied salamander (*D. amphileucus*). Additional research is needed to clarify the geographic distributions and habitat preferences for the individual species
[c]Limited distribution within the central Appalachian region
[d]Limited distribution within the southern Appalachian region
[e]Listed as non-habitat in Mitchell et al. (2006) and optimal in Bailey et al. (2006)
[f]Listed as suitable in Mitchell et al. (2006) and optimal in Bailey et al. (2006)
[g]Listed as optimal in Mitchell et al. (2006) and non-habitat in Bailey et al. (2006)
[h]Listed as optimal in Bailey et al. (2006)
[i]Listed as optimal in Mitchell et al. (2006)
[j]Listed as marginal in Mitchell et al. (2006) and suitable in Bailey et al. (2006)

the scientific literature, we consider spruce-fir to represent suitable rather than optimal habitat for three of these species, northern shovel-nosed salamander (*Desmognathus marmoratus*), spring salamander (*Gyrinophilus porphyriticus*), and eastern newt (*Notophthalmus viridescens*), and marginal rather than optimal habitat for one of these species, southern Appalachian salamander (*Plethodon teyahalee*). The northern shovel-nosed salamander is a highly aquatic species that occupies forest streams over a large elevational gradient (300–1,680 m [990–5,500 ft]; Petranka 1998). Although this species can be found in high elevation red spruce forests, such as White Top Mountain in Virginia and Grandfather Mountain in North Carolina (Pope and Hairston 1947), they are not restricted to this forest type and their distribution appears to be most strongly influenced by aquatic habitat characteristics (Martof 1962). In fact, no literature that we consulted mentioned red spruce or Fraser fir (*Abies fraseri*) in relation to habitat for the northern shovel-nosed salamander. The spring salamander is a semi-aquatic species that occupies forest streams and adjacent terrestrial habitat over a large geographic and elevational (100–2,000 m [330–6,550 ft]) gradient (Petranka 1998). Spring salamanders can be found in spruce-fir forests (Bruce 1972), but similar to the northern shovel-nosed salamander, occurrence is most heavily influenced by aquatic habitat characteristics (Bruce 2003; Lowe 2005; Petranka and Smith 2005). The eastern newt is one of the most widely distributed salamanders in North America and has no elevational restriction (Petranka 1998). It is a wetland breeder with a terrestrial juvenile (red eft) stage and an aquatic adult stage. Eastern newts can be found in spruce-fir forests (Rinehart et al. 2009; Pauley 2015), but there is no evidence that this is a preferred forest type (DeGraaf and Rudis 1990). Southern Appalachian salamanders (*Plethodon teyahalee*) have a restricted distribution, occurring primarily in western North Carolina (Powell et al. 2016). This is a mid-elevation species that is typically found from 640 to 910 m (2,100–2,990 ft; Niemiller and Reynolds 2011), but the species does reach higher elevations in a few localities (Riddell and Sears 2015).

Byers et al. (2010) documented Jefferson salamanders (*Ambystoma jeffersonianum*) and long-tailed salamanders (*Eurycea longicauda*) in red spruce forests in West Virginia, but these species are not considered to be associated with spruce-fir in the PARC guidelines. Pauley (2022) documented only one long-tailed salamander during 40 years of salamander surveys in the high elevation red spruce forests of West Virginia, but he and his graduate students have found several green salamander (*Aneides aeneus*) populations in red spruce forests (Thomas Pauley, Marshall University, oral communication, May 3, 2023). Similarly, Mathews and Echternacht (1984) listed the southern red-backed salamander (*Plethodon serratus*) as known from spruce-fir forests in Great Smoky Mountains National Park, but it is also not considered to be associated with this forest type in the PARC guidelines. Though not spruce-fir forest obligates, PARC guidelines consider four salamanders to be common components therein: spotted salamander (*Ambystoma maculatum*), four-toed salamander (*Hemidactylium scutatum*), eastern red-backed salamander (*Plethodon cinereus*) in the central and southern Appalachians, and red-cheeked salamander (*Plethodon jordani*) in the southern Appalachians. Spotted salamanders and four-toed salamanders are wetland breeders, and thus are likely to be encountered in spruce-fir forest patches adjacent to wetlands (Pauley 2015; Millikin et al. 2019). In contrast, eastern red-backed salamanders and red-cheeked salamanders have a fully terrestrial lifecycle and are widely distributed within spruce-fir forests (DeGraaf and Rudis 1990; Dodd and Dorazio 2004; Kroschel et al. 2014).

We regard seven salamander species in the central and southern Appalachians as either spruce-fir obligates or those for which this forest type is considered optimal habitat. The imitator salamander (*Desmognathus imitator*) has a small geographic distribution at high elevations (900–2,024 m [2,950–6,640 ft]) in eastern Tennessee and western North Carolina (Powell et al. 2016) and has been documented on Great

Smoky, Balsam, and Plott Balsam Mountains (Petranka 1998). Mathews and Echternacht (1984) considered the imitator salamander to be truly indicative of spruce-fir forests in Great Smoky Mountains National Park. Though robust studies are lacking, this salamander is associated with forest seeps and adjacent terrestrial habitats (Tilley 1985; Dodd 2003).

The Allegheny Mountain dusky salamander (*Desmognathus ochrophaeus*; Fig. 6.4) occurs from extreme southwestern Québec to north-central Tennessee (Powell et al. 2016). This is a highly terrestrial species that occupies the largest elevational gradient of any *Desmognathus* species and occupies a wide variety of forest types (Hairston 1949). However, in North Carolina, Tilley (1973) found that Allegheny Mountain dusky salamander annual survival and growth rates were highest in the spruce-fir forest. In the central Appalachians, this is often one of the most abundant species in red spruce forests (Brooks 1948; Pauley 1980; Kroschel et al. 2014).

Historically, pygmy salamanders (*Desmognathus organi*) and northern pygmy salamanders (*Desmognathus wrighti*) were considered to be the same species (*D. wrighti*), with a geographic distribution restricted to the Blue Ridge from southern Virginia to northern Georgia (Harrison 2000). Crespi et al. (2010) concluded that the species are separated by the French Broad River, with pygmy salamanders occurring south and west, and northern pygmy salamanders occurring north and east of the river. These are two of the smallest and most terrestrial species of *Desmognathus*, with no aquatic larval stage (Petranka 1998; Bruce 2011). Although they can be found at elevations < 700 m (2,300 ft; Hocking et al. 2021), these two salamanders are most associated with spruce-fir forests at elevations > 1,600 m (5,250 ft; Hairston 1949; Crespi et al. 2003; Rossell et al. 2018). Occurrence records within

Fig. 6.4 Adult Allegheny Mountain dusky salamander (*Desmognathus ochrophaeus*) observed in a high elevation red spruce (*Picea rubens*) forest in Tucker County, West Virginia. Photo by Donald Brown

their distributions largely correspond to current and former occurrence of spruce-fir forests in the southern Appalachians (Bruce 1977; Harrison 2000; Crespi et al. 2003). Mathews and Echternacht (1984) considered the pygmy salamander to be the other salamander species characteristic of spruce-fir forests in Great Smoky Mountains National Park. Rossell et al. (2018) sampled populations of both species in North Carolina and compared relative abundances among three forest types that occurred at different elevations: spruce-fir (>1,645 m [5,400 ft]), northern hardwood (1,310–1,645 m [4,300–5,400 ft]), and mountain cove (<1,310 m [4,300 ft]). For the combined species data, relative abundance was higher in spruce-fir forest than mountain cove forest, but not significantly different from adjacent northern hardwood forest. However, captures of both species were highest in spruce-fir forest, in contrast to 14 of the 15 additional salamander species encountered. Because the forest types in this study differed by elevation, additional research is needed to determine the relative influences of forest type and elevation on density. Rossell et al. (2018) did not find that microhabitat characteristics differed for occupied locations among the forest types, and only coarse woody debris was a strong predictor of occurrence.

The Cheat Mountain salamander (Fig. 6.5) is restricted to high elevation forests in the Allegheny Mountains portion of the central Appalachians in eastern West Virginia (Pauley 2022). This is the most well-studied amphibian species in relation to associations with red spruce forests (see Sidebar 6.2). Although at present, Cheat Mountain salamanders do not occur exclusively in red spruce forests, presence is strongly associated with this forest type (Dillard et al. 2008a; Pauley 2022), and most contemporary occupied patches probably were red spruce prior to the exploitative logging and subsequent wildfire period (1890–1940) based on post-harvest assessments (Minckler 1945).

Sidebar 6.2 Focal Species—Cheat Mountain Salamander (*Plethodon nettingi*).

The Cheat Mountain salamander was first described in 1938 by Green (1938). At that time, the species was only known to occur at two high elevation localities in eastern West Virginia, Barton Knob (1,351 m [4,432 ft]) and Cheat Bridge (1,084 m [3,556 ft]). Between 1976 and 2016, Thomas K. Pauley surveyed > 1,300 areas in eastern West Virginia to define the geographic distribution of the Cheat Mountain salamander (Pauley 2022). These efforts resulted in the delineation of 81 geographically isolated populations located across a latitudinal and longitudinal distance of approximately 31 km and 84 km (19 mi and 52 mi), respectively, in eastern West Virginia. The lowest documented elevation of occurrence for this species is 805 m (2,640 ft; Pauley and Pauley 1997), and occurrence probability is strongly positively associated with elevation (Dillard et al. 2008b; Kroschel et al. 2014).

In the first publication for the Cheat Mountain salamander, the two known localities of occurrence were described as "originally covered with spruce

Fig. 6.5 Adult female Cheat Mountain salamander (*Plethodon nettingi*) with eggs observed in a high elevation red spruce (*Picea rubens*) forest in Tucker County, West Virginia. Photo by Donald Brown

forest" (Green 1938). Brooks (1948) further described the species' forest type association as "nearly pure stands of red spruce" and "forests in which red spruce is a prominent species." In many occupied areas, red spruce (*Picea rubens*) was replaced by eastern hemlock (*Tsuga canadensis*) and deciduous trees (e.g., yellow birch [*Betula alleghaniensis*]) following extensive clearcutting and wildfires that occurred in the late 1800s and early 1900s (Brooks 1948; Stephenson and Clovis 1983). Pauley (2008) hypothesized that Cheat Mountain salamanders survived the logging and fires in areas that contained large emergent rocks, but were extirpated elsewhere, resulting in the current fragmented distribution. Pauley (2022) recorded contemporary forest type at 195 known-occupied Cheat Mountain salamander sites, and found that 74.3% contained red spruce, with 55.4% of the sites dominated by red spruce. He found that 13.8% of sites did not contain either red spruce or eastern hemlock, indicating that whereas Cheat Mountain salamanders are associated with red spruce and to a lesser extent eastern hemlock, they have persisted in some forest patches that were clearcut and returned as northern hardwood forests.

A few studies have sought to quantitatively characterize Cheat Mountain salamander habitat using observational field data. Pauley (1998) surveyed seven habitat characteristics at two known-occupied sites paired with two unoccupied sites, analyzing each pair separately and found that three of the variables significantly differed between occupied and unoccupied sites, but each variable only differed for one of the site pairs. Air temperature was cooler at one occupied site, soil acidity was lower at one occupied site, and litter mass was higher at one occupied site. Dillard et al. (2008a) surveyed 18 habitat characteristics at 67 known-occupied sites and 37 random sites. They determined that probability of occurrence decreased with soil depth to rock and distance to emergent rocks and increased with bryophyte density and spruce-hemlock density. Brown et al. (2022) quantified soil moisture and soil pH at four occupied sites across their geographic distribution, adjacent forest patches where Cheat Mountain salamanders have not been detected, and nearby forest patches that represented seemingly high-quality habitat but lacked the species. They found that soil moisture was higher at occupied sites, but soil pH did not significantly differ.

Two studies have sought to quantify Cheat Mountain salamander habitat using ecological niche models. Dillard et al. (2008b) used 180 known-occupied locations and random locations within the species' geographic distribution to assess the influence of 13 environmental variables on occurrence probability. A logistic regression analysis indicated that landform and lithology variables were the strongest predictors of occurrence probability, with some additional support for vegetation variables. Specifically, occurrence was positively associated with sandstone, elevation, and red spruce, and negatively associated with slope and southwesterly aspects. A complementary classification tree analysis found that sandstone and annual precipitation were the strongest predictors of occurrence. Another study used 702 known-occupied locations and random background locations within the species' geographic distribution to assess the influence of 11 landscape and 3 climate variables on occurrence probability (Rucker et al. 2025). A random forest analysis found that 10 of the candidate variables were influential for occurrence of Cheat Mountain salamanders, including elevation, existing vegetation, heat load index, canopy cover, soil type, rock type, Euclidean distance to rocky habitat, annual mean temperature, temperature seasonality, and precipitation in the warmest quarter. The estimated relationships generally agreed with previous studies, with occurrence probability positively associated with sandstone, elevation, and red spruce-eastern hemlock, and negatively associated with distance to rocky habitat. The model characterized 6.6%, 23.2%, 12.4%, and 57.8% of area within the known Cheat Mountain salamander distribution with high-quality habitat, moderate-quality habitat, low-quality habitat, and non-habitat, respectively.

In summary, Cheat Mountain salamanders have been considered a red spruce-associated species since their discovery, and research to date has generally supported this association. However, it is not currently known if Cheat Mountain salamanders directly benefit from red spruce. It is plausible that other environmental conditions, such as temperature and moisture (which are strongly correlated with aspect and elevation) and soil/geological characteristics, are influencing occurrence of both species. This may explain why Cheat Mountain salamander populations have persisted in forest patches that were clearcut

> 100 years ago and no longer contain red spruce. Brown et al. (2022) hypothesized that low soil pH found in red spruce and eastern hemlock forests may reduce interspecific competition pressure from other *Plethodon* salamanders, particularly the eastern red-backed salamander (*Plethodon cinereus*), which prefers less acidic soils (Sugalski and Claussen 1997; Wyman and Hawksley-Lescault 1987). If so, Cheat Mountain salamanders may have a competitive advantage in red spruce and eastern hemlock stands as climatic changes increase the habitat quality of high elevation forests for competitor salamanders (Kroschel et al. 2014; Rucker et al. 2022).

The Wehrle's salamander (*Plethodon wehrlei*; Fig. 6.6) occurs from southwestern New York to northwestern North Carolina and northeastern Tennessee, with the largest portion of their distribution occurring in West Virginia (Powell et al. 2016). Wehrle's salamanders occur within, but are not wholly restricted to, high elevations in the central and southern Appalachians (Pauley 1978; Petranka 1998). To our knowledge, forest type associations for the species have not been investigated. In West Virginia, Brooks (1948) noted that Wehrle's salamander was often the most abundant salamander species encountered in mature red spruce forests, and Rucker et al. (2022) observed that it was among the three most common salamander species encountered over three decades in that forest type.

Weller's salamanders (*Plethodon welleri*) have a small, fragmented distribution in the Blue Ridge of the southern Appalachians, from southwestern Virginia (Whitetop Mountain and Mt. Rogers; Hoffman and Kleinpeter 1948) through northwestern North Carolina (Flat Top Mountain, Grandfather Mountain, and Unaka Mountain;

Fig. 6.6 Adult Wehrle's salamander (*Plethodon wehrlei*) observed in a high elevation red spruce (*Picea rubens*) forest in Tucker County, West Virginia. Photo by Donald Brown

Snyder 1946; Thurow 1963) and extreme northeastern Tennessee (Thurow 1963). They are typically found in high elevation spruce-fir forests (Petranka 1998), but also occur in deciduous and eastern hemlock (*Tsuga canadensis*)-dominated forests as low as 700 m (2,300 ft; Thurow 1963; Ronca et al. 2026). Bogert (1952) surveyed several mountains in southwestern Virginia and only encountered this species in spruce-fir on Mt. Rogers.

6.3.2 Amphibians—Anurans

The PARC guidelines consider spruce-fir forests to be optimal and marginal habitat for two and six anuran (i.e., frog and toad) species, respectively (Table 6.1). Spruce-fir is listed as optimal habitat and non-habitat for the American toad (*Anaxyrus americanus*) in the northeastern and southeastern PARC guidelines, respectively. This species is considered a terrestrial habitat generalist, with a preference for open deciduous forests and grasslands (Dodd 2013). American toads occur in red spruce and spruce-fir forests where present in their distribution (Karns 1992; Mazerolle 2004; Pauley 2015). DeGraaf and Rudis (1990) found that American toads used balsam fir (*Abies balsamea*) stands in New Hampshire, but relative abundance was significantly lower than in red maple (*Acer rubrum*) and northern hardwood stands. In Ontario, timber harvesting in red spruce forests did not influence relative abundance of American toads (Thompson et al. 2008). Given this species is a habitat generalist, we consider spruce-fir to be suitable, but not optimal, for the American toad.

Spruce-fir is also listed as optimal habitat and non-habitat for Cope's gray treefrog (*Dryophytes chrysoscelis*) in the northeastern and southeastern PARC guidelines, respectively. This species is considered a terrestrial habitat generalist, with a preference for deciduous forests and grasslands (Dodd 2013). Cope's gray treefrogs have been documented in red spruce forests in West Virginia (Pauley 2015), but to our knowledge, no studies indicate this is a preferred forest type. Byers et al. (2010) listed one additional anuran, northern leopard frog (*Lithobates pipiens*), that has been documented in red spruce forests in West Virginia, but it is not considered to be associated with this forest type in the PARC guidelines. The PARC guidelines do consider two anuran species, wood frog (*Lithobates sylvaticus*) and spring peeper (*Pseudacris crucifer*), to be characteristic species of spruce-fir forests, despite the habitat type not being considered optimal. Both of these species are wetland breeders, and thus are likely to be encountered in forest patches adjacent to wetlands (Pauley 2015; Lambert et al. 2021).

6.3.3 Reptiles—Squamates

The PARC guidelines consider spruce-fir forest to be suitable and marginal habitat for three and four squamate (i.e., snake and lizard) species, respectively, all of which are snakes (Table 6.1). Byers et al. (2010) listed seven additional snake species that have been documented in red spruce forests in West Virginia but are not considered to be associated with the forest type in the PARC guidelines. These include: eastern copperhead (*Agkistrodon contortrix*), common watersnake (*Nerodia sipedon*), smooth greensnake (*Opheodrys vernalis*), central ratsnake (*Pantherophis alleghaniensis*), queensnake (*Regina septemvittata*), Dekay's brownsnake (*Storeria dekayi*), and mountain earthsnake (*Virginia valeriae pulchra*). Two of these species, common watersnake and central ratsnake, have also been documented in spruce-fir forests in Great Smoky Mountains National Park in North Carolina and Tennessee (Mathews and Echternacht 1984). Eastern copperheads and queensnakes were not observed by Thomas Pauley during 40 years of salamander surveys in the high elevation red spruce forests of West Virginia (Thomas Pauley, Marshall University, written communication, February 24, 2024). The PARC guidelines consider four squamates to be characteristic species of spruce-fir forests, despite the habitat type not being considered optimal, including ring-necked snake (*Diadophis punctatus*), eastern milksnake (*Lampropeltis triangulum*), red-bellied snake (*Storeria occipitomaculata*), and common gartersnake (*Thamnophis sirtalis*).

6.3.4 Reptiles—Turtles

The PARC guidelines consider spruce-fir forests to be suitable and marginal habitat for one and four turtle species, respectively (Table 6.1). No turtles are considered to be characteristic species of spruce-fir forests. In Maine and southeastern Canada, alluvial spruce-fir is considered suitable habitat for the wood turtle (*Glyptemys insculpta*; Latham et al. 2023); however, we are unaware of any wood turtle populations occurring in red spruce forests in the central Appalachians, and the species does not occur in the southern Appalachians (Jones and Willey 2021). Eastern box turtles (*Terrapene carolina*) are rarely observed in red spruce forests in West Virginia (Thomas Pauley, Marshall University, written communication, February 24, 2024), but have been found in spruce-fir forests above 1,828 m [6,000 ft; Diggins et al. 2016).

6.4 Birds

Avian (bird) communities of high elevation red spruce and spruce-fir forests in the central and southern Appalachian Mountains are unique and distinctive in their species composition compared to adjacent, lower elevation forests (Rabenold 1978; Clipp et al. 2022). Reflecting the differences in dominant tree species as elevation increases, the bird community composition therein is quantitatively distinct from those of northern hardwood forests, oak-pine forests, and mixed mesophytic forests, which tend to be primarily comprised of birds associated with deciduous tree assemblages (Clipp et al. 2022). Red spruce forests support trailing-edge, breeding populations of boreal bird species (e.g., Blackburnian warbler [*Setophaga fusca*], blue-headed vireo [*Vireo solitarius*], winter wren [*Troglodytes hiemalis*]) more commonly encountered at higher latitudes in New England or southern Canada (Rabenold 1978, 1984; Alsop and Laughlin 1991; Gross 2010). These trailing-edge populations boost both regional diversity and species-level genetic diversity (Hampe and Petit 2005; Ferrari et al. 2018; Merker and Chandler 2020), but they are thought to be at high risk from climate change (Cahill et al. 2014; Rushing et al. 2020), leading to numerous species being listed as local or regional conservation priority species. Southern Appalachian spruce-fir forests also contain endemic subspecies such as Carolina dark-eyed juncos (*Junco hyemalis carolinensis*) and Appalachian black-capped chickadees (*Parus atricapillus practicus*; Rabenold 1984, Rabenold et al. 1998). Despite post-European settlement changes to red spruce and spruce-fir forests, bird assemblages have remained consistent over many decades in the central Appalachians (Stewart and Aldrich 1949; Hall 1984; Clipp et al. 2022) and the southern Appalachians (Brewster 1886; Alsop and Laughlin 1991; Rabenold et al. 1998).

Relative to the boreal bird communities of New England and Canada, the taxonomic composition in central and southern Appalachian red spruce and spruce-fir forests is simpler, with fewer Neotropical migrant songbirds and a notable dominance of species that are residents or short-distance migrants (including altitudinal migrants; Rabenold 1978, 1984). In addition, red spruce forests in the central Appalachians have higher avian richness than spruce-fir forests in the southern Appalachians. To illustrate, in red spruce stands of the Monongahela NF in eastern West Virginia, 45 bird species from 21 families were observed using point count surveys (Clipp et al. 2022), whereas observers counted 32 bird species from 18 families in southern Appalachian spruce-fir forests in North Carolina (Rabenold et al. 1998). Alsop and Laughlin (1991) reported 16 territorial species from 10 families in a virgin spruce-fir stand within the Great Smoky Mountains National Park in Tennessee.

6.4.1 Neotropical Migrant Songbirds

Neotropical migrant songbirds (i.e., long-distance migrants) occur in the U.S. or Canada during the breeding season (approximately late May to mid-July) and spend the non-breeding season in the Caribbean Islands, Central America, or South America. These songbirds comprise a relatively small proportion of central and southern Appalachian red spruce and spruce-fir forest bird communities. Although observers documented 18 Neotropical migrant species during breeding songbird point count surveys in central Appalachian red spruce forests, only three species were considered strongly associated with red spruce: blackburnian warbler, magnolia warbler (*Setophaga magnolia*), and Swainson's thrush (*Catharus ustulatus*; Table 6.2; Clipp et al. 2022). Blackburnian and magnolia warblers, in particular, have a long and consistent history (>70 years) of abundant detections at red spruce survey sites within the central Appalachians (Stewart and Aldrich 1949; Demeo 1999; Kahler and Anderson 2010). In the southern Appalachians, there are fewer breeding Neotropical migrant species; in addition to Blackburnian warblers, observers in spruce-fir forests have recorded relatively low numbers of black-throated green warblers (*Setophaga virens*), black-throated blue warblers (*Setophaga caerulescens*), veeries (*Catharus fuscescens*), and chestnut-sided warblers (*Setophaga pensylvanica*), the latter of which were associated with clearings and disturbed areas (Rabenold 1978, 1992; Alsop and Laughlin 1991).

The Blackburnian warbler (Fig. 6.7) is the primary Neotropical migrant associated with red spruce forests in both the central and southern Appalachians. In West Virginia, multiple studies have linked Blackburnian warblers with forest stands comprised of > 50% red spruce cover of poletimber to sawtimber (13–56 cm [5–22 in] DBH) size (Clipp et al. 2022), and the species' occurrence increases in tandem with those habitats (Bailey and Rucker 2021). Regionally, the Blackburnian warbler is listed as a moderate priority species by the Appalachian Mountains Joint Venture for the Appalachian Mountains Bird Conservation Region (Appalachian Mountains Joint Venture 2019).

In the central Appalachians, magnolia warblers (Fig. 6.7) and Swainson's thrushes are the two other relatively common Neotropical migrants with strong red spruce associations. As one of the most abundant breeding warblers in montane conifer forests of West Virginia, magnolia warblers are associated with understories of dense, young red spruce, eastern hemlock, and eastern white pine (*Pinus strobus*;

Table 6.2 List of breeding songbird species that are strongly associated with red spruce (*Picea rubens*) forests in the central and southern Appalachian Mountains along with the distribution in which they have been documented as indicator species (central Appalachians, southern Appalachians, or both regions) and their migratory status (long-distance migrant [LD], short-distance migrant [SD], or resident [R])

Scientific name	Common name	Appalachia distribution	Migratory status
Catharus guttatus	Hermit thrush	Central	R
Catharus ustulatus	Swainson's thrush	Central	LD
Certhia americana	Brown creeper	Southern	R
Haemorhous purpureus	Purple finch	Central	R
Junco hyemalis	Dark-eyed junco	Both	R
Loxia curvirostra	Red crossbill	Central	R
Poecile atricapillus	Black-capped chickadee	Southern	R
Regulus satrapa	Golden-crowned kinglet	Both	R
Setophaga coronata	Yellow-rumped warbler	Central	SD
Setophaga fusca	Blackburnian warbler	Central	LD
Setophaga magnolia	Magnolia warbler	Central	LD
Sitta canadensis	Red-breasted nuthatch	Both	R
Troglodytes hiemalis	Winter wren	Both	R
Vireo solitarius	Blue-headed vireo	Both	SD

Fig. 6.7 Adult male Blackburnian warbler (*Setophaga fusca*; left; photo by Derek Courtney) and adult male magnolia warbler (*Setophaga magnolia*; right; photo by Hannah Clipp) perched in a mature red spruce (*Picea rubens*) tree and red spruce sapling, respectively, from high elevation sites within the Monongahela NF, West Virginia

Bailey and Rucker 2021). In a recent indicator species analysis identifying individual bird species associated with central Appalachian forest types and cover classes, the magnolia warbler had the second highest indicator value for red spruce forest and the highest indicator value for stands with medium (10–50%) red spruce cover composition (Clipp et al. 2022). Although Swainson's thrushes are the rarest breeding thrush species in the central and southern Appalachians, they are also strongly associated with red spruce-dominated stands. The majority of their trailing-edge breeding population is restricted to high elevation red spruce and mixed forests in the Allegheny Mountains of West Virginia (Bailey and Rucker 2021). In 2000–2001, Kahler and Anderson (2010) recorded Swainson's thrushes only in red spruce stands within their study area in that region. Notably, there is consideration among the ornithological community that the Appalachian Mountain population is a distinctive subspecies in the olive-backed group of Swainson's thrush (*Catharus ustulatus appalachiensis*; Ramos and Warner 1980, Pyle 1997, Mack and Yong 2000). Uncommon Neotropical migrant species in the central Appalachians include Nashville warblers (*Leiothlypis ruficapilla*), northern waterthrushes (*Parkesia noveboracensis*), and olive-sided flycatchers (*Contopus cooperi*). All three species have small, extant breeding populations confined to higher elevations (>1,000 m [3,280 ft]) in the Allegheny Mountains and are found in a few known openings and edges of conifer forests, bogs, and/or shrub wetlands (Hall 1984; Gross 2010), though it is probable that olive-sided flycatchers are regionally extirpated as they are only known to intermittently breed at one location (Bailey and Rucker 2021). The olive-sided flycatcher is a moderate priority species for the Appalachian Mountains Bird Conservation Region and a locally high priority species for conservation in high elevation forests of the southern Blue Ridge Mountains (Hunter 2010).

6.4.2 Short-Distance Migrant and Resident Songbirds

Short-distance migrant songbirds (including altitudinal migrants) are found in the U.S. throughout their full annual cycles, but their breeding grounds tend to be at higher elevations or latitudes than their non-breeding grounds. Similarly, resident songbirds are present year-round in a given area, although individuals may migrate short distances. Combined, these two groups of songbirds comprise the majority of central and southern Appalachian red spruce and spruce-fir forest bird communities (Rabenold 1978, 1984). For breeding songbird point count surveys in red spruce forests of the central Appalachians, observers documented a total of 25 short-distance migrant or resident species (compared to 18 Neotropical migrant species; Clipp et al. 2022). Of those 25 species, 9 were considered to be strongly associated with red spruce forest: dark-eyed junco (*Junco hyemalis*), golden-crowned kinglet (*Regulus satrapa*), winter wren, blue-headed vireo, red-breasted nuthatch (*Sitta canadensis*), yellow-rumped warbler (*Setophaga coronata*), hermit thrush (*Catharus guttatus*), purple finch (*Haemorhous purpureus*), and red crossbill (*Loxia curvirostra*; Table 6.2; Clipp et al. 2022). In the southern Appalachians, dark-eyed juncos, golden-crowned kinglets, winter wrens, red-breasted nuthatches, black-capped chickadees, and brown creepers (*Certhia americana*) comprise the majority of individual birds in spruce-fir forests during the breeding season (Rabenold 1984; Alsop and Laughlin 1991; Rabenold et al. 1998), with the former three species having the densest breeding populations (Alsop and Laughlin 1991, Raynolds et al. 1999). Although most bird species are scarce in winter, a handful of resident species (e.g., dark-eyed juncos, red-breasted nuthatches, black-capped chickadees, brown creepers), along with occasional hardy species such as hairy woodpeckers (*Dryobates villosus*), blue jays (*Cyanocitta cristata*), and common ravens (*Corvus corax*), can be found year-round at the lower boundaries of red spruce forests in the central and southern Appalachians (Rabenold 1984). Winters can also bring irruptive, short-distance migrant finch species, including red crossbills (from populations that breed further north), evening grosbeaks (*Coccothraustes vespertinus*), and pine siskins (*Spinus pinus*; Strong et al. 2015).

Dark-eyed juncos, golden-crowned kinglets, and winter wrens are strongly associated with and abundant within red spruce and spruce-fir forests in the central and southern Appalachians, although dark-eyed juncos may also occur in mesic deciduous forests at lower elevations. At one study site in a spruce-fir forest stand in the southern Appalachians, dark-eyed juncos exhibited the highest breeding density of all species in both 1967 and 1985 surveys (Alsop and Laughlin 1991). Golden-crowned kinglets are year-round residents of high elevation red spruce and mixed forests in the Allegheny Mountains of West Virginia, with a close dependence on red spruce for nesting and foraging (Bailey and Rucker 2021). In a recent indicator species analysis identifying individual bird species associated with central Appalachian forests (Clipp et al. 2022), the golden-crowned kinglet had the highest indicator value for stands dominated by red spruce. Winter wrens are also associated with high elevation forests of the Allegheny Mountains and are likely benefited by maturation of protected red spruce forests and restoration efforts in this region (Bailey and Rucker 2021). Blue-headed vireos (Fig. 6.8) and red-breasted nuthatches are two additional species that are also strongly associated with red spruce and spruce-fir forests. In West Virginia, blue-headed vireo and red-breasted nuthatch populations are highest in red spruce, pine-dominated, mixed, and northern hardwood forests of the Allegheny Mountains; however, blue-headed vireos are not restricted to that region and can be found across much of the eastern and southern portions of the state at lower elevations (Bailey and Rucker 2021).

Yellow-rumped warblers (Fig. 6.8), hermit thrushes, purple finches, and red crossbills are associated with red spruce forests in the central Appalachians, but they occur in limited or low numbers in the southern Appalachians. In West Virginia, yellow-rumped warblers are restricted to breeding in higher elevation forests of pure or mixed red spruce and other montane conifers (Bailey and Rucker 2021). An indicator species analysis in Clipp et al. (2022) showed that the yellow-rumped warbler had the second highest and the purple finch had the highest relative abundance values, respectively, in red spruce-dominated stands. Hermit thrushes tolerate a broader range of habitat within montane northern hardwood and red spruce forests of West Virginia compared to the congeneric Swainson's thrush, but both species are most abundant at high elevations where red spruce is present (Bailey and Rucker 2021). Purple finch breeding records in West Virginia are mainly from the red spruce forest type (Bailey and Rucker 2021), although they will occasionally breed in mixed northern hardwood forests, near bogs, and in riparian areas (Wootton 2020). Red crossbills differ slightly in their life-history traits compared to the other highlighted bird species. They are conifer-obligate finches with bills adapted to extract seeds from conifer cones, and populations in the central and southern Appalachians are linked to conifer seed production and retention, exhibiting nomadic movements driven by food availability (Bailey and Rucker 2021). In the Allegheny Mountains of West Virginia, they are an uncommon breeding bird species that nests in dense conifers and is observed consistently in red

Fig. 6.8 Adult blue-headed vireo (*Vireo solitarius*; left) and adult male yellow-rumped warbler (*Setophaga coronata*; right) from Spruce Knob, an area with high amounts of red spruce (*Picea rubens*) in the Monongahela NF, West Virginia. Photos by Hannah Clipp

spruce. The red crossbill is a high priority species for the Appalachian Mountains Bird Conservation Region and a locally high priority species for conservation in high elevation forests of the southern Blue Ridge Mountains (Hunter 2010).

In the southern Appalachians, black-capped chickadees and brown creepers are associated strongly with spruce-fir forests. In contrast, black-capped chickadees and brown creepers in the central Appalachians are found in red spruce forests, but they are not restricted regionally to the forest type. For example, brown creepers occur more generally in mature mixed coniferous-deciduous forests within West Virginia (Bailey and Rucker 2021). The black-capped chickadee is a moderate priority species for the Appalachian Mountains Bird Conservation Region and a high priority species for high elevation forests of the southern Blue Ridge Mountains (Hunter 2010).

6.4.3 Raptors and Owls

Red spruce and spruce-fir forests provide habitat for several raptor and owl species. Sharp-shinned hawk (*Accipiter striatus*) and American goshawk (*Astur atricapillus*) are the two raptors most strongly associated with red spruce, as they both prefer coniferous forest for nesting (Bailey and Rucker 2021), and both are listed as priority species for the Appalachian Mountains Bird Conservation Region (Appalachian Mountains Joint Venture 2019). In particular, the American goshawk is a rare, apex raptor species of high elevation red spruce and northern hardwood forests in the Allegheny Mountains portion of the central Appalachians, but it does not occur in the southern Appalachians. Although American goshawks will nest in red spruce forests in the central Appalachians, they have become increasingly uncommon in recent years, and they face challenges from nest predation by fishers (*Pekania pennanti*) and mortality from West Nile virus (Bailey and Rucker 2021). Within the past decade, there have been only a few confirmed breeding pairs of American goshawks in West Virginia (Bailey and Rucker 2021).

The primary owl species that occur in central and southern Appalachian red spruce and spruce-fir forests are barred owl (*Strix varia*) and northern saw-whet owl (*Aegolius acadicus*). The latter species is the smallest owl in the region and requires heterogeneous montane coniferous and mixed forests with dense understory structure and well-diversified foliage structure (Milling et al. 1997; Gross 2010). In the central and southern Appalachians, northern saw-whet owls only breed at higher elevations, occupying habitat with large red spruce or eastern hemlock mixed with northern hardwoods (Simpson 1972a, Bailey and Rucker 2021). The northern saw-whet owl is a moderate priority species for the Appalachian Mountains Bird Conservation Region and a high priority species for high elevation forests of the southern Blue Ridge Mountains (Hunter 2010).

6.4.4 Game Birds

Ruffed grouse (*Bonasa umbellus*), American woodcock (*Scolopax minor*), and wild turkeys (*Meleagris gallopavo*) are the only game birds closely associated with red spruce and spruce-fir forests in the Appalachian Mountains. Ruffed grouse are decreasing in abundance throughout all forest types in the central and southern Appalachians due to forest maturation and low abundance of early successional/young forest habitat (Tirpak et al. 2006, Bailey and Rucker 2021). However, they are regularly encountered in red spruce and mixed forests, ecotones, wildlife opening edges, regeneration units, and canopy gaps within surrounding northern hardwood types, as they used these habitats for foraging and mating displays (Gross 2010). Simpson (1972b) found ruffed grouse to be common in spruce-fir on Mt. Mitchell in North Carolina from spring through mid-fall, but largely absent during the winter. The ruffed grouse is a moderate priority species for the Appalachian Mountains Bird Conservation Region.

American woodcock requires wetlands and early successional habitat, and they are also a species with decreasing populations within the eastern U.S. (Gross 2010). In the few areas with abundant wetland resources occurring in red spruce forests, primarily within the central Appalachians, such as Canaan Valley in West Virginia, American woodcock can be common. Goudy et al. (1970) considered Canaan Valley's population of American woodcock, at the time, to be one of the most heavily hunted in the U.S. However, as with ruffed grouse, loss of early successional shrub habitats and development have reduced habitat quality (Steketee et al. 2015; Bailey and Rucker 2021). Meanwhile, wild turkeys are also present and can be locally common in red spruce forests of the central and southern Appalachians (Michael et al. 2015), but the species is limited by hard mast availability and snow depth (Healy and Dickson 1992; McShea et al. 2007), making red spruce forests sub-optimal habitat.

6.5 Mammals

The mammalian biota of high elevation red spruce and spruce-fir forests in the central and southern Appalachians displays a boreal affinity, providing a window to past climatic and landscape conditions that existed across much of the Southeast and mid-Atlantic during the immediate post-Pleistocene and early Holocene. By 10,000 YBP, for example, the post-Pleistocene mammalian mega-fauna that invariably were associated with spruce-fir forests and high elevation open habitats was extinct—whether by the influx of Native Americans or the rapidly changing climate or both (Weigl and Knowles 1995). Still, very few extant mammalian species in present-day red spruce and spruce-fir forest can be considered obligates to that forest type—essentially only the Virginia northern flying squirrel (*Glaucomys sabrinus fuscus*) and the snowshoe hare (*Lepus americanus*) in the central Appalachians, and the

Carolina northern flying squirrel (*Glaucomys sabrinus coloratus*) in the southern Appalachians.

Several species with more boreal, continental distributions occur down the spine of the Appalachians at higher elevations. Some of these species are associated with montane or boreal conifer and mixed forests (e.g., southern red-backed vole [*Myodes gapperi*], woodland jumping mouse [*Napaeozapus insignis*]). Most other species are not restricted to spruce-fir forests, such as masked shrew (*Sorex cinereus*), pygmy shrew (*Sorex hoyi*), water shrew (*Sorex palustris*), hairy-tailed mole (*Parascalops breweri*), red squirrel (*Tamiasciurus hudsonicus*), and fisher. Species with poor cold tolerance or those associated with old field/early successional habitats are uncommon in spruce-fir forests, such as Virginia opossum (*Didelphis virginiana*) and woodchuck (*Marmota monax*), though both have been observed at > 1,900 m (6,230 ft) in Great Smoky Mountains National Park (Linzey 1995). Aptly covered in other chapters, a changing climate from the end of the last glaciation to present, combined with anthropogenic disturbance of the habitat from forest harvesting, burning, and conversion to pastoral uses, as well as post-disturbance succession and forest maturation (for early successional species), has reduced the distributional extent and ecological integrity of these forests and in turn shaped the mammalian community present today.

6.5.1 Insectivores

Of the insectivorous mammals, seven of the eight species of shrew occurring in the region have been documented in red spruce and spruce-fir forests of the central and southern Appalachians (Ford et al. 2006). Based on approximately 38,000 pitfall trap-night surveys performed between 1979 and 2001 within red spruce forests in West Virginia and spruce-fir forests in North Carolina, Tennessee, and southwest Virginia, masked shrews represented 87% of the captures, whereas from a similar level of effort in regional northern hardwood forests, they comprised only 63% of the captures. Ford and Rodrigue (2001) found rock shrews in both red spruce forests and northern hardwood forests with a red spruce component in West Virginia where shaded, moss-covered talus and colluvial rock was present. Kirkland (1977) actually observed a higher abundance of rock shrews in recently cut (<5 years) red spruce stands than in uncut stands that were > 25 years since disturbance. Laerm et al. (2007a) listed streams draining spruce-fir, eastern hemlock-white pine, and high elevation bogs as the preferred habitat for water shrews. On the Ridge and Valley/ Allegheny Front interface in Virginia, Pagels et al. (1998) noted that water shrews required cold, pristine headwater streams. Throughout much of their distribution, water shrews also are associated with American beaver (*Castor canadensis*) ponds (Beneski and Stinson 1987), so occurrence in high elevation ponds in West Virginia is probable. In the southern Appalachians, Linzey (2002) reported an observation of a water shrew in a small bog on Andrew's Bald, North Carolina at around 1,800 m (5,900 ft). For the three mole species in the region, the hairy-tailed mole is the most likely to be encountered in spruce-fir forests (Laerm et al. 2007b), with observations occurring in the Mt. Roger's area in Virginia (Linzey 2021) and Kuwohi Mountains (i.e., Clingman's Dome) in Great Smoky Mountains National Park (Linzey 1995).

6.5.2 Bats

Although considerable bat research has been conducted in the central and southern Appalachians (Britzke et al. 2003; Ford et al. 2005, 2016; Austin et al. 2018; O'Keefe et al. 2019; Rojas et al. 2019; Beilke et al. 2021; Johnson et al. 2021), little work, particularly mist-netting, has purposefully occurred in red spruce or spruce-fir forests. Many records have been chance observations (e.g., a big brown bat [*Eptesicus fuscus*] discovered roosting in Mt. LeConte Lodge in the Great Smoky Mountains National Park at an elevation of 1,920 m [6,300 ft]). Johnson et al. (2021) mist-netted widely on the Monongahela NF, West Virginia from 2003 to 2019, including some sites in red spruce forest, but did not report findings relative to forest cover type or elevation. De La Cruz et al. (2023a, b) created concordance probabilistic distribution maps that included contemporary and historical mist-net capture, day-roost, and acoustic recording records for northern long-eared bats (*Myotis septentrionalis*) and Indiana bats (*Myotis sodalis*) for the Monongahela NF in West Virginia and determined that the cool, high elevation red spruce forests were unsuitable as potential summer day-roosting or foraging habitat. Virginia big-eared bats (*Corynorhinus townsendii virginianus*) inhabit maternity caves and hibernacula in the northern hardwood-red spruce forest transition zone in West Virginia (Stihler 2011) and the Burke's Garden area in Tazewell County, Virginia, which may have been a northern hardwood-red spruce forest prior to European settlement (Hoffman 1950).

Prior to the introduction of white-nose syndrome (WNS; a disease that has decimated populations of many bat species in eastern North America), Francl et al. (2004) acoustically surveyed high elevation (900–1,100 m [2,950–3,610 ft]) wetlands within the red spruce forest zone in and adjacent to Canaan Valley, West Virginia. Although eight species were detected, little brown bats (*Myotis lucifugus*) accounted for 73% of all echolocation files. Post-WNS, Beilke et al. (2021) noted that bat activity attributable to the genus *Myotis* was greater in northern hardwood forests than in spruce-fir forests, in part due to structural clutter in the latter forest type. Similarly, Diggins and Ford (2022) conducted acoustic surveys at seven southern Appalachian *sky island* (1,585–1,920 m [5,200–6,300 ft] in elevation) spruce-fir stands in North Carolina from spring through fall, and found that 53% of bat passes were hoary bats (*Lasiurus cinereus*) and 20% were silver-haired bats (*Lasionycteris noctivagans*). Although these migratory species are present in the Southeast during the non-maternity fall-early spring season, summer activity higher than in the spring and fall migration period therein or as observed in the central Appalachians (Muthersbaugh et al. 2019) suggests that summer foraging and day-roosting indicative of maternity activity is occurring at high elevations in southern Appalachian spruce-fir forests. Only 2% of passes were attributable to bats in the genus *Myotis*, including the eastern small-footed bat (*Myotis leibii*), the threatened northern long-eared bat, and the endangered Indiana bat. Prior to WNS, Kerns and Kerlinger (2004) documented bat mortality at a wind-energy site in Tucker County, West Virginia along Backbone Mountain that was predominantly a high elevation northern hardwood-red spruce community. Even accounting for low detection probability due to rough terrain, bat mortality in the fall and spring migration periods were higher than most wind-energy sites in eastern North America. Eastern red bats (*Lasiurus borealis*) accounted for > 40% of the carcasses recovered, followed by hoary bats (18.5%) and tricolored bats (*Perimyotis subflavus*; 18.3%). No Indiana bats were observed and only six northern long-eared bats were observed (1.3% of total).

6.5.3 Lagomorphs and Rodents

Snowshoe hares occur in red spruce forests, particularly in young regeneration, openings, and forest stands with dense great laurel (*Rhododendron maximum*) and mountain laurel (*Kalmia latifolia*) over 20,000–40,000 ha (49,000–99,000 ac) in the high Allegheny portion of the central Appalachians of West Virginia (Chapman 2007c). Brooks (1955) indicated that snowshoe hares remained in West Virginia in the 1920s to 1940s and were never extirpated but were supplemented in 1949 with individuals from Wisconsin. Until recently, a small number of snowshoe hares persisted in the Laurel Fork area of Highland County, Virginia in dense second-growth thickets of red spruce, yellow birch (*Betula alleghaniensis*), *Rhododendron* spp., mountain laurel, and blackberry (*Rubus* spp.). This population was supplemented by released snowshoe hares from neighboring West Virginia, but post-release survival was low due to predation, averaging only about 24 days (Linzey 2021). Declines in the central Appalachians are attributed to reduced high elevation early successional habitat from forest maturation and climate change, specifically reduced duration and extent of snow cover when snowshoe hare pelage is white, increasing vulnerability to predation (Fies 1991; Michael et al. 1982; Diefenbach et al. 2016). Brooks (1955) noted that even in the first half of the twentieth century, the white winter pelage of snowshoe hares was somewhat of an ecological maladaptation for a region that even in snowy winters often had cycles of melting and limited snow cover.

Whether snowshoe hares historically occurred in the southern Appalachians is subject to debate (Linzey 1995). Merriam (1888) doubted they occurred in the Great Smoky Mountains, whereas Kellogg (1939) reported they were formerly present based on local residents' accounts of "hares" that turned white in the winter along the crest of mountains in spruce-fir forests. Additionally, Campbell et al. (2010) reported a museum record of snowshoe hare from Buncombe County, North Carolina. Although the Appalachian cottontail (*Sylvilagus obscurus*) occurs widely in the Appalachians in early successional habitat and forests with dense, usually ericaceous understories, the species appears to be more common in high elevation red spruce forests in the central Appalachians (Boyce and Barry 2007) and spruce-fir or eastern hemlock-white pine in the southern Appalachians, perhaps in part due to the absence of competition from the eastern cottontail (*Sylvilagus floridanus*; Chapman 2007a). Apodaca et al. (2020) conducted a range-wide assessment of Appalachian cottontails in western North Carolina and found that high elevation populations were associated with shrub and heath balds, but usually in closer proximity to spruce-fir than other forest types. Genetic analyses showed population structuring as a result of their disjunct distribution, and high predicted occupancy was associated with the high elevation *sky islands* in North Carolina where spruce-fir forests occur.

Red squirrels occur at mid to high elevations in the Blue Ridge of the southern Appalachians and the Blue Ridge, Ridge and Valley, and Appalachian Plateau sections of the central Appalachians. Generally, red squirrels require a conifer component, as they feed more heavily on conifer seeds than other sciurids (Linzey 2021). Despite being generally abundant in spruce-fir forests, Laerm and Webster (2007) believed the status of the species more widely in the southern Appalachians was unknown and that stressors to high elevation forests, such as eastern hemlock mortality from the hemlock woolly adelgid (*Adelges tsugae*) and Fraser fir mortality from the balsam woolly adelgid (*Adelges piceae*), are probably a threat to red squirrel populations. Eastern chipmunks occur in spruce-fir forests (Linzey 1995), but are less common due to the paucity of hard mast (Mengak and Laerm 2007b).

Two subspecies of northern flying squirrel (*Glaucomys sabrinus*) occur within spruce-fir forests (Miller 1936; Handley 1953). The Virginia northern flying squirrel occurs in the Allegheny Mountains of West Virginia and Highland County, Virginia (Fig. 6.9), and the Carolina northern flying squirrel occurs in the Blue Ridge portion of the southern Appalachians in western North Carolina, eastern Tennessee, and southwestern Virginia (Weigl et al. 1992; Stihler et al. 1995). Both are Pleistocene relicts that are strongly associated with forests with a high percent stand composition of red spruce, Fraser fir, balsam fir, or eastern hemlock (Weigl et al. 1992; Hughes 2006; Menzel et al. 2006; Ford et al. 2015, 2022; Diggins et al. 2017). Both subspecies were listed as federally endangered in 1985, although the Virginia northern flying squirrel was delisted in 2013 (see Sidebar 6.3). Occupancy probability for northern flying squirrels is highest in mature to old-growth forests with > 30% conifer and occurring above 900 m and 1,350 m (2,950 and 4,430 ft) in the central and southern Appalachians, respectively (Ford et al. 2007, 2015, 2022). Foraging home range size is strongly influenced by habitat quality, such that home ranges in stands with a high red spruce component were < 6 ha (15 ac; Ford et al. 2014; Diggins et al. 2017), whereas home ranges in northern hardwood stands with a small red spruce component were > 50 ha (120 ac; Menzel et al. 2006). Northern flying squirrels are flexible in their denning habits, but dreys (i.e., leaf nests) in conifers and cavities in hardwoods are typically used (Menzel et al. 2004; Diggins et al. 2015, 2017). Though their diet can be varied (Mitchell 2001), northern flying squirrels in the central and southern Appalachians are mycophagists that feed mainly on hypogeal fungi (i.e., truffles) that are associated with red spruce overstories and deep organic soils (Loeb et al. 2000; Diggins et al. 2020).

Fig. 6.9 Adult female Carolina northern flying squirrel (*Glaucomys sabrinus coloratus*) in a spruce-fir forest, William H. Silver Game Land, Haywood County, North Carolina. Photo by Clifton Avery, North Carolina Wildlife Resources Commission

Sidebar 6.3 Focal Species—Northern Flying Squirrel (*Glaucomys sabrinus*).

Two subspecies of northern flying squirrel (*Glaucomys sabrinus*) are strongly associated with red spruce (*Picea rubens*) and spruce-fir forests, the Virginia northern flying squirrel (*G. s. fuscus*) and the Carolina northern flying squirrel (*G. s. coloratus*). Both subspecies were listed as federally endangered in 1985 due to the drastic decline of spruce-fir forests following the exploitative logging period and the threat of climate change to remaining habitat (U.S. Fish and Wildlife Service 1990). The two subspecies are geographically isolated from contiguous populations in the northern Appalachians by approximately 500 and 1,000 km (310 and 620 mi), respectively. Arbogast et al. (2005) found that both Appalachian subspecies had lower levels of genetic variability and the presence of several private alleles compared to continental populations of northern flying squirrels in the northern U.S. and southern Canada.

The reduction of red spruce in the central Appalachians and spruce-fir in the southern Appalachians allowed for the increase of hard mast bearing tree species (i.e., oaks [*Quercus* spp.] and American beech [*Fagus grandifolia*]) that provided a high-energy cacheable food source for the southern flying squirrel (*Glaucomys volans*), an aggressive den site competitor to northern flying squirrels (reviewed in Diggins 2023). Moreover, southern flying squirrels harbor a parasitic nematode (*Strongyloides robustus*) that is deleterious to northern flying squirrels (Wetzel and Weigl 1994). Federal listing resulted in the initiation of long-term monitoring programs using artificial nest-boxes in North Carolina, Virginia, and West Virginia (Fig. 6.10). The monitoring programs have shown that, within their distributions, populations are stable for both subspecies (Reynolds et al. 1999; Ford et al. 2010, 2015). Apparent population stability, coupled with the majority of suitable habitat occurring on protected Federal or State lands, resulted in the Virginia northern flying squirrel being delisted in 2013, although it remains a state endangered animal in Virginia (Ford et al. 2022). Currently, improving northern flying squirrel habitat quality via red spruce planting or midstory release is one of the largest drivers of central and southern Appalachian red spruce restoration efforts (Thomas-Van Gundy and Sturtevant 2014; Rentch et al. 2016).

American beavers occur throughout the central and southern Appalachians, but owing to the small, high gradient streams in southern Appalachian spruce-fir forests, they are uncommon in that setting (Linzey 1995; Mengak and Laerm 2007a). The species was extirpated in Highland County, Virginia and eastern West Virginia by the 1850s, but populations were reestablished in the Allegheny Mountains by the 1940s through reintroduction efforts that began in 1911, and the species occurs in red spruce forests where, unlike the southern Appalachians, larger watercourses and numerous wetland complexes exist (Bailey 1954). Bonner et al. (2009) noted that American beaver-created or maintained wetlands were an important component supporting wetland plant diversity, and presumably, fish, amphibian, and avian diversity in the Canaan Valley area of West Virginia.

No other rodents in the central and southern Appalachians are strongly tied to spruce-fir forests. The small rodent community is dominated by deer mice (*Peromyscus maniculatus*), southern red-backed voles, and woodland jumping mice (Laerm and Castleberry 2007; Laerm and Pagels 2007; Pagels and Laerm 2007b; Linzey 2021). Rock voles (*Microtus chrotorrhinus*) can occur where moss-covered emergent rock and colluvium occur in mesic forests above 900 m (2,950 ft), including in red spruce and spruce-fir forests (Pagels and Laerm 2007a). For this species, Kirkland (1977) found greater abundances in red spruce forests than northern hardwood

Fig. 6.10 A researcher installs a nest-box for Carolina northern flying squirrels (*Glaucomys sabrinus coloratus*) in the Roan Mountain Highlands, Avery County, North Carolina. These artificial den structures are readily used by flying squirrels and constitute a valuable survey and monitoring tool. Photo courtesy of U.S. Fish and Wildlife Service

forests in West Virginia and determined that rock vole abundance increased dramatically in recent red spruce regeneration areas. Francl et al. (2008) observed that current red spruce forests within and surrounding the Canaan Valley area of West Virginia presented a barrier to genetic exchange among high elevation fields and wetlands for the meadow vole (*Microtus pennsylvanicus*), which likely expanded in abundance and distribution following the exploitative logging period. The presence of porcupines (*Erethizon dorsatum*) within historic times in the southern Appalachians is unclear (Linzey 1995). However, for the central Appalachians, sporadic observations in high elevation areas of western Maryland, northwestern Virginia, and northern West Virginia (including Spruce Knob) have been noted since the late 1800s (Goode 1879; Harman and Thoerig 1968). Moncrief and Fies (2020) determined that porcupines had established breeding populations in this part of the central Appalachians and were expanding southward.

6.5.4 Carnivores

No mammalian carnivore in the central or southern Appalachians is strongly tied to spruce-fir forests, but most extant species occur in this habitat type. Although speculative, red spruce forests in West Virginia and spruce-fir forests in the southern Appalachians may have served as localized refugia for species such as the American black bear (*Ursus americanus*) following European settlement and then subsequently after the exploitative logging period, due to their remoteness and low human population density. Throughout its continental distribution, the fisher is associated with boreal forest types, but not exclusively (Chapman 2007b). Miller and Kellogg (1955) believed the species originally ranged down into the southern Appalachians as far south as North Carolina and was probably extirpated by the mid-1800s in the central Appalachians (Linzey 2021). Fishers were reintroduced in two red spruce forest patches in West Virginia in 1969 (Pack and Cromer 1981), and over the subsequent decades have expanded their distribution in West Virginia and have recolonized western Maryland and northwestern Virginia (Moncrief and Fies 2015). Based on the success of the West Virginia reintroduction, a potential fisher reintroduction in Great Smoky Mountains National Park was considered, but ultimately was determined to be a potential threat to the Carolina northern flying squirrel (Weigl 2007).

6.5.5 Ungulates

White-tailed deer (*Odocoileus virginianus*) generally occur at low densities in spruce-fir forests (Linzey 1995), though abundance can be high in localized areas such as Canaan Valley, West Virginia owing to an abundance of improved pastures, early successional wetland, and refugia provided by hunting prohibition on state parks and private lands therein (Michael et al. 2015). Cherefko et al. (2015) noted that white-tailed deer herbivory damage on balsam fir seedlings and saplings in Canaan Valley was severe. Similarly, herbivory on bog Jacob's-ladder (*Polemonium vanbruntiae*), a rare wetland plant, reduced survival and prevented plant recolonization (Flaherty et al. 2018). Despite lower white-tailed deer densities in the red spruce-northern hardwood forest interface in West Virginia, relative browsing pressure on palatable woody browse species can be high in that setting (Campbell et al. 2006). Campbell et al. (2004) did not note any discernible winter "yarding" pattern of white-tailed deer in this habitat type, which has been documented in northern states, despite deep snow and cold temperatures in some winters in the central Appalachians.

Elk (*Cervus canadensis*) were extirpated in the southern Appalachians by the early to mid-1800s (Ganier 1928) and in the central Appalachians by the 1870s in Virginia and West Virginia (Linzey 2021; West Virginia Division of Natural Resources 2021). A reintroduction effort occurred during the 1920s and 1930s in Giles and Bland counties, Virginia, near red spruce habitat at Mountain Lake. The population persisted until 1974 when the combination of forest maturation and crop depredation made their occurrence untenable in the local community (Linzey 2021). Subsequently, elk were successfully reintroduced on the coalfields portion of the Appalachian Plateau on previously surface mined areas in Kentucky, Tennessee, Virginia, and West Virginia, where open grasslands are abundant and the risk of negative interactions with agricultural production is low (Lituma et al. 2021). Interestingly, the West Virginia Division of Natural Resources identified the high elevation areas on the Monongahela NF as the most ecologically suitable habitat for restoration efforts and long-term persistence (Enck and Brown 2005). Similarly, elk were reintroduced to the Cataloochee Valley, North Carolina portion of Great Smoky Mountains National Park in 2001 (Maehr et al. 2007). This small herd has slowly grown in numbers and expanded their distribution in the park as well as adjacent private land (Linehan and Palmer 2014). In 2018, the North Carolina Wildlife Resources Commission purchased 821 ha (2,029 ac) for the William H. Silver Gameland that has high elevation balds, meadows, and mixed red spruce-northern hardwood forests, in part to support expansion of the elk herd (North Carolina Wildlife Resources Commission 2023). Prior to herd expansion in the park, work by Hillard (2013) did not show a strong selection for high elevation habitat types in the Cataloochee Valley, but if the herd continues to grow and expand, potential use of high elevation grassy bald habitats in the spruce-fir forest matrix for foraging habitat and thermal refugia merits investigation.

6.5.6 Feral Swine

Russian wild boar/feral swine (*Sus scrofa*) accidently released from a game preserve on Hooper Bald, North Carolina in the 1920s became established in Great Smoky Mountains National Park by the 1940s (Linzey 1995). As poor thermoregulators, feral swine will often move into higher elevations in the park, including spruce-fir forest during the summer prior to hard mast drop, where their presence has been extremely deleterious to the flora (Bratton 1974). Though the National Park Service has maintained an aggressive control program over the decades, the species' fecundity and high adaptability has allowed it to persist. With additional feral swine releases by the public, the species recently became established in the spruce-fir and grassy balds in the Roan Mountain Highlands of North Carolina and Tennessee (Marquette Crockett, Southern Appalachian Highlands Conservancy, oral communication, 19 Oct 2023).

6.6 Invertebrates

The central and southern Appalachians represent the southern distribution limit for many northern invertebrate species, such as the murky flash-train firefly (*Photinus obscurellus*) and white-spotted sawyer beetle (*Monochamus scutellatus*), due to associations with plants found in high elevation red spruce and spruce-fir forests (Baker 1972; Faust 2017). Invertebrate communities are heavily influenced by local plant communities, relying on them for larval development, oviposition sites, microclimate generation, and habitat for prey. Because red spruce and spruce-fir systems of the central and southern Appalachians contain distinctive plant communities, they also host unique invertebrate assemblages (Niemelä et al. 1996; Bishop et al. 2009). Most documentation of invertebrates in these systems comes from research focusing on tree pest species, but the occurrence of many other invertebrate species can be inferred from the botanical composition of the forest understory. The majority of plants in red spruce and spruce-fir systems are angiosperms, the bulk of which are pollinated by insects. For example, blueberries (*Vaccinium* spp.), an important food supply for vertebrates, rely on pollination by bumble bees (*Bombus* spp.) to produce fruit, and rare flowers in these forest ecosystems, such as orchids, typically depend on bees or flies for pollination.

The high elevation streams and wetlands associated with red spruce and spruce-fir forests also harbor uncommon invertebrates such as the superb jewelwing (*Calopteryx amata*; Paulson and Dunkle 2021). Cool, acidic, nutrient poor waters can be inhospitable to more common, generalist species found at lower elevations, allowing specialist species such as the superb jewelwing to remain extant. High elevation streams are associated with higher stream gradients, increased scour, reduced catchment areas, and increased shading, resulting in smaller amounts of fine particulate matter (Cormack 1949; Vannote et al. 1980; Arbuckle and Downing 2002). They are also typically characterized by lower pH and reduced fish diversity (see Sect. 6.2). These conditions are unsuitable for many filter feeding invertebrates, such as mussels, which are impacted by limited food, substrates that are not conducive to burrowing, and reduced diversity of larval (glochidial) host fish (Arbuckle and Downing 2002; Modesto et al. 2018). Instead, aquatic invertebrate communities contain higher proportions of predatory invertebrates and *shredder* invertebrates (which feed on coarse organic matter, such as twigs, leaves, and needles, that fall into the streams) than streams at lower elevations (Vannote et al. 1980).

Pest insect associations with red spruce are well studied, as understanding the impacts these organisms have on red spruce ecosystems is critical for forest management. Important insect pests of red spruce include eastern spruce gall adelgid (*Adelges abietis*), pine leaf adelgid (*Pineus pinifoliea*), eastern spruce beetle (*Dendroctonus rufipennis*), European spruce sawfly (*Diprion hercyniae*), yellowheaded spruce sawfly (*Pikonema alaskensis*), and spruce budworm (*Choristoneura fumiferana*; Potter et al. 2019). These insects can damage host trees in various ways (e.g., defoliation, sap feeding), and the significance of damage ranges from minor to potentially lethal. Although these pest species are not exclusively red spruce specialists and may utilize various hosts (typically other *Picea*), red spruce is a potential host and may be the primary host, within the central and southern Appalachians.

Studies of non-pest insects and other invertebrates in red spruce and spruce-fir forests of the central and southern Appalachians are scarce. Nonetheless, notable non-pest invertebrates that associate with red spruce include numerous high elevation wetland species (Byers et al. 2007, 2010), the spruce-fir moss spider (*Microhexura montivaga*), stoneflies (*Plecoptera* spp.), and several land snails. While this chapter emphasizes major pest species, select non-pest insects are included where the literature supports their inclusion.

6.6.1 *Insecta: Coleoptera*

6.6.1.1 Eastern Spruce Beetle (*Dendroctonus Rufipennis*)

The eastern spruce beetle is a major pest of red spruce, preferentially boring into weak or windthrown spruce as these hosts make excellent nesting sites and represent a source of nutrition without the need to battle host defenses. During outbreak years, eastern spruce beetles become substantially more aggressive and will attack standing trees in numbers significant enough to overwhelm host defenses (Wallin and Raffa 2004; Fig. 6.11). This species may have up to three generations per year. Population size fluctuates intermittently, with peak densities capable of killing large numbers of spruce over a few years, followed by many years of relatively low impact.

Several organisms are strongly associated with the eastern spruce beetle. The beetle acts as a vector for multiple fungi including *Leptographium abietinum* (Six and Bentz 2003) and *Ceratocystis rufipenni* (Wingfield et al. 1997). Additionally, Cardoza et al. (2008) listed eight mites and six nematodes carried by eastern spruce beetles. *Leptographium abietinum* has a context-dependent mutualism with the eastern spruce beetle, as the fungi can weaken host defenses and synthesize sterols that are important to the beetle, but with a tradeoff of fewer eggs produced (Cardoza et al. 2008). *Ceratocystis rufipenni* is a particularly virulent fungus and represents a serious threat to infected trees (Six and Bracewell 2015). The nature of its relationship with the eastern spruce beetle is less clear, but presumably the fungus weakens the host tree, making invasion easier. While the specific nature of the phoretic mite and nematode associations described by Cardoza et al. (2008) are not fully known, it is known that

Fig. 6.11 Adult eastern spruce beetle (*Dendroctonus rufipennis* [left]) and an example of the potential for extensive spruce mortality caused by this species (white spruce [*Picea glauca*] in south-central Alaska [right]; photos by USDA Forest Service Forest Health Protection program)

one mite species, *Histiogaster arborsignis*, feeds on *L. abietinum* and likely transfers spores to eastern spruce beetles.

6.6.2 *Insecta: Hemiptera*

6.6.2.1 Eastern Spruce Gall Adelgid/Pineapple Gall Adelgid (*Adelges Abietis*)

The eastern spruce gall adelgid is an introduced species from Europe which typically does only minor damage, primarily aesthetic, to host trees (Fig. 6.12). As an entirely female species, reproduction is by parthenogenesis. Because it is a poor flier, dispersal potential is low. This often leads to individual trees being densely covered in galls while nearby trees have relatively few galls. Because gall abundance and volume are directly related to production of nymphal gallicolae, any factors that impact these gall parameters will also impact eastern spruce gall adelgid population size. Stems with larger, more numerous galls will have a larger number of gallicolae nymphs present, and these nymphs will be larger and possess greater fecundity than smaller individuals (Sopow and Quiring 1998, McKinnon et al. 1999). Gall density is typically highest on intermediate-sized mid-crown shoots or on upper-crown shoots after crown closure (Fidgen et al. 1994). Gall abundance and volume may be impacted by tree stress. For example, drought-stressed trees may show greater gall abundance, though reduced gall volume.

6.6.2.2 Balsam Woolly Adelgid (*Adelges Piceae*)

Balsam woolly adelgid is an exotic invasive insect that invaded North America in the early twentieth century and is a serious pest of trees in the *Abies* genus. The

Fig. 6.12 Example of pineapple-shaped galls caused by feeding activity of exotic invasive eastern spruce gall adelgids (*Adelges abietis*); photo by USDA Forest Service Forest Health Protection program

species has expanded to many of North America's fir populations, including the Appalachian Mountains (Montgomery and Havill 2014; Hrinkevich et al. 2016). Despite its specific epithet, this insect does not feed on *Picea* species. Balsam woolly adelgid has caused extensive fir mortality within its invasion distribution, resulting in altered stand composition of spruce-fir ecosystems (Ragenovich and Mitchell 2006). Notably, this includes extensive die-offs of Fraser fir in Great Smoky Mountains National Park in the 1960s and 1980s (Dull et al. 1988; Alsop and Laughlin 1991). Rapid reductions in canopy cover associated with fir die-offs can result in increased understory temperatures, drier soil conditions, and replacement of bryophyte, herbaceous, and shrub layers with more sun-tolerant plants, such as *Rubus* spp. (Potter et al. 2005; McManamay et al. 2011).

North American fir species respond to balsam woolly adelgid feeding by creating dense wood growth, causing reduced nutrient availability and subsequently a distinctive progression of foliage discoloration; first yellow, then red, and finally brown (Ragenovich and Mitchell 2006). Prolonged infestation can also cause curled crowns (Amman and Speers 1965). This species produces 2–4 generations per year, depending on weather conditions. While there is some evidence of sexual reproduction, creation of males generally requires an additional host tree species, which is not present in North America and is suspected to have gone extinct in its native range during a European interglacial period. Consequently, most reproduction is parthenogenic (Havill et al. 2021), with females producing more than 100 eggs that hatch into a mobile stage called crawlers. Crawlers feed in protected bark crevices on the trunk, near nodes, and bud bases (Varty 1956). Once feeding begins, they become immobile for the rest of their lives, secreting a protective, waxy coating that results in a distinctive woolly appearance. As the insect matures, this woolly coating increases in volume, allowing severe infestations to be identified from a distance.

Adults lay eggs underneath this woolly layer at the end of their lives. Generations are not synchronous and all life stages can be found on trees from April to November (Sidebottom 2019). However, crawlers are the only life stage to survive the winter.

6.6.2.3 Pine Leaf Adelgid (*Pineus Pinifoliae*)

The pine leaf adelgid has a complex life cycle which occurs on two different host plants and requires 2 years to complete (Drooz 1985). The primary host is either red spruce or black spruce (*Picea mariana*), and the secondary host is eastern white pine. In stands where both spruce species are present, pine leaf adelgids show a preference for red spruce (Howse and Dimond 1965). This alternation of host-dependence has important implications for population size. Dimond and Bishop (1968) showed that the relative proportions of primary and secondary hosts are an important predictor of population size. Because these insects are relatively poor fliers (Lowe 1966), dispersal capability is limited. Thus, population size is also likely limited by distance between primary and secondary hosts. Forest stands that possess a mixture of primary and secondary hosts in close proximity have the greatest potential for large population sizes.

Pine leaf adelgids feed from the host tree using piercing/sucking mouthparts. Populations tend to be highly synchronized, with nearly all individuals being on either the primary or secondary host in a given year (Lowe 1966, Dimond and Allen 1974). These insects infrequently outbreak, and the conelike galls of gallicolae do relatively little damage to the primary host. It is the secondary host, eastern white pine, that suffers significant damage due to exule feeding, particularly in the mid-crown (Lowe 1966). This damage may result in tree mortality in some instances (Dimond 1974). Feeding by exules creates a diagnostic growth pattern in the secondary host, with alternating reduced needle length and reduced shoot growth being a direct result of the pine leaf adelgid's 2-year life cycle (DeBoo et al. 1964).

6.6.3 *Insecta: Hymenoptera*

6.6.3.1 European Spruce Sawfly (*Gilpinia Hercyniae*)

The European spruce sawfly is an invasive pest of red spruce that may have up to three generations per year. Populations of European spruce sawflies are kept in check primarily by a virus (*Lefavirales*: Baculoviridae) mistakenly introduced alongside several biological control agents for the species in the 1930s (Morris 1958; Entwistle et al. 1983). There is evidence that small mammals consume cocoons of European spruce sawflies, with some areas seeing 40–50% of cocoons damaged by small mammals (Morris 1942). Tree mortality occurs gradually over outbreak periods, with the most mortality seen in host trees suffering losses of all old foliage and 75% of new foliage (Morris 1958). For example, Morris (1958) pointed to the work of

Fig. 6.13 Eggs (left) and feeding larvae (right) of the yellowheaded spruce sawfly (*Pikonema alaskensis*); photos by USDA Forest Service Forest Health Protection program

Reeks and Barter (1951) on the 1920s through 1930s outbreak of European spruce sawflies in Gaspé, Québec. In that study, tree mortality took 7 years to develop and another 5 years to reach peak loss rate. Reeks and Barter (1951) also found that trees under severe feeding stress experienced prolonged (e.g., post-infestation) reduction of growth in annual rings and tree height.

6.6.3.2 Yellowheaded Spruce Sawfly (*Pikonema Alaskensis*)

Yellowheaded spruce sawflies feed heavily on young needles of host spruce trees and may cause either reduced growth or outright mortality of host trees (Houseweart and Kulman 1976; Fig. 6.13). Younger trees and those growing in the open are preferred oviposition sites, and thus are at greatest risk of damage (Rose et al. 1994). This preference makes the yellowheaded spruce sawfly particularly damaging in spruce plantations (Bartelt et al. 1982). For example, Morse and Kulman (1984) investigated factors causing mortality of plantation white spruce (*Picea glauca*) in Minnesota. They attributed 65% of white spruce mortality in 5–12-year-old trees to yellowheaded spruce sawflies. Outbreaks occur frequently and may persist over several years (Katovich et al. 1995).

6.6.4 Insecta: Lepidoptera

6.6.4.1 Spruce Budworm (*Choristoneura Fumiferana*)

Spruce budworm densities are typically low and relatively little damage occurs to host trees in most years (Fig. 6.14). However, because spruce budworm population density is strongly associated with the extent of defoliation (Anderson and Sturtevant

Fig. 6.14 Larval (left) and adult (right) spruce budworms (*Choristoneura* sp.); photos by USDA Forest Service Forest Health Protection program

2011), these insects are capable of severely damaging or killing their hosts during outbreak years. These outbreaks also impact forest floor nutrient availability (Paré et al. 1993) and synergistically interact with other forest disturbances (e.g., fire) to restructure local environments (Bergeron and Leduc 1998). Population outbreaks occur approximately every 30 years, although this varies throughout the insect's range. For example, in New Brunswick, population outbreaks occur approximately every 35 years (Royama 1984).

Population density has two important effects on spruce budworm populations: triggering of dispersal behavior and negative impacts on body size and fecundity of adult moths (Bauce et al. 1994, Bakthavatsalam et al. 2016). Female spruce budworms are capable of autodetection (i.e., the detection of their own pheromones; Ross et al. 1979). While it may not seem intuitive that females would need this ability, it is likely that autodetection triggers dispersal behavior in females when population density is high (Bakthavatsalam et al. 2016). When densities are high, it is possible for new shoots to be totally consumed, forcing larvae to feed from older needles. Adult spruce budworm fecundity and body size are reduced in this circumstance (Bauce and Carisey 1996), and autodetection-triggered dispersal at high densities may alleviate this issue. When moths find themselves in heavily defoliated stands post-dispersal, these individuals may engage in a second dispersal flight.

The microsporidium (*Nosema fumiferanae*) is a lethal pathogen of spruce budworms that is transmitted between individuals orally or passed from mother to offspring through the egg (Eveleigh et al. 2007). This pathogen thrives when foliage nitrogen content is high (Bauer and Nordin 1988), which is optimal for spruce budworm development (Shaw and Little 1977). Consequently, those areas where spruce budworm populations outbreak may experience heavy transmission of this pathogen and subsequent steep population decline.

6.6.5 *Arachnida: Araneae*

6.6.5.1 Spruce-Fir Moss Spider (*Microhexura montivaga*)

The spruce-fir moss spider is the world's smallest tarantula, a mygalomorph spider with a body size ranging from 0.25 to 0.40 cm (0.10–0.16 in; Fig. 6.15). This spider is federally listed as endangered, one of four arachnids listed under the Endangered Species Act (Fridell 1995). It is endemic to six spruce-fir *sky islands* in the southern Appalachians (Coyle 1981, 1997, 2009; U.S. Fish and Wildlife Service 2001; Hedin et al. 2015). Spruce-fir moss spider habitat occurs under dense canopies of spruce-fir forests underneath bryophyte mats found on north-facing emergent rocky outcrops, cliffs, and colluvial boulders (Coyle 1997, 2009; U.S. Fish and Wildlife Service 2001). The main threat to spruce-fir moss spider populations is the loss of bryophyte mats through desiccation or disturbance. Bryophyte mats provide the spiders with relatively stable microclimates (Seaborn and Catley 2016). Shade-adapted bryophytes are sensitive to loss of overstory cover (Proctor 2000; Nelson and Halpern 2005); therefore, reduced canopy cover following balsam woolly adelgid-induced dieback events can dry out bryophyte mats. Climate change may also impact habitat through increased drought events or alteration in the patterns of cloud immersion events, an important source of moisture for high elevation bryophyte communities (Norris 1974). In the 1980s, population declines were observed at several sites via bryophyte mat desiccation events linked to drought and balsam woolly adelgid-induced dieback events (Coyle 2004, 2009). Since the rocky outcrops upon which bryophyte mats occur are a patchy resource and the types of moss and bryophytes favored by the spider are also patchily distributed, loss of habitat at one rocky outcrop could lead to extirpation at that site. Spruce-fir restoration activities that impact the overstory or directly impact bryophyte mats may impact spruce-fir moss spider habitat (Diggins and Ford 2020).

6.7 Wildlife Considerations for Management and Restoration Efforts

Forest management and restoration actions can have immediate and delayed positive and negative effects on wildlife (e.g., Thompson et al. 2003; Harper et al. 2016; Irwin et al. 2018). Species differ in habitat requirements, spatial and temporal activity patterns, and tolerance to ecological disturbances. Thus, it is not possible to make singular management recommendations for the benefit of all wildlife species. Rather, we suggest that it is incumbent on land and wildlife managers to consider the specific details of management actions during the development of site-specific management plans. Major considerations for wildlife include: (1) what species are present or likely to be present; (2) what action is to be conducted and what are all of its subcomponents (e.g., if timber is harvested, access road ground disturbance impacts would be a part

Fig. 6.15 Spruce-fir moss spider (*Microhexura montivaga*) documented on Roan Mountain. Photo by Gary Peeples

of the actions to consider); (3) when will the action occur; and (4) how much of the action will occur.

Understanding what species are present is perhaps the single most important consideration with regard to wildlife, as this may guide all subsequent discussions for conservation measures. The taxa lists included in this chapter provide useful guidance for identifying species likely to occur in a management unit, but obtaining species occurrence records for the area and performing site-specific wildlife surveys is optimal to validate occurrence. Identifying the action and all relevant subcomponents also is important, as this can help determine *what* species are likely to be affected and will help understand *how* species may be affected. Action timing is also important as some effects may be avoided entirely if the action is conducted at a time when species are not present in the disturbance area. How much of the action is to be performed will in large part determine the scope for potential effects, including the likelihood of direct impacts, how many individuals of each species may be impacted, and the potential for indirect effects on habitat quality.

Potential negative effects include mortality and injury as a direct result of management actions (e.g., crushing animals with equipment or falling trees), noise disturbances, and temporary habitat degradation (e.g., micro-site desiccation from overstory removal). For example, when conducting tree felling for red spruce release, the use of chainsaws could disturb nesting birds and cause nest abandonment (Ortega 2012), and the act of felling trees could impact microclimate quality for amphibians or spruce-fir moss spider (Diggins and Ford 2020; Brown et al. 2022). In this simple example, potential impacts to nesting birds can be avoided by conducting the tree removal outside of the nesting season, impacts to amphibians could be reduced by limiting the intensity of the tree removal and retaining coarse woody debris on the forest floor, and impacts to spruce-fir moss spider could be reduced by avoiding canopy gap creation treatments around rocky outcrops.

The wildlife species benefiting from red spruce restoration will change with age of the restored forests. For example, Stewart and Aldrich (1949) determined that Blackburnian warblers, magnolia warblers, golden-crowned kinglets, and dark-eyed juncos were the most common bird species in a mature second-growth red spruce stand in West Virginia, whereas only magnolia warblers and dark-eyed juncos were present in a young red spruce stand. In general, habitat quality will increase with stand age for most of the species of conservation concern associated with red spruce, such as Cheat Mountain salamanders, Virginia northern flying squirrels, and northern saw-whet owls (Simpson 1972a; Ford et al. 2022; Pauley 2022). As discussed in Sect. 7.2, benefits for downstream fish communities will also increase with maturity of the restored red spruce stands. Thus, from a wildlife perspective, red spruce restoration is largely a proactive management strategy to improve habitat quality in the long term.

References

Ahn S, De Steiguer JE, Palmquist RB et al (2000) Economic analysis of the potential impact of climate change on recreational trout fishing in the southern Appalachian Mountains: an application of a nested multinomial logit model. Clim Change 45:493–509

Alsop FJ III, Laughlin TF (1991) Changes in the spruce-fir avifauna of Mt. Guyot, Tennessee, 1967–1985. J Tenn Acad Sci 66:207–209

Amman GD, Speers CF (1965) Balsam woolly aphid in the southern Appalachians. J For 63:18–20

Amman GD, Laughlin TF (1991) Changes in the spruce-fir avifauna of Mt. Guyot, Tennessee, 1967–1985. J Forest 63:18–20

Anderson DP, Sturtevant BR (2011) Pattern analysis of eastern spruce budworm *Choristoneura fumiferana* dispersal. Ecography 34:488–497

Apodaca JJ, Diggins CA, Erb L (2020) Distribution, habitat preferences, and landscape genetics of Appalachian cottontail (*Sylvilagus obscurus*) in western North Carolina. Report WM-0323 to the North Carolina Wildlife Resource Commission, Raleigh, North Carolina

Appalachian Mountains Joint Venture (2019) Priority species. https://amjv.org/priority-species/. Accessed 26 Mar 2024

Arbogast BS, Browne RA, Weigl PD et al (2005) Conservation genetics of endangered flying squirrels (*Glaucomys*) from the Appalachian Mountains of eastern North America. Anim Conserv 8:123–133

Arbuckle KE, Downing JA (2002) Freshwater mussel abundance and species richness: GIS relationships with watershed land use and geology. Can J Fish Aquat Sci 59:310–316

Austin LV, Silvis A, Ford WM et al (2018) Bat activity following restoration prescribed burning in the central Appalachian upland and riparian habitats. Nat Areas J 38:183–195

Bailey RW (1954) Status of beaver in West Virginia. J Wildl Manag 18:184–190

Bailey RS, Rucker CB (eds) (2021) The second atlas of breeding birds in West Virginia. Pennsylvania State University Press, University Park, Maryland

Bailey MA, Holmes JN, Buhlmann KA et al (2006) Habitat management guidelines for amphibians and reptiles of the southeastern United States. Partners in Amphibian and Reptile Conservation Technical Publication HMG-2, Montgomery, Alabama

Baker WL (1972) Eastern forest insects. USDA Forest Service Miscellaneous Publication No. 1175, Washington, DC

Bakthavatsalam N, Vinutha J, Ramakrishna P et al (2016) Autodetection in *Helicoverpa armigera* (Hubner). Curr Sci 110:2261–2267

Ball BA, Kominoski JS, Adams HE et al (2010) Direct and terrestrial vegetation-mediated effects of environmental change on aquatic ecosystem processes. Bioscience 60:590–601

Bartelt RJ, Jones RL, Kulman HM (1982) Evidence for a multicomponent sex pheromone in the yellowheaded spruce sawfly. Journal of Chemical Ecology 8:83–94

Bauce É, Carisey N (1996) Larval feeding behaviour affects the impact of staminate flower production on the suitability of balsam fir trees for spruce budworm. Oecologia 105:126–131

Bauce É, Crépin M, Carisey N (1994) Spruce budworm growth, development and food utilization on young and old balsam fir trees. Oecologia 97:499–507

Bauer LS, Nordin GL (1988) Nutritional physiology of the eastern spruce budworm, *Choristoneura fumiferana*, infected with *Nosema fumiferanae*, and interactions with dietary nitrogen. Oecologia 77:44–50

Beachy CK, Bruce RC (1992) Lunglessness in plethodontid salamanders is consistent with the hypothesis of a mountain stream origin: a response to Ruben and Boucot. Am Nat 139:839–847

Beilke EA, Blakey RV, O'Keefe JM (2021) Bats partition activity in space and time in a large, heterogeneous landscape. Ecol Evol 11:6513–6526

Beneski JT Jr, Stinson DW (1987) Sorex palutris. Mamm Species 296:1–6

Bergeron Y, Leduc A (1998) Relationships between change in fire frequency and mortality due to spruce budworm outbreak in the southeastern Canadian boreal forest. Journal of Vegetation Science 9:492–500

Bishop DJ, Majka CG, Bondrup-Nielsen S et al (2009) Deadwood and saproxylic beetle diversity in naturally disturbed and managed spruce forests in Nova Scotia. ZooKeys 22:309–340

Bogert CM (1952) Relative abundance, habitats, and normal thermal levels of some Virginian salamanders. Ecology 33:16–30

Bonner JL, Anderson JT, Rentch JS et al (2009) Vegetative composition and community structure associated with beaver ponds in Canaan valley, West Virginia, USA. Wetlands Ecol Manage 17:543–554

Boyce KA, Barry RE (2007) Seasonal home range and diurnal movements of *Sylvilagus obscurus* (Appalachian cottontail) at Dolly Sods, West Virginia. Northeast Nat 14:99–110

Bratton SP (1974) The effect of the European wild boar (*Sus scrofa*) on the high-elevation vernal flora in Great Smoky Mountains National Park. Bull Torrey Bot Club 101:198–206

Brewster W (1886) An ornithological reconnaissance in western North Carolina. Auk 3:94–112

Britzke ER, Harvey MJ, Loeb SC (2003) Indiana bat, *Myotis sodalis*, maternity roosts in the southern United States. Southeast Nat 2:235–242

Brooks M (1948) Notes on the Cheat Mountain salamander. Copeia 1948:239–244

Brooks M (1955) An isolated population of the Virginia varying hare. J Wildl Manag 19:54–61

Brown DJ, Rucker LE, Johnson C et al (2022) Microhabitat associations for the threatened Cheat Mountain salamander in relation to early-stage red spruce restoration areas. J Fish Wildl Manag 13:68–80

Bruce RC (1972) Variation in the life cycle of the salamander *Gyrinophilus porphyriticus*. Herpetologica 28:230–245

Bruce RC (1977) The pygmy salamander, *Desmognathus wrighti* (Amphibia, Urodela, Plethodontidae), in the Cowee Mountains, North Carolina. J Herpetol 11:246–247

Bruce RC (2003) Ecological distribution of the salamanders *Gyrinophilus* and *Pseudotriton* in a southern Appalachian watershed. Herpetologica 59:301–310

Bruce RC (2011) Community assembly in the salamander genus *Desmognathus*. Herpetol Monogr 25:1–24

Byers EA, Vanderhorst JP, Streets BP (2007) Classification and conservation assessment of high elevation wetland communities in the Allegheny Mountains of West Virginia. West Virginia Division of Natural Resources Report, Elkins, West Virginia

Byers EA, Vanderhorst JP, Streets BP (2010) Classification and conservation assessment of upland red spruce communities in West Virginia. West Virginia Division of Natural Resources Report, Elkins, West Virginia

Cahill AE, Aiello-Lammens ME, Fisher-Reid CM et al (2014) Causes of warm-edge range limits: systematic review, proximate factors and implications for climate change. J Biogeogr 41:429–442

Campbell TA, Laseter BR, Ford WM et al (2004) Movements of female white-tailed deer (*Odocoileus virginianus*) in relation to timber harvests in the central Appalachians. For Ecol Manage 199:371–378

Campbell TA, Laseter BR, Ford WM et al (2006) Abiotic factors influencing deer browsing in West Virginia. North J Appl for 23:20–26

Campbell JW, Mengak MT, Castleberry SB et al (2010) Distribution and status of uncommon mammals in the southern Appalachian Mountains. Southeast Nat 9:275–302

Cardoza YJ, Moser JC, Klepzig KD et al (2008) Multipartite symbioses among fungi, mites, nematodes, and the spruce beetle, *Dendroctonus rufipennis*. Environ Entomol 37:956–963

Chaplin SJ, Gerrard RA, Watson HM et al (2000) The geography of imperilment: targeting conservation toward critical biodiversity areas. In: Stein BA, Kutner LS, Adams JS (eds) Precious heritage: the status of biodiversity in the United States. Oxford University Press, New York, New York, pp 159–199

Chapman BR (2007a) Appalachian cottontail *Sylvilagus obscurus*. In: Trani-Griep M, Ford WM, Chapman BR (eds) Land managers guide to mammals of the south. The Nature Conservancy, Durham, North Carolina, pp 243–246

Chapman BR (2007b) Fisher *Martes pennati*. In: Trani-Griep M, Ford WM, Chapman BR (eds) Land managers guide to mammals of the south. The Nature Conservancy, Durham, North Carolina, pp 486–489

Chapman BR (2007c) Snowshoe hare *Lepus americanus*. In: Trani-Griep M, Ford WM, Chapman BR (eds) Land managers guide to mammals of the south. The Nature Conservancy, Durham, North Carolina, pp 226–229

Cherefko C, Fridley C, Medsger J et al (2015) White-tailed deer and balsam woolly adelgid effects on balsam fir in Canaan Valley. Southeast Nat 14:218–231

Clayton JL, Dannaway ES, Menendez R et al (1998) Application of limestone to restore fish communities in acidified streams. North Am J Fish Manag 18:347–360

Clingerman JW (2008) Statewide analysis of brook trout (*Salvelinus fontinalis*) population status and reach-scale conservation priorities in West Virginia watersheds. West Virginia University, Morgantown, West Virginia, Thesis

Clipp HL, Brown DJ, Rota CT et al (2022) Distinct forest bird communities are strongly associated with red spruce-northern hardwood ecosystems in central Appalachia, USA. Ecol Ind 135:108568

Cormack RGH (1949) A study of trout streamside cover in logged-over and undisturbed virgin spruce woods. Can J Res 27:78–95

Coyle FA (1981) The mygalomorph spider genus *Microhexura* (Araneae, Dipluridae). Bull Am Mus Nat Hist 170:64–75

Coyle FA (1997) Status survey of the endangered spruce-fir moss spider, *Microhexura montivaga* Crosby & Bishop, on Mount LeConte. Report to the U.S. Fish and Wildlife Service, Asheville, North Carolina

Coyle FA (2004) Status survey of the endangered spruce-fir moss spider, *Microhexura montivaga* in the Great Smoky Mountains National Park. Report to the Great Smoky Mountains National Park, Asheville, North Carolina

Coyle FA (2009) Status survey of the spruce-fir moss spider, *Microhexura montivaga*. Report to the U.S. Fish and Wildlife Service, Asheville, North Carolina

Crespi EJ, Rissler LJ, Browne RA (2003) Testing Pleistocene refugia theory: phylogeographical analysis of *Desmognathus wrighti*, a high-elevation salamander in the southern Appalachians. Mol Ecol 12:969–984

Crespi EJ, Browne RA, Rissler LJ (2010) Taxonomic revision of *Desmognathus wrighti* (Caudata: Plethodontidae). Herpetologica 66:283–295

De La Cruz JL, Ford WM, Jones SC et al (2023a) Distribution of summer habitat for the Indiana bat on the Monongahela National Forest, West Virginia. J Southeast Assoc Fish Wildl Agencies 10:125–134

De La Cruz JL, Ford WM, Jones SC et al (2023b) Distribution of northern long-eared bat summer habitat on the Monongahela National Forest, West Virginia. J Southeast Assoc Fish Wildl Agencies 10:114–124

DeBoo R, Dimond J, Lowe J (1964) Impact of pine leaf aphid, *Pineus pinifoliae* (Chermidae) on its secondary host, eastern white pine. Can Entomol 96:765–772

DeGraaf RM, Rudis DD (1990) Herpetofaunal species composition and relative abundance among three New England forest types. For Ecol Manage 32:155–165

Delcourt PA, Delcourt HR (1998) Paleoecological insights on conservation of biodiversity: a focus on species, ecosystems, and landscapes. Ecol Appl 8:921–934

Demeo TE (1999) Forest songbird abundance and viability at multiple scales on the Monongahela National Forest, West Virginia. Dissertation, West Virginia University, Morgantown, West Virginia

Diefenbach DR, Rathbun SL, Vreeland JK et al (2016) Evidence for range contraction of snowshoe hare in Pennsylvania. Northeast Nat 23:229–248

Diggins CA (2023) Anthropogenically-induced range expansion as an invasion front in native species: an example in North American flying squirrels. Front Ecol Evol 11:1096244

Diggins CA, Ford WM (2020) Effects of surveying for the federally endangered spruce-fir moss spider (*Microhexura montivaga* Crosby & Bishop) on its bryophyte habitat. Southeast Nat 20:77–91

Diggins CA, Ford WM (2022) Seasonal activity patterns of bats in high-elevation conifer sky islands. Acta Chiropterologica 24:91–101

Diggins CA, Kelly CA, Ford WM (2015) Atypical den use of Carolina northern flying squirrels (*Glaucomys sabrinus coloratus*) in the southern Appalachian Mountains. Southeast Nat 14:44–49

Diggins CA, Higdon S, Ford WM (2016) High elevation record for eastern box turtle in the southern Appalachians. Herpetol Rev 47:454

Diggins CA, Silvis A, Kelly CA et al (2017) Home range, den selection and habitat use of Carolina northern flying squirrels (*Glaucomys sabrinus coloratus*). Wildl Res 44:427–437

Diggins CA, Castellano MA, Ford WM (2020) Hypogeous, sequestrate fungi (genus *Elaphomyces*) found at small-mammal foraging sites in high-elevation conifer forests of West Virginia. Northeast Nat 27:N40–N47

Dillard LO, Russell KR, Ford WM (2008a) Site-level habitat models for the endemic, threatened Cheat Mountain salamander (*Plethodon nettingi*): the importance of geophysical and biotic attributes for predicting occurrence. Biodivers Conserv 17:1475–1492

Dillard LO, Russell KR, Ford WM (2008b) Macrohabitat models of occurrence for the threatened Cheat Mountain salamander, *Plethodon nettingi*. Appl Herpetol 5:201–224

Dimond JB (1974) Sequential surveys for the pine leaf chermid, *Pineus pinifoliae*. University of Maine Life Sciences and Agriculture Experiment Station Technical Bulletin 68, Orono, Maine

Dimond J, Allen D (1974) Sampling populations of pine leaf chermid *Pineus pinifoliae* (Homoptera: Chermidae): III. Neosistentes on white pine. Canadian Entomologist 106:509–518

Dimond JB, Bishop RH (1968) Susceptibility and vulnerability of forests to the pine leaf aphid *Pineus pinifoliae* (Fitch) (Adelgidae). University of Maine Agricultural Experiment Station Bulletin 658, Orono, Maine

Dodd CK Jr (2013) Frogs of the United States and Canada. Johns Hopkins University Press, Baltimore, Maryland

Dodd CK Jr, Dorazio RM (2004) Using counts to simultaneously estimate abundance and detection probabilities in a salamander community. Herpetologica 60:468–478

Dodd CK Jr (2003) Monitoring amphibians in Great Smoky Mountains National Park. U.S. Geological Survey Circular 1258, Tallahassee, Florida

Downey DM, French CR, Odom M (1994) Low cost limestone treatment of acid sensitive trout streams in the Appalachian Mountains of Virginia. Water Air Soil Pollut 77:49–77

Drooz A (1985) Insects of eastern forests. USDA Forest Service Miscellaneous Publication No. 1426, Washington, DC

Dull CW, Ward JD, Brown HD et al (1988) Evaluation of spruce and fir mortality in the southern Appalachian Mountains. USDA Forest Service Southern Region Protection Report R8-PR 13, Atlanta, Georgia

Eastern Brook Trout Joint Venture (2018) Eastern brook trout roadmap to conservation. https://easternbrooktrout.org/about/reports/eastern-brook-trout-roadmap-to-conservation-2018/view. Accessed 11 May 2023

Enck JW, Brown TL (2005) Social feasibility of restoring elk to West Virginia. Report to the West Virginia Division of Natural Resources, Cornell University, Ithaca, New York

Entwistle PF, Adams PHW, Evans HF et al (1983) Epizootiology of a nuclear polyhedrosis virus (Baculoviridae) in European spruce sawfly (*Gilpinia hercyniae*): spread of disease from small epicentres in comparison with spread of baculovirus diseases in other hosts. J Appl Ecol 20:473–487

Etnier DA, Starnes WC (1993) The fishes of Tennessee. University of Tennessee Press, Knoxville, Tennessee

Eveleigh ES, Lucarotti CJ, McCarthy PC et al (2007) Occurrence and effects of *Nosema fumiferanae* infections on adult spruce budworm caught above and within the forest canopy. Agric For Entomol 9:247–258

Faust LF (2017) Fireflies, glow-worms, and lightning bugs: identification and natural history of the fireflies of the eastern and central United States and Canada. University of Georgia Press, Athens, Georgia

Ferrari BA, Shamblin BM, Chandler RB et al (2018) Canada warbler (*Cardellina canadensis*): novel molecular markers and a preliminary analysis of genetic diversity and structure. Avian Conservation and Ecology 13:8

Fidgen JG, Teerling CR, McKinnon ML (1994) Intra- and inter-crown distribution of the eastern spruce gall adelgid, *Adelges abietis* (L.), on young white spruce. Can Entomol 126:1105–1110

Fies ML (1991) Snowshoe hare. In: Terwilliger K (ed) Virginia's endangered species. McDonald and Woodward Publishing Company, Newark, New Jersey, pp 576–578

Flaherty KL, Grafton WN, Anderson JT (2018) White-tailed deer florivory influences the population demography of *Polemonium vanbruntiae*. Plant Biosyst 152:453–463

Ford WM, Rodrigue JL (2001) Soricid abundance in partial overstory removal harvests and riparian areas in an industrial forest landscape of the central Appalachians. For Ecol Manage 152:159–168

Ford WM, Menzel MA, Rodrigue JL et al (2005) Relating bat species presence to simple habitat measures in a central Appalachian forest. Biol Cons 126:528–539

Ford WM, Kelly CA, Rodrigue JL et al (2014) Late winter and early spring home range and habitat use of the endangered Carolina northern flying squirrel in western North Carolina. Endanger Species Res 2:73–82

Ford WM, Evans AM, Odom RH et al (2015) Predictive habitat models derived from nest-box occupancy for the endangered Carolina northern flying squirrel in the southern Appalachians. Endanger Species Res 27:131–140

Ford WM, Silvis A, Rodrigue JL et al (2016) Deriving habitat models for northern long-eared bats from historical detection data: a case study using the Fernow Experimental Forest. J Fish Wildl Manag 7:86–98

Ford WM, Diggins CA, De La Cruz JL et al (2022) Distribution probability of the Virginia northern flying squirrel in the high Allegheny Mountains. J Southeast Assoc Fish Wildl Agencies 9:168–175

Ford WM, McCay TS, Menzel MA et al (2006) Influence of elevation and forest type on shrew community assemblage and species distribution in the central and southern Appalachian Mountains. In: Merritt JF, Churchfield S, Hutterer R et al (eds) Advances in the biology of the shrews II. International Society of Shrew Biologists Special Publication No. 1, New York, New York, p 303–315

Ford WM, Mertz KM, Menzel JM et al (2007) Winter home range and habitat use of the Virginia northern flying squirrel (*Glaucomys sabrinus fuscus*). USDA Forest Service Northern Research Station Research Paper NRS-4, Newtown Square, Pennsylvania

Ford WM, Moseley KR, Stihler CW et al (2010) Area occupancy and detection probabilities of the Virginia northern flying squirrel (*Glaucomys sabrinus fuscus*) using nest-box surveys. In: Rentch JS, Schuler, TM (eds) Proceedings from the conference on the ecology and management of high elevation forests in the central and southern Appalachian Mountains. USDA Forest Service Northern Research Station General Technical Report NRS-P-64, Newtown Square, Pennsylvania, p 39–47

Francl KE, Ford WM, Castleberry SB (2004) Bat activity in central Appalachian wetlands. Georg J Sci 62:87–94

Francl KE, Glenn TC, Castleberry SB et al (2008) Genetic relationship of meadow vole (*Microtus pennsylvanicus*) populations in central Appalachian wetlands. Can J Zool 86:344–355

Fridell JA (1995) Endangered and threatened wildlife and plants: spruce-fir moss spider determined to be endangered. Fed Reg 60:6968–6973

Ganier AF (1928) The wildlife of Tennessee. J Tenn Acad Sci 3:10–22

Goode GB (1879) The occurrence of the Canada porcupine in West Virginia. Proc US Natl Mus 1:264–265

Goudy WH, Kletzly RC, Rieffenberger JC (1970) Characteristics of a heavily hunted woodcock population in West Virginia. Trans N Amn Wildl Nat Resour Conf 35:183–195

Green NB (1938) A new salamander, *Plethodon nettingi*, from West Virginia. Ann Carnegie Mus 27:295–299

Gregory SV, Swanson FJ, McKee WA et al (1991) An ecosystem perspective of riparian zones: focus on links between land and water. Bioscience 41:540–551

Gross DA (2010) Pennsylvania boreal conifer forests and their bird communities: past, present, and potential. In: Rentch JS, Schuler, TM (eds) Proceedings from the conference on the ecology and management of high elevation forests in the central and southern Appalachian Mountains. USDA Forest Service Northern Research Station General Technical Report NRS-P-64, Newtown Square, Pennsylvania, pp 48–73

Habera J, Moore S (2005) Managing southern Appalachian brook trout: a position statement. Fisheries 30:10–20

Hairston NG (1949) The local distribution and ecology of the plethodontid salamanders of the southern Appalachians. Ecol Monogr 19:47–73

Hall GA (1984) A long-term bird population study in an Appalachian spruce forest. Wilson Bull 96:228–240

Hampe A, Petit RJ (2005) Conserving biodiversity under climate change: the rear edge matters. Ecol Lett 8:461–467

Handley CO Jr (1953) A new flying squirrel from the southern Appalachian Mountains. Proc Biol Soc Wash 66:191–194

Harman DM, Thoerig T (1968) Occurrence of the porcupine *Erethizon dorsatum* and the nutria *Myocastor coypus bonariensis* in western Maryland. Chesap Sci 9:138–139

Harper CA, Ford WM, Lashley MA et al (2016) Fire effects on wildlife in the central hardwoods and Appalachian regions, USA. Fire Ecology 12:127–159

Harrison JR III (2000) *Desmognathus wrighti* king pygmy salamander. Cat Am Amphib Reptil 704(1–704):7

Havill NP, Griffin BP, Andersen JC et al (2021) Species delimitation and invasion history of the balsam woolly adelgid, *Adelges* (*Dreyfusia*) *piceae* (Hemiptera: Aphidoidea: Adelgidae), species complex. Syst Entomol 46:186–204

Healy WM, Dickson JG (1992) Population influences. Environment 1:129–143
Hedin M, Carlson D, Coyle F (2015) Sky island diversification meets the multispecies coalescent – divergence in the spruce-fir moss spider (*Microhexura monitvaga*, Araneae, Mygalomorphae) on the highest peaks of southern Appalachia. Mol Ecol 24:3467–3484
Herlihy AT, Kaufmann PR, Church MR et al (1993) The effects of acidic deposition on streams in the Appalachian Mountain and Piedmont region of the Mid-Atlantic United States. Water Resour Res 29:2687–2703
Hillard EM (2013) Elk (*Cervus elaphus* L.) habitat selection in Great Smoky Mountains National Park. Thesis, Western Carolina University, Cullowhee, North Carolina
Hocking DJ, Crawford JA, Peterman WE et al (2021) Abundance of montane salamanders over an elevational gradient. Ecol Evol 11:1378–1391
Hoffman RL (1950) Records of *Picea* in Virginia. Castanea 15:55–58
Hoffman RL, Kleinpeter HI (1948) A collection of salamanders from Mount Rogers, Virginia. J Wash Acad Sci 38:106–108
Houseweart MW, Kulman HM (1976) Life tables of the yellowheaded spruce sawfly, *Pikonema alaskensis* (Rohwer) (Hymenoptera: Tenthredinidae) in Minnesota. Environmental Entomology 5:859–867
Howse GM, Dimond JB (1965) Sampling populations of pine leaf adelgid *Pineus pinifoliae* (Fitch): I. The gall and associated insects. Can Entomol 97:952–961
Hrinkevich KH, Progar RA, Shaw DC (2016) Climate risk modelling of balsam woolly adelgid damage severity in subalpine fir stands of western North America. PLoS ONE 11:e0165094
Hudy M, Downey DM, Bowman DM (2000) Successful restoration of an acidified native brook trout stream through mitigation with limestone sand. North Am J Fish Manag 20:453–466
Hudy M, Thieling TM, Gillespie N et al (2008) Distribution, status, and land use characteristics of subwatersheds within the native range of brook trout in the eastern United States. North Am J Fish Manag 28:1069–1085
Hughes RS (2006) Home ranges of the endangered Carolina northern flying squirrel in the Unicoi Mountains of North Carolina. Proc Annu Conf Southeast Assoc Fish Wildl Agencies 60:19–24
Hunter WC (2010) Bird conservation issues in high-elevation (red spruce-Fraser fir-northern hardwood) forests of the southern Blue Ridge. In: Rentch JS, Schuler, TM (eds) Proceedings from the conference on the ecology and management of high elevation forests in the central and southern Appalachian Mountains. USDA Forest Service Northern Research Station General Technical Report NRS-P-64, Newtown Square, Pennsylvania, p 212
Irwin LL, Riggs RA, Verschuyl JP (2018) Reconciling wildlife conservation to forest restoration in moist mixed-conifer forests of the inland northwest: a synthesis. For Ecol Manage 424:288–311
Jenkins RE, Burkhead NM (1994) Freshwater fishes of Virginia. American Fisheries Society, Bethesda, Maryland
Jenkins CN, Van Houtan KS, Pimm SL et al (2015) US protected lands mismatch biodiversity priorities. Proc Natl Acad Sci 112:5081–5086
Johnson C, Brown DJ, Sanders C et al (2021) Long-term changes in occurrence, relative abundance, and reproductive fitness of bat species in relation to arrival of white-nose syndrome in West Virginia, USA. Ecol Evol 11:12453–12467
Jones MT, Willey LL (eds) (2021) Biology and conservation of the wood turtle. Northeast Association of Fish and Wildlife Agencies, Petersburgh, New York
Kahler HA, Anderson JT (2010) Factors influencing avian communities in high-elevation southern Allegheny Mountain forests. In: Rentch JS, Schuler, TM (eds) Proceedings from the conference on the ecology and management of high elevation forests in the central and southern Appalachian Mountains. USDA Forest Service Northern Research Station General Technical Report NRS-P-64, Newtown Square, Pennsylvania, pp 94–103
Karas N (1997) Brook trout. Lyons & Burford, New York, New York
Karns DR (1992) Effects of acidic bog habitats on amphibian reproduction in a northern Minnesota peatland. J Herpetol 26:401–412

Katovich SA, McCullough DG, Haack RA (1995) Yellowheaded spruce sawfly–its ecology and management. USDA Forest Service North Central Forest Experiment Station General Technical Report NC-179, St. Paul, Minnesota
Kazyak DC, Lubinski BA, Kulp MA et al (2022) Population genetics of brook trout in the southern Appalachian Mountains. Trans Am Fish Soc 151:127–149
Kellogg R (1939) Annotated list of Tennessee mammals. Proc US Natl Mus 86:245–303
Kennedy PP (1853) The Blackwater chronicle. Refield Publishing, New York, New York
Kerns J, Kerlinger P (2004) A study of bird and bat collision fatalities at the Mountaineer Wind Energy Center, Tucker County, West Virginia: annual report for 2003. Curry & Kerlinger, LLC Report, Frostburg, Maryland
Kirkland GL Jr (1977) Responses of small mammals to the clearcutting of northern Appalachian forests. J Mammal 58:600–609
Kozak KH (2017) What drives variation in plethodontid salamander species richness over space and time? Herpetologica 73:220–228
Kroschel WA, Sutton WB, McClure CJW et al (2014) Decline of the Cheat Mountain salamander over a 32-year period and the potential influence of competition from a sympatric species. J Herpetol 48:415–422
Laerm J, Castleberry SB (2007) Deer mouse *Peromyscus maniculatus*. In: Trani-Griep M, Ford WM, Chapman BR (eds) Land managers guide to mammals of the south. The Nature Conservancy, Durham, North Carolina, pp 337–341
Laerm J, Pagels JF (2007) Woodland jumping mouse *Napaeozapus insignis*. In: Trani-Griep M, Ford WM, Chapman BR (eds) Land managers guide to mammals of the south. The Nature Conservancy, Durham, North Carolina, pp 426–429
Laerm J, Webster WD (2007) Red squirrel *Tamiasciurus hudsonicus*. In: Trani-Griep M, Ford WM, Chapman BR (eds) Land managers guide to mammals of the south. The Nature Conservancy, Durham, North Carolina, pp 422–425
Laerm J, Ford WM, Chapman BR (2007a) American water shrew *Sorex palustris*. In: Trani-Griep M, Ford WM, Chapman BR (eds) Land managers guide to mammals of the south. The Nature Conservancy, Durham, North Carolina, pp 109–112
Laerm J, Ford WM, Chapman BR (2007b) Hairy-tailed mole *Parascalops breweri*. In: Trani-Griep M, Ford WM, Chapman BR (eds) Land managers guide to mammals of the south. The Nature Conservancy, Durham, North Carolina, pp 117–119
Lambert M, Drayer AN, Leuenberger W et al (2021) Evaluation of created wetlands as amphibian habitat on a reforested surface mine. Ecol Eng 171:106386
Latham SR, Sirén APK, Reitsma LR (2023) Space use and resource selection of wood turtles (*Glyptemys insculpta*) in the northeastern part of its range. Can J Zool 101:20–31
Linehan K, Palmer D (2014) Landowner views of elk in western North Carolina. Report to the North Carolina Wildlife Resources Commission, Raleigh, North Carolina
Linzey DW (1995) Mammals of the Great Smoky Mountains—1995 update. J Elisha Mitchell Sci Soc 111:1–81
Linzey DW (2002) Significant new mammal records from the Great Smoky Mountains National Park, Tennessee-North Carolina. J N C Acad Sci 118:91–96
Linzey DW (2021) The mammals of Virginia. McDonald and Woodward Publishing Company, Newark, New Jersey
Lituma CM, Cox JJ, Spear S et al (2021) Terrestrial wildlife in the post-mined Appalachian landscape: status and opportunities. In: Zipper C, Skousen J (eds) Appalachia's coal-mined landscapes. Springer, Cham, Switzerland, pp 135–166
Loeb SC, Tainter FH, Cázares E (2000) Habitat associations of hypogeous fungi in the southern Appalachians: implications for the endangered northern flying squirrel (*Glaucomys sabrinus coloratus*). Am Midl Nat 144:286–296
Lowe JH Jr (1966) Biology and dispersal of *Pineus pinifoliae* (Fitch). Dissertation, Yale University, New Haven, Connecticut

Lowe WH (2005) Factors affecting stage-specific distribution in the stream salamander *Gyrinophilus porphyriticus*. Herpetologica 61:135–144
MacDonald LH, Coe D (2007) Influence of headwater streams on downstream reaches in forested areas. Forest Science 53:148–168
Mack DE, Yong W (2000) Swainson's thrush (*Catharus ustulatus*), ver 1.0. In: Poole AF, Gill FB (eds) Birds of the world. Cornell Lab of Ornithology, Ithaca, New York
Maehr DS, Cox JJ, Larkin JL (2007) Elk *Cervus elaphus*. In: Trani-Griep M, Ford WM, Chapman BR (eds) Land managers guide to mammals of the south. The Nature Conservancy, Durham, North Carolina, pp 562–532
Marschall EA, Crowder LB (1996) Assessing population responses to multiple anthropogenic effects: a case study with brook trout. Ecol Appl 6:152–167
Martof BS (1962) Some aspects of the life history and ecology of the salamander *Leurognathus*. Am Midl Nat 67:1–35
Mathews RC, Echternacht AC (1984) Herpetofauna of the spruce-fir ecosystem in the southern Appalachian Mountain regions, with emphasis on the Great Smoky Mountains National Park. In: White PS (ed) The southern Appalachian spruce-fir ecosystem: its biology and threats. National Park Service Uplands Field Research Laboratory Research/Resources Management Report SER-71, Gatlinburg, Tennessee, pp 155–167
Mazerolle MJ (2004) Amphibian road mortality in response to nightly variations in traffic intensity. Herpetologica 60:45–53
McClurg SE, Petty JT, Mazik PM et al (2007) Stream ecosystem response to limestone treatment in acid impacted watersheds of the Allegheny Plateau. Ecol Appl 17:1087–1104
McDonnell TC, Sloat MR, Sullivan TJ et al (2015) Downstream warming and headwater acidity may diminish coldwater habitat in southern Appalachian Mountain streams. PLoS ONE 10:e0134757
McKinnon ML, Quiring DT, Bauce E (1999) Influence of tree growth rate, shoot size and foliar chemistry on the abundance and performance of a galling adelgid. Funct Ecol 13:859–867
McManamay RH, Resler LM, Campbell JB, McManamay RA (2011) Assessing the impacts of balsam woolly adelgid (*Adelges piceae* Ratz.) and anthropogenic disturbance on the stand structure and mortality of Fraser fir [*Abies fraseri* (Pursh) Poir.] in the Black Mountains, Noth Carolina. Castanea 76:1–19
McNeill D (1877) The last forest: tales of the Allegheny woods. McClain Printing Company, New York, New York
McShea WJ, Healy WM, Devers P et al (2007) Forestry matters: decline of oaks will impact wildlife in hardwood forests. J Wildl Manag 71:1717–1728
Menendez R (1976) Chronic effects of reduced pH on brook trout (*Salvelinus fontinalis*). J Fish Board Can 33:118–123
Mengak MT, Laerm J (2007a) American beaver *Castor canadensis*. In: Trani-Griep M, Ford WM, Chapman BR (eds) Land managers guide to mammals of the south. The Nature Conservancy, Durham, North Carolina, pp 252–257
Mengak MT, Laerm J (2007b) Eastern chipmunk *Tamias striatus*. In: Trani-Griep M, Ford WM, Chapman BR (eds) Land managers guide to mammals of the south. The Nature Conservancy, Durham, North Carolina, pp 417–421
Menzel JM, Ford WM, Edwards JW et al (2004) Nest tree use by the endangered Virginia northern flying squirrel in the central Appalachian Mountains. Am Midl Nat 151:355–368
Menzel JM, Ford WM, Edwards JW et al (2006) Home range and habitat use of the vulnerable Virginia northern flying squirrel *Glaucomys sabrinus fuscus* in the central Appalachian Mountains, USA. Oryx 40:204–210
Merker SA, Chandler RB (2020) Identifying global hotspots of avian trailing-edge population diversity. Glob Ecol Conserv 22:e00915
Merriam CH (1888) Remarks on the fauna of the Great Smoky Mountains; with description of a new species of red-backed mouse (*Evotomys carolinensis*). Am J Sci 36:458–460
Michael ED, Hahn BL, Hasen HJ (1982) Response of deer, hare, and grouse to whole-tree harvesting in central Appalachia. Proc Annu Conf Southeast Fish Wildl Agencies 36:627–633

Michael ED, Brown SL, Brown WS (2015) Historic game harvests in Canaan Valley and Tucker County, West Virginia. Southeast Nat 14:382–404

Miller GS Jr (1936) A new flying squirrel from West Virginia. Proc Biol Soc Wash 49:143–144

Miller GS Jr, Kellogg R (1955) List of North American recent mammals. United States National Museum Bulletin 205, Washington, DC

Millikin AR, Woodley SK, Davis DR et al (2019) Habitat characteristics in created vernal pools impact spotted salamander water-borne corticosterone levels. Wetlands 39:803–814

Milling TC, Rowe MP, Cockerel BL et al (1997) Population densities of northern saw-whet owls (*Aegolius acadicus*) in degraded boreal forests of the southern Appalachians. In: Duncan JR, Johnson DH, Nicholls TH (eds) Biology and conservation of owls in the northern hemisphere: 2nd international symposium. USDA Forest Service North Central Forest Experiment Station General Technical Report NC-190, St. Paul, Minnesota

Minckler LS (1945) Reforestation in the spruce type in the southern Appalachians. J Forest 43:349–356

Mitchell D (2001) Spring and fall diet of the endangered West Virginia northern flying squirrel (*Glaucomys sabrinus fuscus*). Am Midl Nat 146:439–443

Mitchell JC, Breisch AR, Buhlmann KA (2006) Habitat management guidelines for amphibians and reptiles of the northeastern United States. Partners in Amphibian and Reptile Conservation Technical Publication HMG-3, Montgomery, Alabama

Modesto V, Ilarri M, Souza AT et al (2018) Fish and mussels: importance of fish for freshwater mussel conservation. Fish Fish 19:244–259

Moncrief ND, Fies ML (2015) Report of first specimens of *Pekania pennanti* (fisher) from Virginia. Northeast Nat 22:N31–N34

Moncrief ND, Fies ML (2020) Recent records and range expansion of *Erethizon dorsatum* (North American porcupine) in Virginia. Northeast Nat 27:N21–N27

Montgomery ME, Havill NP (2014) Balsam woolly adelgid. In: Van Driesche R, Reardon R (eds) The use of classical biological control to preserve forests in North America. USDA Forest Service Forest Health Technology Enterprise Team Report FHTE-2013–2, Morgantown, West Virginia, pp 9–19

Morris RF (1958) A review of the important insects affecting the spruce - fir forests in the Maritime provinces. For Chron 34:159–189

Morris RF (1942) Preliminary notes on the natural control of the European spruce sawfly by small mammals. Can Entomol 74:176–202

Morse BW, Kulman HM (1984) Plantation white spruce mortality: estimates based on aerial photography and analysis using a life-table format. Can J For Res 14:195–200

Muthersbaugh MS, Ford WM, Powers KE et al (2019) Activity patterns in regional and long-distance migrant bat species during the fall and spring along ridgelines in the central Appalachians. J Fish Wildl Manag 10:180–195

Nelson CR, Halpern CB (2005) Short-term effects of timber harvest and forest edges on ground-layer mosses and liverworts. Can J Bot 83:610–620

Nicholson, KE (committee chair) (2025) Scientific and standard English names of amphibians and reptiles of North America north of Mexico, with comments regarding confidence in our understanding. 9th edn. Society for the Study of Amphibians and Reptiles

Niemelä J, Haila Y, Punttila P (1996) The importance of small-scale heterogeneity in boreal forests: variation in diversity in forest-floor invertebrates across the succession gradient. Ecography 19:352–368

Niemiller ML, Reynolds RG (2011) The amphibians of Tennessee. University of Tennessee Press, Knoxville, Tennessee

Norris DH (1974) Bryoecology of the Appalachian spruce–fir zone. Dissertation, University of Tennessee, Knoxville, Tennessee

North Carolina Wildlife Resources Commission (2023) Elk—North Carolina wildlife profiles. https://www.ncwildlife.org/Portals/0/Learning/documents/Profiles/Mammals/Elk-Species-Profile-FINAL.pdf. Accessed 20 Jan 2023

O'Keefe JM, Pettit JL, Loeb SC et al (2019) White-nose syndrome dramatically altered the summer bat assemblage in a temperate southern Appalachian forest. Mamm Biol 98:146–153

Olson DM, Dinerstein E (2002) The global 200: priority ecoregions for global conservation. Ann Mo Bot Gard 89:199–224

Ortega CP (2012) Effects of noise pollution on birds: a brief review of our knowledge. Ornithol Monogr 74:6–22

Pack JC, Cromer JI (1981) Reintroduction of fisher in West Virginia. In: Chapman JA, Pursley D (eds) Worldwide furbearer conference proceedings, vol 2. Frostburg State College. Frostburg, Maryland, pp 1431–1442

Pagels JF, Smock LA, Sklarew SH (1998) The water shrew, *Sorex palustris* Richardson (Insectivora: Soricidae), and its habitat in Virginia. Brimleyana 25:120–134

Pagels JF, Laerm J (2007a) Rock vole *Microtus chrotorrhinus*. In: Trani-Griep M, Ford WM, Chapman BR (eds) Land managers guide to mammals of the south. The Nature Conservancy, Durham, North Carolina, pp 276–279

Pagels JF, Laerm J (2007b) Southern red-backed vole *Clethrionomys gapperi*. In: Trani-Griep M, Ford WM, Chapman BR (eds) Land managers guide to mammals of the south. The Nature Conservancy, Durham, North Carolina, pp 272–275

Paré D, Bergeron Y, Camiré C (1993) Changes in the forest floor of Canadian southern boreal forest after disturbance. J Veg Sci 4:811–818

Pauley TK (1978) Plants as indicators of occurrence of two sympatric *Plethodon* species. Bull Md Herpetol Soc 14:29–35

Pauley TK (1980) Amphibians and reptiles of the West Virginia mountains above 975 meters. W Va Acad Sci 50:84–92

Pauley BA (1998) The use of emergent rocks as refugia for the Cheat Mountain salamander. Thesis, Marshall University, Huntington, West Virginia, Plethodon nettingi Green

Pauley TK (2008) The Appalachian inferno: historical causes for the disjunct distribution of *Plethodon nettingi* (Cheat Mountain salamander). Northeast Nat 15:595–606

Pauley TK (2015) Amphibians in the Canaan Valley drainage. Southeast Nat 14:314–322

Pauley TK (2022) Forty-years of field notes: the Cheat Mountain salamander (*Plethodon nettingi*). Proc W Va Acad Sci 94:1–37

Pauley BA, Pauley TK (1997) Range and distribution of the Cheat Mountain salamander, *Plethodon nettingi*: an update. Proc W Va Acad Sci 69:3–4

Paulson DR, Dunkle SW (2021) A checklist of North American Odonata including English name, etymology, type locality, and distribution. https://www.odonatacentral.org/public/media/uploads/files/NA_Odonata_Checklist_2021_update.pdf. Accessed 26 Mar 2024

Petranka JW (1998) Salamanders of the United States and Canada. Smithsonian Books, Washington, DC

Petranka JW, Smith CK (2005) A functional analysis of streamside habitat use by southern Appalachian salamanders: implications for riparian forest management. For Ecol Manage 210:443–454

Petty JT, Lamothe PJ, Mazik PM (2005) Spatial and seasonal dynamics of brook trout populations inhabiting a central Appalachian watershed. Trans Am Fish Soc 134:572–587

Petty JT, Fulton JB, Strager MP et al (2010) Landscape indicators and thresholds of stream ecological impairment in an intensively mined Appalachian watershed. J N Am Benthol Soc 29:1292–1309

Pickering J, Kays R, Meier A et al (2002) The Appalachians. In: Mittermeier RA, Mittermeier CG, Gil PR et al (eds) Wilderness: Earth's last wild places. Conservation International, pp 458–467

Pope CH, Hairston NG (1947) The distribution of *Leurognathus* a southern Appalachian genus of salamanders. Fieldiana Zool 31:155–162

Poplar-Jeffers IO, Petty JT, Anderson JT et al (2009) Culvert replacement and stream habitat restoration: implications from brook trout management in an Appalachian watershed, U.S.A. Restor Ecol 17:404–413

Potter KM, Escanferla ME, Jetton RM et al (2019) Important insect and disease threats to United States tree species and geographic patterns of their potential impacts. Forests 10:304

Potter KM, Frampton J, Sidebottom J (2005) Impacts of balsam woolly adelgid on the southern Appalachian spruce-fir ecosystem and the North Carolina Christmas tree industry. In: Onken B, Reardon R (eds). Proceedings of the Third Symposium on Hemlock Woolly Adelgid in the Eastern United States, USDA Forest Service Forest Health Technology Enterprise Team, Morgantown, West Virginia, pp 25–41

Powell R, Conant R, Collins JT (2016) Peterson field guide to reptiles and amphibians of eastern and central North America, 4th edn. Houghton Mifflin Harcourt, New York, New York

Proctor MCF (2000) Physiological ecology. In: Shaw AJ, Goffinet B (eds) Bryophyte biology. Cambridge University Press, Cambridge, England, pp 225–247

Pyle P (1997) Identification guide to North American birds: a compendium of information on identifying, ageing, and sexing "near-passerines" and passerines in the hand. Slate Creek Press, Bolinas, California

Pyron RA, Beamer DA (2022) Nomenclatural solutions for diagnosing 'cryptic' species using molecular and morphological data facilitate a taxonomic revision of the black-bellied salamanders (Urodela, *Desmognathus 'quadramaculatus'*) from the southern Appalachian Mountains. Bionomina 27:1–43

Rabenold KN (1978) Foraging strategies, diversity, and seasonality in bird communities of Appalachian spruce-fir forests. Ecol Monogr 48:397–424

Rabenold KN, Fauth PT, Goodner BW et al (1998) Response of avian communities to disturbance by an exotic insect in spruce-fir forests of the southern Appalachians. Conserv Biol 12:177–189

Rabenold KN (1984) Birds of Appalachian spruce-fir forests: dynamics of habitat-island communities. In: White PS (ed) The southern Appalachian spruce-fir ecosystem: its biology and threats. National Park Service Southeast Region Research/Resources Management Report SER-71, Atlanta, Georgia, pp 168–186

Rabenold KN (1992) Bird communities of Great Smoky Mountains National Park: establishing monitoring protocol and measuring effects of disturbance. Report to the National Park Service, Gatlinburg, Tennessee

Ragenovich IR, Mitchell RG (2006) Balsam woolly adelgid. USDA Forest Service Forest Insect & Disease Leaflet 118, Washington, DC

Ramos MA, Warner DA (1980) Analysis of North America subspecies of migrant birds wintering in Los Tuxtlas, southern Veracruz, Mexico. In: Keast A, Morton ES (eds) Migrant birds in the Neotropics: ecology, behavior, distribution, and conservation. Smithsonian Institution Press, Washington, DC, pp 173–180

Ray S, Pool J (2018) Once degraded, West Virginia's Monongahela National Forest is restored to glory. World Resources Institute, Washington, DC. https://www.wri.org/insights/once-degraded-west-virginias-monongahela-national-forest-restored-glory. Accessed 20 Jan 2023

Reeks WA, Barter GW (1951) Growth reduction and mortality of spruce caused by the European spruce sawfly, *Gilpinia hercyniae* (HTG.) (Hymenoptera: Diprionidae). For Chron 27:140–156

Rentch JS, Ford WM, Schuler TS et al (2016) Release of suppressed red spruce using canopy gap creation—ecological restoration in the central Appalachians. Nat Areas J 36:29–37

Reynolds RJ, Pagels JF, Fies ML (1999) Demography of northern flying squirrels in Virginia. Proc Southeast Assoc Fish Wildl Agencies 53:340–349

Rhodes BM, Barton CD (2024) Comparing the response of red spruce plantings on legacy coal mines and old-field restoration sites in the West Virginia highlands. Nat Areas J 44:65–75

Richardson JS (2019) Biological diversity in headwater streams. Water 11:366

Riddell EA, Sears MW (2015) Geographic variation of resistance to water loss within two species of lungless salamanders: implications for activity. Ecosphere 6:86

Rinehart KA, Donovan TM, Mitchell BR et al (2009) Factors influencing occupancy patterns of eastern newts across Vermont. J Herpetol 43:521–531

Rissler LJ, Smith WH (2010) Mapping amphibian contact zones and phylogeographical break hotspots across the United States. Mol Ecol 19:5404–5416

Rojas VG, Loeb SC, O'Keefe JM (2019) False-positive occupancy models produce less-biased occupancy estimates for a rare and elusive bat species. J Mammal 100:212–222

Ronca R, Winterton E, Davenport JM (2026) Bridging data gaps: evidence-based population assessment for an endemic amphibian of conservation concern. J Wildl Manag 2026:e70216
Rose AH, Lindquist OH, Syme P (1994) Insects of eastern spruces, fir and hemlock. Canadian Forest Service, Ottawa, Canada
Ross RJ, Palaniswamy P, Seabrook WD (1979) Electroantennograms from spruce budworm moths (*Choristoneura fumiferana*) (Lepidoptera: Tortricidae) of different ages and for various pheromone concentrations. Can Entomol 111:807–816
Rossell CR Jr, Haas IC, Williams LA et al (2018) Comparison of relative abundance and microhabitat of *Desmognathus organi* (northern pygmy salamander) and *Desmognathus wrighti* (southern pygmy salamander) in North Carolina. Southeast Nat 17:141–154
Royama T (1984) Population dynamics of the spruce budworm *Choristoneura fumiferana*. Ecol Monogr 54:429–462
Rucker LE, Brown DJ, Watson MB et al (2022) Long-term occupancy dynamics of the threatened Cheat Mountain salamander and its competitors in relation to linear habitat fragmentation. For Ecol Manage 505:119847
Rucker LE, Brown DJ, Strager MP et al (2025) Projected future changes in the geographic distributions of the threatened *Plethodon nettingi* and a potential competitor. Endanger Species Res 57:103–118
Rushing CS, Royle JA, Ziolkowski DJ Jr et al (2020) Migratory behavior and winter geography drive differential range shifts of eastern birds in response to recent climate change. Proc Natl Acad Sci 117:12897–12903
Seaborn T, Catley K (2016) Abiotic microhabitat parameters of the spruce–fir moss spider, *Microhexura monitvaga* Crosby and Bishop (Araneae: Dipluridae). Southeast Nat 15:16–75
Shaw GG, Little CHA (1977) Natural variation in balsam fir foliar components of dietary importance to spruce budworm. Can J for Res 7:47–53
Shingleton MV, Brown WS (2010) Wild trout restoration in acidified streams. In: Carline RF, LoSapio C (eds) Conserving wild trout. Proceedings of the wild trout X symposium, Bozeman, Montana, pp 304–310
Sidebottom J (2019) Balsam woolly adelgid Christmas tree notes. https://content.ces.ncsu.edu/balsam-woolly-adelgid#section_heading_4932. Accessed 16 Dec 2024
Simbeck DJ (1990) Distribution of the fishes of the Great Smoky Mountains National Park. University of Tennessee, Knoxville, Tennessee, Thesis
Simpson MB Jr (1972a) Annotated checklist of the birds of Mt. Mitchell State Park, North Carolina. J Elisha Mitchell Sci Soc 88:244–251
Simpson MB Jr (1972b) The saw-whet owl population of North Carolina's southern Great Balsam Mountains. Chat 36:39–47
Six DL, Bracewell R (2015) *Dendroctonus*. In: Vega FE, Hofstetter RW (eds) Bark beetles: biology and ecology of native and invasive species. Academic Press, London, England, pp 305–350
Six DL, Bentz BJ (2003) Fungi associated with the North American spruce beetle, *Dendroctonus rufipennis*. Can J for Res 33:1815–1820
Snyder RC (1946) *Plethodon welleri* from Flat Top Mountain, North Carolina. Copeia 1946:174
Sopow SL, Quiring DT (1998) Body size of spruce-galling adelgids is positively related to realized fecundity in nature. Ecol Entomol 23:476–479
Stauffer JR, Boltz JM, White LR (1995) The fishes of West Virginia. Proc Acad Natl Sci Phila 146:1–389
Steele MA, Powell RA (1999) Biogeography of small mammals in the southern Appalachians: patterns of local and regional abundance and distribution. In: Eckerlin RP (ed) Proceedings of the Appalachian biogeography symposium. Virginia Museum of Natural History Special Publication 7, Martinsville, Virginia, pp 155–165
Steketee AK, Wood PB, Gregg ID (2015) American woodcock habitat changes in Canaan Valley and environs. Southeast Nat 14:331–343
Stephenson SL, Clovis JF (1983) Spruce forests of the Allegheny Mountains in central West Virginia. Castanea 48:1–12

Stewart RE, Aldrich JW (1949) Breeding bird populations in the spruce region of the central Appalachians. Ecology 30:75–82

Stihler CW, Wallace JW, Michael ED et al (1995) Range of *Glaucomys sabrinus fuscus*, a federally endangered subspecies of the northern flying squirrel, in West Virginia. Proc W Va Acad Sci 67:13–20

Stihler CW (2011) Status of the Virginia big-eared bat (*Corynorhinus townsendii virginianus*) in West Virginia: twenty-seven years of monitoring cave roosts. In: Loeb SC, Lacki MJ, Miller DA (eds) Conservation and management of eastern big-eared bats: a symposium. USDA Forest Service Southern Research Station General Technical Report SRS-145, Asheville, North Carolina, pp 75–84

Strong C, Zuckerberg B, Betancourt JL et al (2015) Climatic dipoles drive two principal modes of North American boreal bird irruption. Proc Natl Acad Sci 112:E2795–E2802

Sugalski MT, Claussen DL (1997) Preference for soil moisture, soil pH, and light intensity by the salamander, *Plethodon cinereus*. J Herpetol 31:245–250

Thomas-Van Gundy MA, Sturtevant BR (2014) Using scenario modeling for red spruce restoration planning in West Virginia. J Forest 112:457–466

Thompson ID, Baker JA, Ter-Mikaelian M (2003) A review of the long-term effects of post-harvest silviculture on vertebrate wildlife, and predictive models, with an emphasis on boreal forests in Ontario, Canada. For Ecol Manage 177:441–469

Thompson ID, Baker JA, Jastrebski C et al (2008) Effects of post-harvest silviculture on use of boreal forest stands by amphibians and marten in Ontario. For Chron 84:741–747

Thurow GR (1963) Taxonomic and ecological notes on the salamander, *Plethodon welleri*. Univ Kans Sci Bull 44:87–108

Tilley SG (1973) Life histories and natural selection in populations of the salamander *Desmognathus ochrophaeus*. Ecology 54:3–17

Tilley SG (1985) *Desmognathus imitator* imitator salamander. Cat Am Amphib Reptil 359(1–359):2

Tirpak JM, Giuliano WM, Miller CA et al (2006) Ruffed grouse population dynamics in the central and southern Appalachians. Biol Cons 133:364–378

U.S. Fish and Wildlife Service (2001) Endangered and threatened wildlife and plants; designation of critical habitat for the spruce-fir moss spider. Fed Reg 66:35547–35566

U.S. Fish and Wildlife Service (1990) Appalachian northern flying squirrels (*Glaucomys sabrinus fuscus* and *Glaucomys sabrinus coloratus*) recovery plan. U.S. Fish and Wildlife Service, Hadley, Massachusetts

Vannote RL, Minshall GW, Cummins KW et al (1980) The river continuum concept. Can J Fish Aquat Sci 37:130–137

Varty I (1956) *Adelges* insects of silver firs. Her Majesty's Stationary Office Forestry Commission Bulletin No. 26, Edinburgh, Scotland

Wallin KF, Raffa KF (2004) Feedback between individual host selection behavior and population dynamics in an eruptive herbivore. Ecol Monogr 74:101–116

Weigl PD (2007) The northern flying squirrel (*Glaucomys sabrinus*): a conservation challenge. J Mammal 88:897–907

Weigl PD, Knowles TW (1995) Megaherbivores and southern Appalachian grass balds. Growth Chang 26:365–382

Weigl PD, Knowles TW, Boyton AC (1992) The distribution and ecology of the northern flying squirrel *Glaucomys sabrinus coloratus* in the southern Appalachians. North Carolina Wildlife Resources Commission Nongame and Endangered Wildlife Program, Raleigh, North Carolina

West Virginia Division of Natural Resources (2021) West Virginia elk management plan. West Virginia Division of Natural Resources, Charleston, West Virginia

Wetzel EJ, Weigl PD (1994) Ecological implications for flying squirrels (*Glaucomys* spp.) of effects of temperature on the in vitro development and behavior of *Strongyloides robustus*. Am Midl Nat 131:43–54

Wingfield MJ, Harrington TC, Solheim H (1997) Two species in the *Ceratocystis coerulescens* complex from conifers in western North America. Can J Bot 75:827–834

Wootton JT (2020) Purple finch (*Haemorhous purpureus*), ver 1.0. In: Poole AF, Gill FB (eds) Birds of the world. Cornell Lab of Ornithology, Ithaca, New York

Wyman RL, Hawksley-Lescault DS (1987) Soil acidity affects distribution, behavior, and physiology of the salamander *Plethodon cinereus*. Ecology 68:1819–1827

Yetter E, Chhin S, Brown JP (2021) Sustainable management of central Appalachian red spruce. Sustainability 13:10871

Zurbuch PE (2015) Historic fishery of the Blackwater River. Southeast Nat 14:276–296

Zurbuch PE, Menendez R, Clayton JL (1996) Limestone neutralization of Dogway Fork, West Virginia, by means of a rotary-drum system. Restor Ecol 4:206–219

Chapter 7
Climate Change Implications and Adaptation Solutions

Stephen R. Keller, Matthew C. Fitzpatrick, Susanne Lachmuth, Danika Mosher, Thibaut Capblancq, Kevin M. Potter, Elizabeth A. Byers, and John R. Butnor

7.1 Historical and Contemporary Climate Change Responses of Red Spruce

7.1.1 The Climatic Niche of Red Spruce

Aspects of red spruce's climatic niche and its relation to the biogeography of the species are covered in Chap. 1. Our emphasis here is on red spruce's interaction with its biophysical environment in terms of a *changing* climate, what this might illuminate about the nature of its climate sensitivities, and how it may respond to future climate change. We approach this from an interdisciplinary viewpoint that integrates perspectives across biological, temporal, and spatial scales not often discussed together in the same work. We first briefly summarize key features of red spruce's climate preferences and tolerances, as they relate to the ensuing discussion of its realized and potential responses to climate changes, past, present, and future.

S. R. Keller (✉)
University of Vermont, Department of Plant Biology, Burlington, VT, USA
e-mail: stephen.keller@uvm.edu

M. C. Fitzpatrick · S. Lachmuth
University of Maryland Center for Environmental Science, Appalachian Lab, Frostburg, MD, USA
e-mail: matt.fitzpatrick@umces.edu

S. Lachmuth
e-mail: susanne.lachmuth@zalf.de

D. Mosher
USDA Forest Service, Southern Research Station, Eastern Forest Environmental Threat Assessment Center, Research Triangle Park, Durham, NC, USA
e-mail: danika.mosher@usda.gov

D. J. Brown et al. (eds.), *Ecology and Restoration of Red Spruce Ecosystems in the Central and Southern Appalachians*, https://doi.org/10.1007/978-3-032-19616-3_7

The Appalachian Mountains span a diverse range of environmental conditions in terms of precipitation patterns, humidity, mean and extreme temperatures, and length of growing season. The climatic niche of red spruce has been described as cool, moist, and temperate, as reflected by its distribution at mid to high elevations within the Appalachians and along the cool, humid coastal areas of Maine and the Atlantic Canadian provenances (Fig. 7.1; Day 2000; Dumais and Prévost 2007). Red spruce is not, as it is sometimes described, a boreal tree species or one that is preferentially found in the coldest microclimates available within the geography it inhabits. Perhaps the most easily defined and visible aspect of red spruce's climatic niche is that its distribution generally exhibits a latitude-elevation relationship, with central and southern Appalachian spruce forests confined to the highest elevations in the region, typically above 1,000 m and 1,600 m (3,280 ft and 5,250 ft), respectively (Cogbill and White 1991, Byers et al. 2010). In their analysis of latitude-elevation relationships of Appalachian spruce-fir forests, Cogbill and White (1991) found the spruce-fir/deciduous ecotone to closely follow a mean July isotherm of 17.1 °C (62.8 °F). High temperatures (>32 °C [90 °F]) during the growing season inhibit photosynthesis of red spruce seedlings and saplings and can cause irreversible foliar injury in mature trees (Day 2000; Dumais and Prévost 2007). This is consistent with red spruce favoring environments with low vapor pressure deficits, characterized by moderate temperatures and high humidity (Day 2000). Red spruce is also sensitive to extreme cold and shows moderate levels of mid-winter freezing hardiness among conifers, even compared to co-occurring balsam fir (*Abies balsamea*; Strimbeck et al. 2007, 2015). Red spruce foliage is especially susceptible to winter freezing injury as a result of dehardening that occurs during prolonged temperatures >0 °C (32 °F), followed by rapid return of sub-freezing temperatures (Schaberg 2000). Detailed descriptions of red spruce's physiological tolerances as they relate to climate are presented in Sect. 7.2.

In comparison to the northern part of its range, the climate inhabited by red spruce in the central and southern Appalachians is characterized by milder summer and winter temperatures, longer frost-free period and growing season length, lower

T. Capblancq
University of Grenoble, Alpine Ecology Laboratory, Grenoble, France
e-mail: thibaut.capblancq@univ-grenoble-alpes.fr

K. M. Potter
USDA Forest Service, Southern Research Station, Eastern Forest Environmental Threat Assessment Center, Research Triangle Park, Durham, NC, USA
e-mail: kevin.potter@usda.gov

E. A. Byers
West Virginia Division of Natural Resources (retired), Elkins, WV, USA
e-mail: appalachian.ecology@gmail.com

J. R. Butnor
USDA Forest Service, Northern Research Station, Burlington, VT, USA
e-mail: john.butnor@usda.gov

Fig. 7.1 High-elevation red spruce community after an ice storm in Tucker County, West Virginia. Photo by Matthew Fitzpatrick

amplitude of mean monthly temperature variation across the year, enhanced precipitation, lower snowfall, and frequent orographic-derived cloud cover (Cogbill and White 1991). Red spruce reaches its largest size and has its greatest growth potential in the southern portion of the range (King and Stupka 1950). The persistence of red spruce in the southern extent of the range is likely dependent on ecophysiological adaptations to cloud forest conditions that mitigate climatic conditions near the physiologically determined limits of its range. In the context of a warming climate, the climatic niche of red spruce suggests that as temperatures increase, southern portions of its range may become too warm, especially when combined with increasing frequency of drought and associated increases in vapor pressure that, collectively, are likely to decrease climatic suitability and could lead to extirpation of populations at both local and regional scales. In contrast, portions of the northern range currently limited by severe winter cold are likely to increase in suitability as long as sufficient moisture remains available. Detailed projections of the future climatic niche of red spruce are discussed in Sect. 7.3.

7.1.2 The Paleoecological Perspective: What Does the Pollen Record Tell Us About the Response of Spruce to Climate Change?

The paleoecological record of the last ice age provides a unique opportunity to examine how species have responded to rapid climate change and thereby can provide a glimpse of what might be expected in the future, even if the analogy between climate change in the past and that expected in the future is less than perfect. Fortunately, the ice age history of spruce is relatively well documented, with the important caveat that, due to limitations in assigning fossil material to species, most inferences are limited to the genus level (*Picea*; unless otherwise noted, we use the term spruce in this section to refer to all species in the genus). Nonetheless, available fossil evidence suggests spruce responded relatively quickly to climate changes during the Last Glacial Maximum (approximately 22,000 YBP) and has largely tracked its climatic niche in the process. It therefore is not surprising that the predictions of change in the distribution of spruce from ecological niche models (ENM), which attempt to fit a statistical model to the climatic niche, tend to be in good agreement with observed changes in spruce abundance inferred from fossil pollen records (Veloz et al. 2012), and therefore lend support to the use of ENMs to forecast the potential response of spruce to future climatic changes. In this section, we first summarize relevant past climatic trends and then discuss how spruce appears to have responded to warming according to changes in fossil pollen abundance through time, as well as some limited evidence of the responses of red spruce specifically.

7.1.2.1 Climatic and Vegetation Changes Since the Last Glacial Maximum

The Quaternary period, which began 2.58 million YBP, was characterized by colder temperatures than the present and encompassed about 30 cycles of glaciation with expanded ice sheets in the northern hemisphere (Marshall 2009). Whereas cooling typically occurred gradually over periods of 50,000–100,000 years, the interglacials, which were comparatively short at 10,000–20,000 years, were characterized by relatively rapid climate warming and associated environmental changes that were more disruptive to vegetation than the cooling periods (Davis 1981). The last glaciation started approximately 100,000 YBP, and by the Last Glacial Maximum about 22,000 YBP, the Laurentide ice sheet extended into central Illinois, northern Pennsylvania, and southern New England. The Last Glacial Maximum terminated with a sudden steep temperature rise beginning about 16,000 YBP. The interglacial reached its warmest period during the Holocene Climatic Optimum at 8,500–5,000 YBP. From a conservation (and modeling) perspective, the paleoecological record makes it clear that during these periods of climatic upheaval, plant taxa responded individually and opportunistically to Quaternary environmental changes rather than migrating as intact forest communities (Williams et al. 2004). Existing plant communities disassembled as individual taxa migrated at different rates and along different routes such that forest communities rarely exhibited constant species composition for more than 2,000–3,000 years at a time during the interglacials (Davis 1981).

Spruce pollen is abundant in the fossil record and suggests that *Picea* was among the most responsive to changes in climate and also among the first trees to colonize recently deglaciated regions (Watts 1979; Davis 1981; Williams et al. 2004). *Picea*'s propensity for rapid migration in response to climate change helps explain the relative stability of its niche through time, suggesting that changes in the distribution of spruce were closely linked to changes in climate and that the genus was able to largely track rapid climate changes (Veloz et al. 2012). The late Quaternary vegetation history of North America was characterized by rapid north–south, but also east–west shifts in plant distributions (Williams et al. 2004). The center of abundance of spruce pollen over the past 22,000 YBP has shifted several times between eastern and western North America according to pollen records (Williams et al. 2004). Range shifts of red spruce likely remained restricted to the eastern part of the continent but still were rapid and substantial according to modeling (Thibaut Capblancq and Susanne Lachmuth et al. unpublished manuscript).

7.1.2.2 Geographic Distribution of Spruce During the Last Glacial Maximum

The eastern Laurentide Ice Sheet was fringed by tundra in the continental interior (Jackson et al. 2000), and alpine communities also covered the Appalachian ridges up to several hundred km south of the glacial limit (Watts 1979; Davis 1981). Multiple lines of evidence suggest spruce occurred in the cold, dry, and windy conditions near the glacial margin as spruce logs were found to have been overridden by the oscillating ice front while still growing in Ohio (Watts 1979 and citations therein). Moreover, spruce macrofossils in Pennsylvania 60 km (37 mi) south of the ice sheet and at 200 m (656 ft) elevation indicate the proximity of the tree limit (Watts 1979), which was formed by open spruce forest or spruce tundra along most of the Laurentide and Cordilleran ice sheets (Watts 1979; Williams et al. 2004). Macrofossils analyzed by Jackson and Overpeck (2000) also confirmed that the occurrence of spruce in the continental interior ranged from very close to the ice margin (in particular in the Ohio Valley) south into east-central Louisiana. However, given the current distribution of red spruce and the species' sensitivity to extreme cold, it is likely populations close to the ice sheets were white spruce (*Picea glauca*) and/or black spruce (*Picea mariana*).

In the South, vegetation appears to have remained relatively stable from 22,000 to 13,500 YBP. In this period, a spruce refugium stretched from the Great Plains, where spruce codominated with larch (Davis 1981), to the eastern Coastal Plain (Jackson et al. 2000; Williams et al. 2004). In the lower Mississippi Valley, spruce pollen reached maximum abundance, likely reflecting the presence of Critchfield's spruce (*Picea critchfieldii*), which also occurred in Georgia (Jackson et al. 2000). Despite its seemingly high abundance in the Southeast in the past, Critchfield's spruce is now extinct and represents the only documented plant extinction from the last ice age in North America, though the cause of its extirpation remains unresolved. When still extant, Critchfield's spruce was associated with oak (*Quercus* spp.) and other nonboreal species, a plant assemblage with no close modern analog (Jackson et al. 2000). Spruce abundance was low in the Carolinas, Georgia, and eastern Tennessee and the genus was absent from the central and southern Appalachian highlands (Davis 1981), even though these areas were not glaciated (White and Cogbill 1992). All members of the genus *Picea* were absent from the central and southern Appalachians and the absence of red spruce specifically is supported by both pollen records (Delcourt and Delcourt 1988) and ENM hindcasts and demogenetic simulations (Thibaut Capblancq and Susanne Lachmuth et al., unpublished manuscript). This is also consistent with Davis (1981), who described the distribution of *Picea* in eastern North America as a belt stretching from the Mississippi Valley to the East Coast, excluding the mountainous region of the southern Appalachians. According to Watts (1979), red spruce occurred along the southern edge of this belt for its entire east–west extent. ENM hindcasts and demogenetic simulations (Thibaut Capblancq and Susanne Lachmuth et al., unpublished manuscript) delineate this belt as the glacial refugium of red spruce, which was largely maintained in its east–west extent during northward migration until approximately 15,000 YBP.

7.1.2.3 Rapid Shifts in Spruce Distribution During the Late Pleistocene and Early Holocene

With the onset of warming around 15,000 YBP, which probably increased the length of the growing season (Watts 1979), spruce proved to be an aggressive pioneer as it rapidly spread northward into the tundra of the Great Lakes Region (Davis 1981). In New England, the earliest records of spruce date to between 16,500 and 14,000 YBP (Davis and Jacobson 1985), with recalibrated dates in Lindbladh et al. (2003). Lindbladh et al. (2007) suggested that around 14,000–13,000 YBP a transition from open white spruce forest-tundra to closed black spruce-dominated forest occurred. Lindbladh et al. (2003, 2007), who applied a palynological method to distinguish red spruce from white spruce and black spruce (Lindbladh et al. 2002), also suggested that red spruce was largely absent from New England before 12,000 YBP during the immediate post-glacial in which climate was characterized by high seasonality and extreme low winter minimum temperatures. While it remains difficult to distinguish whether the increase in spruce pollen after 12,000 YBP was primarily red spruce or if black spruce was also increasing in abundance at this time, the timing aligns well with ENM hindcasts (Thibaut Capblancq and Susanne Lachmuth et al., unpublished manuscript), which suggest that red spruce rapidly colonized this region between 12,000 and 10,000 YBP.

Evidence suggests that climate initially hindered red spruce colonization of higher elevations in the central and southern Appalachians (Watts 1979; Davis 1981), and that the expansion front was likely formed mostly by white spruce, followed by black spruce. Watts (1979) claimed to have distinguished red spruce pollen from other *Picea* species and documented a sudden massive increase of red spruce around 11,000 YBP at a site in the Appalachians of southern Pennsylvania (Crider's Pond, 40° N), where members of the *Picea* genus occurred as early as 15,200 YBP. Watts (1979) also found red spruce and no other members of the genus in a sample dated to 11,100 YBP from the southcentral Appalachians (Potts Mountain, Virginia, 37.5° N).

Around 12,500 YBP during the Pleistocene-Holocene transition, cold, dry, and windy Pleistocene climates were replaced by warmer and more humid climates (White and Cogbill 1992). In the south, lowland spruce populations declined (Williams et al. 2004) and the open spruce forest and forest tundra were replaced with species-rich conifer-hardwood forests (Watts 1979). At a continental scale and genus level, peak abundances of spruce pollen occurred in the eastern interior until 14,000 YBP, then shifted to western Canada as a consequence of asymmetrical retreat of the Canadian ice sheet and did not move back to eastern Canada until around 7,000 YBP (Williams et al. 2004).

7.1.2.4 Near Loss of Midcontinent Spruce During the Holocene Climatic Optimum and Its Sudden Reappearance Afterward

During the Holocene Climatic Optimum, spruce nearly disappeared entirely from its midcontinent belt when temperatures peaked between 9,000 and 4,000 YBP (White and Cogbill 1992; Lindbladh et al. 2003). In the Appalachians, spruce-fir forests shifted upwards by 200–500 m (656–1,640 ft; White and Cogbill 1992). White and Cogbill (1992) suggested that red spruce may have survived this time period in damp lowlands, on steep rocky slopes, and/or in the northeastern Maritime region, after which time spruce suddenly re-emerged and presumably expanded to the Appalachian slopes ca. 2,000 YBP. Similarly, Lindbladh et al. (2003) and citations therein suggest a disappearance of spruce from the lowlands in New England during the Holocene Climatic Optimum followed by an increase in abundance 1,000–500 YBP, probably indicative of red spruce. Further south, Lindbladh et al. (2003) also noted that "the presence of a continuous *Picea* pollen curve (identified as *P. rubens*) at a high-altitude site (Cranberry Glades) in the Appalachian Mountains of West Virginia (Watts 1979) raises the possibility that the species survived the generally warm and dry middle Holocene climate in cool, moist microhabitats at suitable altitudes there."

7.1.3 Contemporary Evidence of Red Spruce Responses to Climate Change

There have been mixed reports of how the distribution and abundance of red spruce in natural forested stands have responded to contemporary climate trends over the last century. Some of this complexity appears to reflect the confounding of the climate signal, which has generally been progressively warmer, with other temporal changes such as recovery from past land-use history and atmospheric pollution (e.g., acid rain). Initial evidence of red spruce range contraction came from upslope shifts of the deciduous/coniferous ecotone in Vermont between 1964 and 2004 (Beckage et al. 2008). The position of this ecotone is highly correlated with mid-summer temperatures (Hamburg and Cogbill 1988; Cogbill and White 1991). However, later studies documented widespread regional trends of a subsequent downslope movement of red spruce and associated conifer forests throughout New England (Foster and D'Amato 2015; Wason and Dovciak 2017), including some of the same areas where upslope movement was previously documented (Verrico et al. 2020). This re-colonization of lower elevations has principally been attributed to recovery from historical anthropogenic land-use history, which negatively impacted low-elevation red spruce, and reductions in atmospheric pollution since passage of the Clean Air Act (Wason and Dovciak 2017; Verrico et al. 2020).

Other evidence of growth responses of red spruce to contemporary climate comes from dendroecological studies. Here again, the response to climate change is complicated. In western Massachusetts, where red spruce is at the southern margin of its distribution in New England, climate warming trends over the past century resulted in reductions in annual growth rings, particularly associated with the previous year's July maximum temperature (Ribbons 2014). In contrast, positive associations between annual growth rings and summer temperatures (both current and previous year's) were reported elsewhere in New England, despite a similar climate trend of regional warming (Wason et al. 2017; Kosiba et al. 2018). In the central Appalachians, tree-ring chronologies of red spruce from three stands in West Virginia showed evidence of recovery from historical atmospheric pollution and increased growth linked to higher CO_2 concentration and warmer April temperatures, the latter likely reflecting a positive phenological response to climate warming (Mathias and Thomas 2018). Another dendroclimatic study of red spruce from 18 sites throughout the Monongahela NF in West Virginia also found tree-ring growth was positively associated with warmer April temperatures, as well as warmer autumn temperatures (Yetter et al. 2021). However, this same study documented negative growth responses to warming summer temperatures, and climate change projections forecasted an overall reduction in growth in response to warm-temperature stress under future climates (Yetter et al. 2021). These projections are in line with most red spruce ENMs for this region, which generally predict a range contraction in response to increasing temperatures (summarized in Sect. 7.3.1.1).

7.2 Ecological, Physiological, and Evolutionary Perspectives on Climate Adaptation in Red Spruce Ecosystems

Here we summarize what is known about the responses of red spruce ecosystems to climate and climate change, looking across scales including the level of communities of interacting species and the physiological scale of tree stress response. We finish this section with a summary of recent research on red spruce local adaptation to climate, and the traits and genes that have been discovered as underlying this adaptation. Our intent is to impart a detailed understanding of the current distribution of climate-adapted diversity in red spruce, as a preface to Sect. 7.3, which explores model predictions of how this diversity is likely to interact with a changing climate over the next century.

7.2.1 Red Spruce Communities, Species Interactions, and Climate Change

Red spruce communities, and the unique biodiversity they contain, occupy the highest-elevation habitats and the coolest climatic niches in the central and southern Appalachians and are clearly at risk due to climate change. Warming and possible drying of habitats will negatively impact many species that are adapted to the cool, moist conditions prevailing within the red spruce ecosystem. Extreme events, including floods, droughts, and severe storms will also place new stresses on species struggling to adapt to changing conditions.

The majority of foundational tree species in the red spruce ecosystem are predicted to decline as the climate changes, based on modeling by the USDA Forest Service (Table 7.1; Iverson et al. 2012; Peters et al. 2020). As these foundational tree species decline, cascading effects on specialist species are anticipated. For example, the Cheat Mountain salamander (*Plethodon nettingi*; Fig. 7.2) depends upon the moist, cool, mossy, rock-strewn forest floor of red spruce forests, and with a home range of just a few m^2, has very limited ability to disperse (Pauley 2022). Virginia northern flying squirrels (*Glaucomys sabrinus fuscus*) depend on red spruce and red spruce-northern hardwood forests with abundant snags, mature trees, lichens, and mosses (Menzel et al. 2006; Ford et al. 2022). Certain snails, such as the Spruce Knob threetooth (*Triodipsis picea*), live in moist or wet areas associated with calcium-poor soils common to the red spruce ecosystem. This species accumulates calcium in its shell, which upon its death becomes a key available nutrient for birds and small mammals (Dourson and West Virginia Division of Natural Resources 2015). Larvae of the salt-and-pepper looper moth (*Syngrapha rectangula*) feed on the needles of red spruce, balsam fir (*Abies balsamea*), and eastern hemlock (*Tsuga canadensis*), and these larvae in turn form a key food source for birds (Opler et al. 2010).

Table 7.1 Modeled climate change adaptability (biological and disturbance factors) and capability (adaptability, predicted range shift, and abundance) scores for common dominant trees in the red spruce ecosystem

Common name	Scientific name	Adaptability	Capability	Traits that increase adaptability	Traits that decrease adaptability
Red spruce	*Picea rubens*	Low	Poor	Environmental habitat specificity; shade tolerance	Fire topkill; seedling establishment
Balsam fir/Fraser fir	*Abies balsamea/ Abies Fraseri*	Low	Fair	Shade tolerance	Insect pests; fire topkill; drought
Eastern hemlock	*Tsuga canadensis*	Low	Poor	Shade tolerance	Insect pests; drought
Yellow birch	*Betula alleghaniensis*	Medium	Fair	Dispersal	Fire topkill; insect pests; disease
Red maple	*Acer rubrum*	High	Very good	Seedling establishment, environment habitat specificity, edaphic specificity, shade tolerance, dispersal	None
American beech	*Fagus grandifolia*	Medium	Good	Shade tolerance	Insect pests; fire topkill
Yellow buckeye	*Aesculus flava*	High	Poor	Shade tolerance	Drought; seedling establishment; fire topkill; environment habitat specificity; dispersal

Adapted from Peters et al. (2020) and Iverson et al. (2012)

The relative risk of local extirpation of plant and animal species due to climate change was assessed by Byers and Norris (2011) for 185 taxa in West Virginia using NatureServe's Climate Change Vulnerability Index (Young et al. 2011). Thirty-seven of these taxa are associated with red spruce ecosystems. Climate change vulnerability assessments involve describing the severity and scope of exposure that species experience and combining this with species' sensitivity and capacity to adapt to climate change (Foden et al. 2019). Inputs to the assessment are downscaled projections of future climate, natural history data for each species, and distributional data for the species and its habitat. Of the 37 red spruce-associated species assessed, thirteen are globally rare, twelve are rare in West Virginia, and twelve are common (Table 7.2).

All of the globally rare species assessed were determined to be vulnerable to climate change, including two species, Cheat Mountain salamander and glade spurge

Fig. 7.2 Adult Cheat Mountain salamander (*Plethodon nettingi*) in a high-elevation red spruce (*Picea rubens*) forest in Tucker County, West Virginia. Photo by Donald Brown

(*Euphorbia purpurea*), which scored *Extremely Vulnerable*. Of the state-rare species assessed, half (six) were determined to be vulnerable to climate change, five were presumed stable, and one was predicted to increase. Of the common species, eight were presumed stable, one was predicted to increase, and the following were determined to be vulnerable to climate change: red spruce, black cherry (*Prunus serotina*), sugar maple (*Acer saccharum*), and brook trout (*Salvelinus fontinalis*).

Several factors were key to influencing climate change vulnerability in this species assemblage, including natural barriers to dispersal (mountaintop location with low-elevation barriers to the north), physiological thermal niche, physiological hydrologic niche, and historical hydrologic niche. For amphibians and some plants, dispersal ability was a strong limiting factor. Thirteen of the 37 species assessed, including red spruce, black cherry, and many of the more mobile taxonomic groups such as birds and butterflies, were predicted to shift their range northward and disappear from West Virginia.

On the positive side, red spruce communities have several key characteristics that are likely to impart resilience in the face of climate stress and other stresses. For one, they occur on large tracts of relatively well-connected landscapes. The red spruce ecosystem as a whole, including both forested and wetland communities, contains high levels of biodiversity, which means that even with some extinctions, there may be enough connected threads of the ecosystem left to function. The southwest to northeast orientation of the mountains, and the connectivity of natural habitats between low and high elevations, may allow some rapidly dispersing species to migrate northward or, in some cases, upward as the climate envelope shifts. Regardless of what happens to the individual species in the red spruce ecosystem, the landscape itself is

Table 7.2 Climate change vulnerability assessment of selected species inhabiting red spruce (*Picea rubens*) ecosystems in West Virginia (WV)

Common name	Scientific name	Conservation status	Climate change vulnerability	Relationship to red spruce ecosystems
Alder flycatcher	*Empidonax alnorum*	WV vulnerable	Highly vulnerable; range may shift out of WV	Associate; also occurs in other habitats
Allegheny crayfish	*Orconectes obscurus*	Apparently secure	Presumed stable	Associate; also occurs in other habitats
Allegheny woodrat	*Neotoma magister*	Globally vulnerable	Moderately vulnerable	Associate; also occurs in other habitats
American woodcock	*Scolopax minor*	WV vulnerable	Presumed stable	Associate; also occurs in other habitats
Appalachian brook crayfish	*Cambarus bartonii cavatus*	Secure	Presumed stable	Associate; also occurs in other habitats
Bear Creek slitmouth snail	*Stenotrema simile*	Globally imperiled	Moderately vulnerable	Associate; also occurs in other habitats
Big River crayfish	*Cambarus robustus*	Apparently secure	Presumed stable	Associate; also occurs in other habitats
Black cherry	*Prunus serotina*	Secure	Moderately vulnerable; range may shift out of WV	Associate; also occurs in other habitats
Black dash	*Euphyes conspicua*	WV critically imperiled	Moderately vulnerable; range may shift out of WV	Associate; also occurs in other habitats
Blackgum	*Nyssa sylvatica*	Secure	Increase likely	Associate; also occurs in other habitats
Blue crawfish	*Cambarus monongalensis* (mountains)	Globally vulnerable	Moderately vulnerable	Associate; also occurs in other habitats
Bog copper	*Lycaena epixanthe*	WV critically imperiled	Moderately vulnerable; range may shift out of WV	Obligate (wetlands)

(continued)

Table 7.2 (continued)

Common name	Scientific name	Conservation status	Climate change vulnerability	Relationship to red spruce ecosystems
Brook trout	*Salvelinus fontinalis*	Secure	Highly vulnerable	Associate; also occurs in other habitats
Cheat Mountain salamander	*Plethodon nettingi*	Globally imperiled	Extremely vulnerable	Close associate
Elk River crayfish	*Cambarus elkensis*	Globally imperiled	Moderately vulnerable	Associate; also occurs in other habitats
Fisher	*Martes pennanti*	WV vulnerable	Presumed stable; range may shift out of WV	Associate; also occurs in other habitats
Glade spurge	*Euphorbia purpurea*	Globally vulnerable	Extremely vulnerable	Obligate (wetlands)
Monongahela Barbara's-buttons	*Marshallia pulchra*	Globally imperiled	Moderately vulnerable	Associate; also occurs in other habitats
Mountain earthsnake	*Virginia valeriae pulchra*	Globally vulnerable	Highly vulnerable; range may shift out of WV	Associate; also occurs in other habitats
New River crayfish	*Cambarus chasmodactylus*	WV vulnerable	Presumed stable	Associate; also occurs in other habitats
Northern goshawk	*Accipiter gentilis*	WV critically imperiled	Presumed stable; range may shift out of WV	Associate; also occurs in other habitats
Northern saw-whet owl	*Aegolius acadicus* (breeding)	WV imperiled	Moderately vulnerable; range may shift out of WV	Close associate
Northern saw-whet owl	*Aegolius acadicus* (wintering)	WV vulnerable	Increase likely	Associate; also occurs in other habitats
Olive-sided flycatcher	*Contopus cooperi*	WV critically imperiled	Presumed stable; range may shift out of WV	Obligate (openings and wetlands)

(continued)

Table 7.2 (continued)

Common name	Scientific name	Conservation status	Climate change vulnerability	Relationship to red spruce ecosystems
Pink-edged sulphur	*Colias interior*	WV critically imperiled	Moderately vulnerable; range may shift out of WV	Close associate (openings)
Pitch pine	*Pinus rigida*	Secure	Presumed stable	Associate; also occurs in other habitats
Red maple	*Acer rubrum*	Secure	Presumed stable	Associate; also occurs in other habitats
Red spruce	*Picea rubens*	Apparently secure	Highly vulnerable; range may shift out of WV	Obligate
Rock crawfish	*Cambarus carinirostris*	Secure	Presumed stable; range may shift out of WV	Associate; also occurs in other habitats
Sanborn's crayfish	*Orconectes sanbornii*	Apparently secure	Presumed stable	Associate; also occurs in other habitats
Shriver's frilly orchid	*Platanthera shriveri*	Globally critically imperiled	Highly vulnerable	Associate; also occurs in other habitats
Southern rock vole	*Microtus chrotorrhinus carolinensis*	Globally vulnerable	Moderately vulnerable	Associate; also occurs in other habitats
Southern water shrew	*Sorex palustris puctulatus*	Globally vulnerable	Moderately vulnerable	Associate; also occurs in other habitats
Spruce Knob threetooth snail	*Triodopsis picea*	Globally vulnerable	Moderately vulnerable	Close associate
Sugar maple	*Acer saccharum*	Secure	Moderately vulnerable	Associate; also occurs in other habitats

(continued)

Table 7.2 (continued)

Common name	Scientific name	Conservation status	Climate change vulnerability	Relationship to red spruce ecosystems
Swainson's thrush	*Catharus ustulatus*	WV critically imperiled	Moderately vulnerable; range may shift out of WV	Close associate
Virginia northern flying squirrel	*Glaucomys sabrinus fuscus*	Globally imperiled	Highly vulnerable	Close associate

Adapted from Byers and Norris (2011)

likely to continue to support important conservation targets for the foreseeable future (Byers et al. 2010).

7.2.2 *Physiological Responses to Seasonal and Annual Variation in Climate*

7.2.2.1 Gas Exchange and Environmental Conditions

Photosynthetic rates in red spruce follow ambient air temperature, achieving the highest rates in midsummer, declining in the autumn, and reducing to zero under freezing conditions. Red spruce trees are able to efficiently photosynthesize over a wide optimum temperature range (16–32 °C [61–90 °F]), with sharp declines at temperatures >32 °C (90 °F; Vann et al. 1994; Day 2000). Specific rates are dependent on leaf nutrition, hydration status, and genetics, though the photosynthetic response to air temperature is very predictable (Fig. 7.3). The species is very responsive to warming in the winter and can photosynthesize any time the air temperature is above freezing for a few hours (Schaberg et al. 1998). Winter photosynthesis is limited by water availability in the foliage and inhibition from the accumulation of carbohydrates (Schaberg 2000), so it is not a primary means of carbon capture in northern regions, but the long season with temperatures permissive for photosynthesis is likely relevant to growth in the southern end of the range. The rapid response of photosynthesis to increasing temperature may confer growth advantages over deciduous trees and, in mixed forests, permit enhanced carbon capture in the understory before spring leaf-out. However, the warm conditions that permit photosynthesis also lead to a reduction in cold tolerance and a vulnerability to injury if there is a rapid return to very cold temperatures (Schaberg 2000).

Red spruce is characterized by comparatively low stomatal conductance or degree of stomatal opening, which relates to gas exchange and water loss. This results in low to moderate growth rates and conservation of water in dry conditions. Photosynthesis and stomatal conductance in red spruce are less sensitive to low vapor pressure deficit than other conifers (Day 2000), enabling them to capture carbon during periods of high humidity (Reinhardt and Smith 2008). It has been hypothesized that elevation, temperature, and cloud cover combine to define where red spruce is abundant across its range (Cogbill and White 1991). In studies of high and low-elevation saplings near Mt. Mitchell, North Carolina, Berry et al. (2019) found that immersion in clouds resulted in higher leaf water potential and rates of photosynthesis, especially in the afternoon, even though lower light conditions prevailed. Immersion in clouds also helped to maintain leaf hydration which allowed for higher photosynthetic rates throughout the day. Evidence that cloud levels are rising from the central to northern Appalachians (but not southern Appalachians) could exacerbate drought stress for

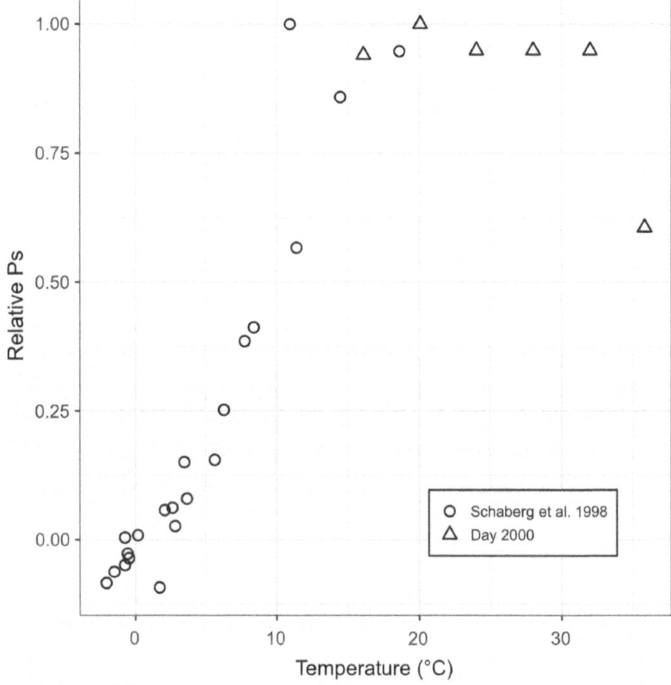

Fig. 7.3 Relative red spruce (*Picea rubens*) photosynthesis (Ps) where the maximum value is normalized to 1 (per study) follows a predictable pattern from negative to near zero at air temperatures of 0 °C (32 °F), and rises to an optimum range from 16 to 32 °C (61–90 °F), before declining

red spruce (Richardson et al. 2003). Climate change scenarios that lead to diminished cloud levels and high vapor pressure deficit would likely be detrimental to productivity.

7.2.2.2 Light Response

Red spruce is a late-successional species with adaptations to low-light environments. Compared to its sister species black spruce, red spruce has lower quantum yield or efficiency in bright light and greater thermal energy dissipation efficiency, which offer a greater degree of photoprotection and resulting photoinhibition (Major et al. 2003). In low light, red spruce exhibits a higher photochemical efficiency than other conifers allowing it to take advantage of fleeting sunflecks, though light saturation of photosynthesis can occur at levels as low as 250 μmol m^{-2} m s^{-1} or 1/8 that of full sunlight in the understory (Alexander et al. 1995). Seedlings grown under 50% shade cloth experienced light-saturated photosynthesis at levels between 600 and 1,000 μmol m^{-2} m s^{-1} (J. Butnor, unpublished data). Seedlings and advance-regeneration understory trees can be stressed by sudden exposure to bright light as may occur with clearcutting (Dumais and Prévost 2007). For this reason, gradually increasing light through multiple entries or gap creation is recommended for releasing suppressed red spruce in the understory (Rentch et al. 2016).

7.2.2.3 Response to High Temperatures and Drought

Based on the distribution of the species on the landscape, red spruce is favored in cool, moist environments and microsites (Koo et al. 2014a, b). On a physiological level, red spruce is unable to efficiently photosynthesize at temperatures above 32 °C (90 °F; Day 2000) and severe cellular damage has been reported for temperatures between 32 and 40 °C (90 and 104 °F; Vann et al. 1994). Red spruce's low stomatal conductance limits transpirational cooling. Air temperatures at elevations where red spruce occurs in the southern Appalachians seldom exceed these thresholds. For example, a series of data loggers on Mt. Mitchell (North Carolina) that were installed across eight elevations from 1,024 to 2,027 m (3,360–6,650 ft) found the maximum hourly air temperature at the summit never surpassed 21.2 °C (70.2 °F) during a 3-year period (2018–2020), while the lowest elevation experienced a maximum temperature of 29.8 °C (85.6 °F; Table 7.3). However, seed beds in full sunlight and exposed surfaces after disturbances can easily exceed lethal temperatures (Gray and Spies 1997), and leaf temperatures in full light can be several degrees warmer than ambient air, emphasizing the species' need for cool establishment conditions.

Drought affects the growth patterns of most woody plants, including red spruce, though relatively low stomatal gas exchange limits water loss during dry conditions (Day 2000). Red spruce is noted for shallow rooting and requiring ample moisture for seedling germination and establishment (Blum 1990), which likely are responsible for its drought vulnerability. Prior-season heat and drought stress have been correlated with growth declines and susceptibility to injury from cold and other abiotic stressors (Kosiba et al. 2018).

7.2.2.4 Freezing Injury

Populations adapted to cold climates typically have greater cold tolerance, though seasonal acclimation to cold temperatures determines the susceptibility to injury at specific times (Bigras and Colombo 2013). Acclimation is a complex biochemical process where changes in intracellular carbohydrates, solutes, proteins (particularly dehydrins), antioxidants, and lipid membranes are initiated by environmental cues and function to protect cells from freezing (Kalberer et al. 2006; Strimbeck et al. 2008, 2015). Tree species from warm temperate mountains may be injured by winter temperatures between −25 and −35 °C (−13 and −31 °F), while true boreal species can survive temperatures in excess of −80 °C (−112 °F; Strimbeck and Schaberg 2009).

Table 7.3 Mean, maximum, minimum, January (mean), and August (mean) air temperature (temp) and maximum vapor pressure deficit (VPD; kilopascals [kPa]) statistics based on weather station data collected hourly from eight elevations at Mt. Mitchell, North Carolina, 2018–2020 (J. Butnor and C. Maier, USDA Forest Service, Southern Research Station, unpublished data)

Elevation (m [ft])	Mean temp (°C [°F])	Maximum temp (°C [°F])	Minimum temp (°C [°F])	January temp (°C [°F])	August temp (°C [°F])	VPD (kPa)
2,027 (6,650)	6.3 (43.3)	21.2 (70.2)	−22.2 (−8.0)	−3.4 (25.9)	13.8 (56.8)	1.7
1,914 (6,280)	7.2 (45.0)	22.6 (72.7)	−21.3 (−6.3)	−2.3 (27.9)	14.5 (58.1)	1.9
1,807 (5,928)	8.0 (46.4)	23.4 (74.1)	−20.2 (−4.4)	−1.3 (29.7)	15.2 (59.4)	2.0
1,699 (5,574)	8.3 (46.9)	23.0 (73.4)	−18.9 (−2.0)	−1.0 (30.2)	15.4 (59.7)	1.9
1,543 (5,062)	9.4 (48.9)	23.8 (74.8)	−17.4 (0.7)	−0.1 (31.8)	16.7 (62.1)	1.9
1,377 (4,518)	10.6 (51.1)	28.0 (82.4)	−17.4 (0.7)	0.8 (33.4)	17.8 (64.0)	2.6
1,217 (3,993)	11.5 (52.7)	28.3 (82.9)	−18.0 (−0.4)	1.4 (34.5)	18.7 (65.7)	2.8
1,024 (3,360)	11.9 (53.4)	29.8 (85.6)	−17.0 (1.4)	2.1 (35.8)	18.8 (65.8)	3.3

Red spruce foliage is only marginally cold-tolerant (e.g., −35 to −50 °C [−31 to −58 °F] in Vermont mid-winter conditions), compared to the sympatric species balsam fir, which can survive freezing in liquid nitrogen after acclimation (DeHayes et al. 1990; Schaberg et al. 2000; Strimbeck et al. 2015). These temperatures may seem unrealistic in the warming central and southern Appalachians, but extensive evaluation of red spruce cold tolerance has only been done on material acclimated to the temperatures of the northern Appalachians, where extreme low temperatures may be 10–20 °C colder than in the southern portion of the range. A red spruce provenance trial planted near Colebrook, New Hampshire that included collections from North Carolina, West Virginia, and Pennsylvania did not find strong relationships between climate of origin and either cold tolerance or winter injury for these southern provenances (DeHayes et al. 1990). Mid-winter cold hardiness of red spruce may be reduced by as much as 14 °C after a thaw of ≥3 days, opening a window for freezing injury (Strimbeck et al. 1995), after which trees may require several weeks to return to pre-thaw cold hardiness. This highlights the tradeoff between the risk of freezing injury versus the benefit of additional carbon gain (Schaberg 2000), especially for productivity of understory red spruce before deciduous trees leaf out in the spring.

A variety of abiotic factors have been shown to affect red spruce physiology in ways that impact susceptibility to cold injury. During the severe red spruce declines in the 1980s, symptoms analogous to winter injury of foliage were commonly observed and related to air pollution and acid rain (Johnson and Siccama 1983; DeHayes et al. 1990; Vann et al. 1992). Low-pH conditions were found to alter the mobility of Al ions in the soil, which affected Ca nutrition in red spruce, leading to winter injury under otherwise nonlethal temperatures (Shortle et al. 1997; DeHayes et al. 1999; Driscoll et al. 2001).

7.2.2.5 Physiological Component of Recent Gains in Productivity

Over the past decade, there have been numerous reports of increases in general health and radial growth, combined with downslope migration of red spruce (Foster and D'Amato 2015; Kosiba et al. 2018; Mathias and Thomas 2018), and the last major winter injury event was in 2003 (Lazarus et al. 2004). Dramatic reductions in atmospheric deposition of nitrate and sulfate associated with red spruce decline have helped ameliorate winter injury, and warmer temperatures outside of the typical growing season have increased photosynthesis and growth (Kosiba et al. 2018). Between 1980 and 2020, the concentration of CO_2 in the atmosphere has increased by more than 20% (National Oceanic and Atmospheric Administration 2023), making photosynthesis more efficient by increasing the differential between atmospheric CO_2 concentration and CO_2 inside the leaf, leading to increased water-use efficiency. Using stable isotopes, Mathias and Thomas (2018) found that red spruce growth more than doubled in the central Appalachians since 1989, an effect they attributed largely to reduced sulfate deposition and increased atmospheric CO_2.

7.2.3 Evidence and Nature of Adaptation to Climate in Red Spruce

7.2.3.1 Local Adaptation to Climate in Common Garden Studies

Local adaptation is the product of an evolutionary process by which natural selection favors individuals with genetically based trait values that maximize fitness (e.g., survival, growth, reproduction) under local environmental conditions. A strict definition of local adaptation posits that locally adapted genotypes have higher fitness in their native environment than genotypes originating from other environments (Kawecki and Ebert 2004). Local adaptation to climate is commonly observed in forest trees (Savolainen et al. 2007), often studied using provenance trials or common gardens, and has been shown to be present in the genus *Picea* in relation to different climatic drivers. For example, white spruce is locally adapted to the frequency of extreme drought events (Depardieu et al. 2021), Sitka spruce (*Picea sitchensis*) and black spruce are locally adapted to the length of the growing season (Mimura and Aitken 2010; Rossi 2015), and Norway spruce (*Picea abies*) shows variation in both height and diameter associated with source-locality genetic make-up (Milesi et al. 2019).

Two recent studies showed that red spruce is no exception to the ubiquity of local adaptation to climate among spruce species (Capblancq et al. 2022; Prakash et al. 2022). Prakash et al. (2022) conducted a common garden study (Fig. 7.4) in which seed was sampled from >300 mother trees representing 65 geographically diverse localities throughout the geographic distribution of red spruce (Fig. 7.5). This sampling captured variation naturally present across three geographically and genetically differentiated ancestry groups: (1) the northern Appalachians, (2) the Allegheny and Pocono plateaus of Pennsylvania, and (3) the central and southern Appalachians (Capblancq et al. 2020a). Seedlings from each family (>5,000 total) were planted into three common gardens located along a latitudinal gradient in Burlington, Vermont; Frostburg, Maryland; and Asheville, North Carolina. By monitoring vegetative phenology (spring bud break and autumn bud set) and height growth over 2 years, Prakash et al. (2022) showed the presence of significant heritable genetic variation in these traits (H^2 from 0.075–0.408), as well as strong plasticity in trait expression across the different common garden sites. Growing plants from different source climates together under common garden conditions revealed significant genetic differentiation in phenology among the ancestry groups, with seedlings from the central and southern Appalachians exhibiting the latest bud break and bud set across all garden sites, while seedlings from the northern ancestry group had the earliest phenology (Prakash et al. 2022). This somewhat counterintuitive result (i.e., southern genotypes with delayed phenology relative to northern genotypes under common garden conditions) is a frequent observation in other tree species and reflects the warmer and more variable winter temperatures commonly encountered in southern regions that select for genotypes that require a large number of growing degree days to accumulate before breaking dormancy (Thibault et al. 2020). Phenology traits also showed a high degree of plasticity in response to climatic differences between the garden sites, occurring earlier in the growing season at the North Carolina garden for both bud break and bud set relative to the two more northern gardens. Interestingly, plasticity in bud break was significantly correlated with climate of origin, with southern seedlings expressing higher bud break plasticity compared to northern seedings (Prakash et al. 2022). Recent analyses have demonstrated that the genetic differentiation among ancestry groups in bud break and bud set phenology, as well as in bud break plasticity, reflect local adaptation of these traits to climate (A. Prakash and S. Keller, unpublished manuscript).

Evidence of locally adaptive bud phenology at a finer spatial scale within the southern Appalachian region comes from a common garden greenhouse study using 32 seed families collected along replicated altitudinal gradients in Great Smoky Mountains National Park, Tennessee and Mt. Mitchell State Park, North Carolina (Butnor et al. 2019). Seedlings from high-elevation sources set bud up to 10 days earlier than low-elevation sources, but no clinal variation was observed for bud break (Butnor et al. 2019). The accelerated bud set of high-elevation sources closely paralleled the earlier bud set observed for high-latitude sources by Prakash et al. (2022),

Fig. 7.4 Red spruce (*Picea rubens*) common garden experiment at the University of Maryland Center for Environmental Science, Appalachian Lab in Frostburg, Maryland. Photo by Matthew Fitzpatrick

confirming that this trait serves as a local adaptation to growing season length in red spruce.

A further demonstration of local adaptation of red spruce populations to their source climates comes from Capblancq et al. (2022). Using the same experimental seedlings reported in Prakash et al. (2022), Capblancq et al. (2022) analyzed height growth over two years of growth in the common gardens and related these to the difference between source climate and the climate of the common garden (a.k.a., climate transfer distance). They showed a highly significant negative correlation between the magnitude of climate transfer distance and seedling height growth in the common gardens over 2 years, thus demonstrating local adaptation in red spruce linked to gradients of temperature and precipitation (Fig. 7.6). Further, because the climate of the garden sites was generally warmer and drier than most of the source climates for the seedlings (especially the North Carolina site), the climate transfer analysis by Capblancq et al. (2022) provides a much-needed experimental assessment of how red spruce will respond to a warming climate, using a space-for-time proxy for climate change. However, this experiment only addressed seedling performance and may or may not be indicative of fitness at later life stages.

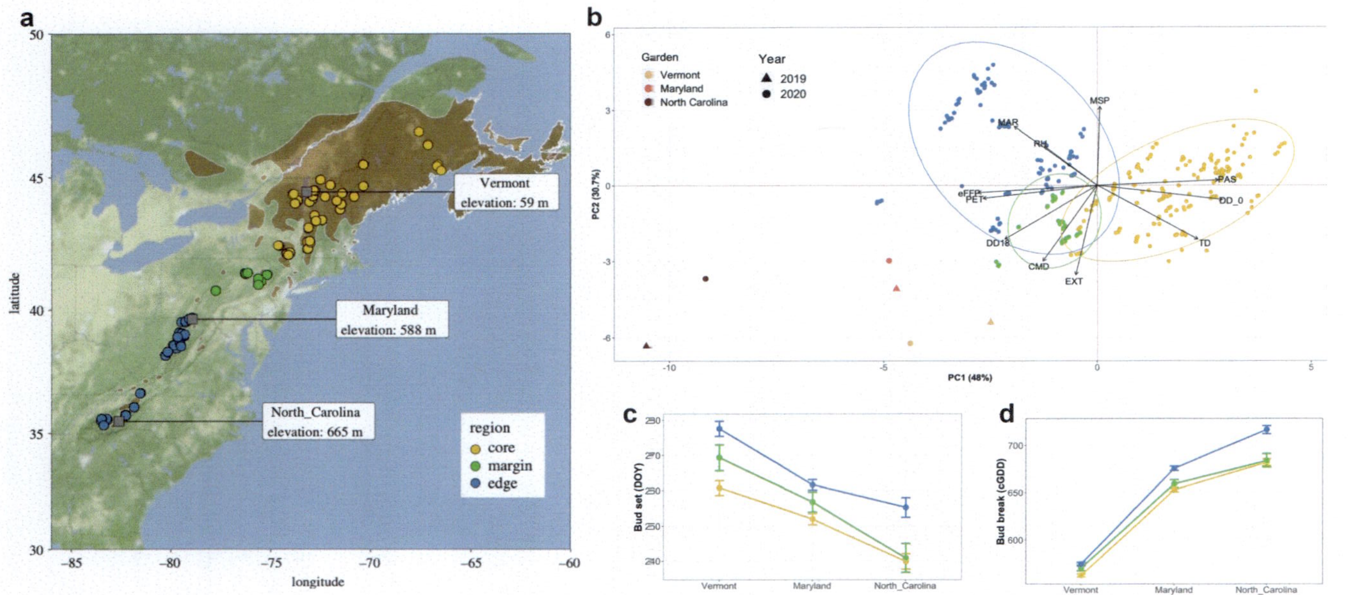

Fig. 7.5 Regional differentiation in quantitative genetic variation in phenology traits along climatic gradients in red spruce (*Picea rubens*). **a** Genetic ancestry groups sampled in the northern Appalachians (yellow, *core*), Allegheny and Pocono plateaus (green, *margin*), and central and southern Appalachians (blue, *edge*) are genetically differentiated at the molecular level (Capblancq et al. 2020a). Seeds from each of these populations were grown as seedlings and planted into three common garden locations (gray squares) in Vermont, Maryland, and North Carolina. **b** The genetic ancestry groups inhabit different climatic zones, based on a principal component analysis of 11 climate variables: mean summer precipitation (MSP), precipitation as snow (PAS), degree-days per year <0 °C (32 °F; DD_0), temperature continentality (TD), extreme summer temperature (EXT), climate moisture deficit (CMD), degree-days per year >18 °C (64 °F; DD_18), potential evapotranspiration (PET), estimated frost-free period (eFFP), mean annual solar radiation (MAR), and mean annual relative humidity (RH). **c** Mean bud set versus day of year (DOY) in 2019 by genetic ancestry group and common garden site. **d** Mean bud break versus cumulative growing degree days >0 °C (32 °F; cGDD) in 2020. Values represent means ± SE. Figures used with permission of The Royal Society (U.K.), from Prakash et al. (2022); permission conveyed through Copyright Clearance Center, Inc.

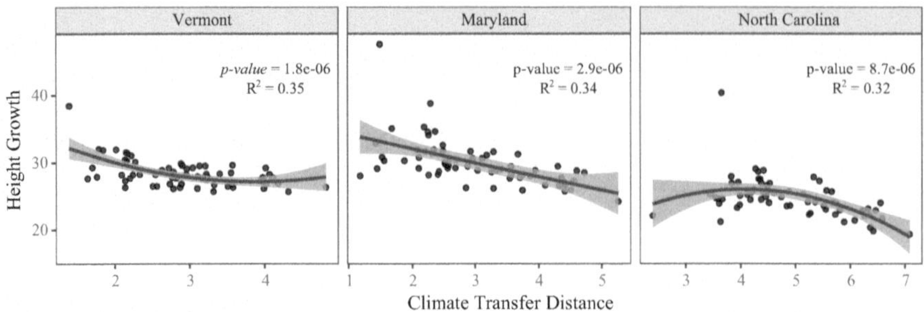

Fig. 7.6 Association between climate transfer distance based on a principal component analysis of 11 climate variables (see Fig. 7.5) and red spruce (*Picea rubens*) seedling fitness represented by mean height growth per source locality at the three common garden sites. The significance of the association was tested in each garden using a quadratic regression; the resulting p-values for the quadratic term and R^2 are shown on each panel. Lines show model predictions and the gray areas show the 95% confidence intervals. Figure used with permission of John Wiley & Sons, from Capblancq et al. (2022); permission conveyed through Copyright Clearance Center, Inc.

Other common garden studies with evidence of genetically based climate adaptation come from red spruce provenance trials established by foresters in the mid-twentieth century, which are now mature enough to provide informative data on adult tree responses. These trials were located in the northern part of the range (Morgenstern et al. 1981; Wilkinson 1990), but since many incorporated at least a few central or southern Appalachian seed sources, they provide valuable evidence of genetic differentiation in quantitative traits between southern and northern populations and evidence for local adaptation to their respective climates. For example, Li et al. (2020) used height and diameter measurements on mature trees established in six historical provenance trials in eastern Canada's Acadian forests to evaluate if red spruce is likely to persist under climate change in this region. The trials were established in the early 1960s in New Brunswick, Nova Scotia, Prince Edward Island, and Newfoundland by the Canadian Forest Service and its collaborators, and each included trees from the same 30 provenances. Except for the two provenances from West Virginia, the locations of origin were all located near the northern range edge or fell within the northern range core in New Brunswick, Nova Scotia, and Maine. Remeasurements from 2013 revealed that temperature-related climate variables played a greater role than precipitation in the growth response of red spruce to climate transfer distance between provenance origin and trial site (Li et al. 2020). Interestingly, under these northern climates, trees showed a significant positive growth response to climate transfer into warmer temperatures, in particular to increases in the length of the growing season, although climate transfers into shorter growing seasons decreased tree size, likely as a result of frost damage (Li et al. 2020). Trees from provenances with a shorter growing season were smaller in size and benefited more from warming than southern provenances, which in turn reacted negatively to cooling.

The findings of Li et al. (2020) indicate that red spruce is cold-limited at its northern range edge, and might at least temporarily benefit from climate warming in these regions, and imply that northward assisted migration of southern populations might help mitigate climate change effects. The positive effect of mean annual temperatures did not peak, probably due to the fact that none of the experimental sites had a warmer climate than the southernmost West Virginia provenances. Mean annual precipitation had no significant effect, and the positive effects of increased moisture availability (measured, for example, as climatic moisture deficit) were weak but increased the explanatory power of the temperature-related variables. Due to the location of the experiments in the Canadian Maritime provinces, trees received abundant precipitation and were likely not water-limited. In conclusion, Li et al. (2020) provided evidence of the importance of growing season length and other temperature-related climatic conditions for red spruce growth and highlighted the role that phenotypic plasticity and local adaptation may play in its climate change response. However, the study design, which included a limited number of southern provenances and no trial sites in the central or southern Appalachians, limits our ability to draw conclusions regarding how climate warming or increased drought might affect growth of adult red spruce in these regions.

7.2.3.2 Gene-Environment Association Studies Reveal Genes Under Climatic Selection

Local adaptation in red spruce has also been investigated using genomics, which allowed the identification of genes linked to climatically driven divergent selection and revealed the distribution of adaptive genetic variation across the species' range (Fig. 7.7). Three main interconnected axes of adaptation to climate are described for temperate and boreal coniferous species: drought tolerance (Moran et al. 2017), cold hardiness (Howe et al. 2003; Chang et al. 2021), and phenological timing of growth and dormancy (Gyllenstrand et al. 2007). In red spruce, genes involved in all three adaptive pathways were found to be under selection using genome scan methods (Capblancq et al. 2022). The identified candidates included, for example, the FPA gene which regulates the phenology of flowering time in *Arabidopsis thaliana* (Schomburg et al. 2001) and could control the same process in red spruce in response to temperature variation across the latitudinal range or along altitudinal gradients. Various genes associated with drought and heat tolerance were also identified as important potentially adaptive genes in red spruce, especially including genes involved in the abscisic acid pathway. Abscisic acid is known to initiate responses to drought stress (Hamanishi and Campbell 2011), with high abscisic acid concentration acting as a signal for plants to close their stomata and prevent water loss (Moran et al. 2017). Interestingly, some drought resistance genes were also shown associated with pathogen resistance in Norway spruce (Capador-Barreto et al. 2021) and in red spruce, with the gene At4g33300 identified as one of the most important candidate genes.

The geographic distribution of candidate adaptive alleles is particularly interesting in red spruce. Climatically driven divergent selection has resulted in a clear differentiation between southern and northern ancestry groups (Fig. 7.8). Climate-associated

Fig. 7.7 Red spruce (*Picea rubens*) cuttings collected for DNA extraction and sequencing for use in genomic analyses (photo by Matthew Fitzpatrick)

alleles, which make up just a small proportion of the overall diversity within the genome, show strong clinal differentiation of central and southern Appalachian populations compared to populations in the northeastern part of the range. The climate-adaptive alleles contained within the central and southern Appalachian gene pool make this region very important for future conservation and management plans, particularly in the context of assisted migration (Aitken and Bemmels 2016). The presence of warm-adapted alleles in the southern portion of red spruce's range suggests this region could be an important source of seed for assisted migration strategies and highlights the importance of preserving in situ populations that contain unique functional but also genetic diversity.

In addition to adaptive genetic divergence, there is a clear difference in the neutral genetic composition of the central and southern Appalachians from the fragmented

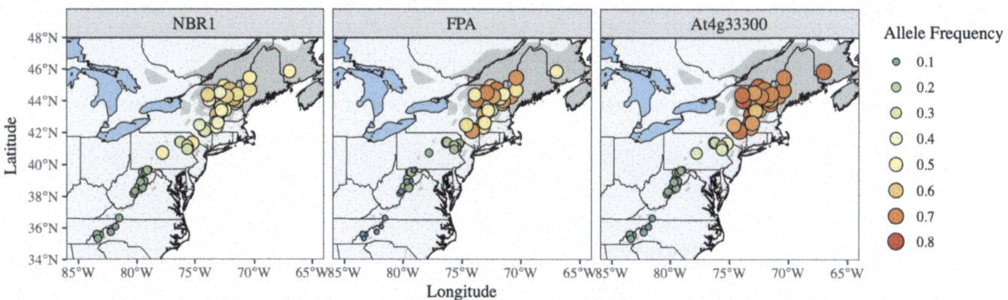

Fig. 7.8 Among-population variation in red spruce (*Picea rubens*) climate-adaptive alleles for three candidate genes identified through gene-environment association analysis. Figure used with permission of John Wiley & Sons, from Capblancq et al. (2022); permission conveyed through Copyright Clearance Center, Inc.

midlatitude region in Pennsylvania and the northeastern region in New England and Canada (Capblancq et al. 2020a). While the estimated genetic difference is relatively low between these regional populations ($F_{ST} < 0.03$), it points to a lack of gene flow along the latitudinal gradient and thus a poor natural dispersal capacity.

In conclusion, recent work on the genetics of climate-adaptive variation has shown that red spruce forests of the central and southern Appalachians are an important pool of adaptive genetic diversity that could become a key component of future genetically informed management programs. However, these forests are already adapted to the warmest portion of the species' climatic niche, which may present constraints to adaptation under warmer future climates. Maintaining high connectivity and gene flow among forest patches will likely be an important consideration for facilitating red spruce persistence in the southern part of its range.

7.3 Adaptation and Management of Red Spruce in a Dynamic Climate

7.3.1 Future Forecasts: What Do Models Suggest About the Future of Red Spruce?

While not without shortcomings, forecasting models offer one of the best methods available to anticipate what the future might hold for natural systems. To date, numerous studies have used models to forecast how red spruce could be impacted by expected changes in climate over the next several decades. While most studies have focused on modeling red spruce's response at the species level, recent studies have begun to consider local adaptation to climate and population-level responses (e.g., Lachmuth et al. 2024). Despite differences in techniques and data, these studies generally depict a gloomy future for red spruce in the central and southern Appalachians. However, recent studies incorporating genomic data and local adaptation to climate suggest that some populations could persist through this century. In this section, we summarize studies that have used models to forecast the potential impacts of climate on red spruce and what these studies suggest in light of important uncertainties, caveats, and assumptions.

7.3.1.1 Species-Level Models

Numerous studies have used various implementations of ENMs to forecast changes in the potential distribution of red spruce in response to climate change. ENMs are empirical models that combine observations of species occurrence (e.g., presence-absence records) with abiotic covariates that describe important habitat characteristics. Habitat predictors typically include measures of precipitation and temperature and, occasionally, soils or other abiotic correlates (Elith and Leathwick 2009). Once fitted, these models can be forecasted to scenarios of future climate to estimate potential changes in distribution, abundance, or habitat suitability (Guisan and Thuiller 2005).

To describe future climate conditions, modeling studies use forecasts from general circulation models. These forecasts, typically generated as part of assessment reports from the Intergovernmental Panel on Climate Change, span a range of possible emission scenarios and different periods in the future. Best modeling practices dictate that studies should consider climate forecasts from multiple general circulation models (to capture variation between models) and emission scenarios (to explore how different assumptions about future emissions impact forecasts). The typical periods to which models are forecasted are mid-century (2050) and end-of-century (2080). Studies therefore often report results for multiple models and time periods.

Model forecasts based on the worst-case emission scenarios typically project near-complete loss of red spruce habitat from the southern Appalachians by the end of the twenty-first century (Iverson et al. 2008; Stanton 2009; Beane and Rentch 2015; Walter et al. 2017; Yetter et al. 2021). Under the best-case emission scenarios, models generally also forecast overall declines in habitat suitability in the central and southern Appalachians, but the magnitude and extent of these declines tend to be not as severe (e.g., Mosher 2020).

Regions where models suggest red spruce could maintain suitable habitats, especially into the late twenty-first century and regardless of the severity of expected climate change, represent potential climate refugia and areas to focus restoration efforts. For the southern Appalachians, potential refugial locations include Big Bald, Beech Mountain, Plott Balsam Mountain, Sugar Mountain, and Elk Knob (Mosher 2020). For West Virginia, refugial areas reside primarily in Pocahontas and parts of Randolph and Pendleton counties, which contain the Monongahela NF (Beane and Rentch 2015). The broad agreement in forecasted declines in habitat suitability across studies that used different methods, general circulation models, and emission scenarios suggests conservation and restoration efforts in the region will be challenging under ongoing climate change (Walter et al. 2017).

Given their relative simplicity, computational efficiency, and modest data requirements, ENMs are the class of models most often used to forecast potential changes in species distributions in response to climate change. However, these correlative models often lack incorporation of important biological processes such as plasticity, evolution, and dispersal. Mechanistic models offer an alternative to purely empirical ENMs that can incorporate aspects of biological processes into forecasts and a few such models have been developed for red spruce. For example, Rentch et al. (2007) used the NE-TWIGS growth model and the Forest Vegetation Simulator to forecast changes in red spruce abundance in 10-year intervals from 2005 to 2106. The simulations showed initial, early increases in abundance after 10–20 years, followed by declines starting around 2050, and that areas above 1,035 m (3,395 ft) supported the highest total basal area. Prasad et al. (2020) combined predictions from ENMs with a dispersal simulation in an effort to forecast actual changes in occurrence of 25 eastern North American tree species rather than simply potential shifts in habitat suitability. The study provided summaries of the amount of habitat expected to be lost and gained in the U.S. and Canada. Even under moderate emissions, Prasad et al. (2020) reported that red spruce would colonize only 36% of newly suitable habitats, nearly all of which were in Canada, while red spruce habitat would be lost in most other regions. Koo et al. (2011a, 2011b) applied a mechanistic model for red spruce habitat in a series of papers. For example, Koo et al. (2011b) generated submodels that fed into a primary model in order to fully describe the distribution of red spruce based on available and plausible data. This method aimed to assess both direct and indirect, and within- and across-scale, interactions to more accurately portray red spruce habitat. The simulations suggested that at lower elevations (1,450–1,700 m [4,757–5,577 ft]) red spruce exhibits more complex interactions with the environment. Empirical studies by the same group were then able to connect temperature, precipitation, and pollution with tree-ring growth (Koo et al. 2011a). The combination of increased acid rain, temperature, and cloud frequencies negatively impacted suitability the most at higher elevations (Koo et al. 2014a, b). Pollution had an overall negative effect on red spruce regardless of elevation, but the magnitude of impact was less at higher elevations and importantly had more of an impact than did climate change.

7.3.1.2 Population-Level

Species-level ENMs do not take intra-specific differentiation into account and therefore do not consider how local adaptation to climate will influence how different populations could respond to climate change (Aitken et al. 2008). As a promising tool for closing this research gap, the concept of genomic offset (Fitzpatrick and Keller 2015; Capblancq et al. 2020b) allows the incorporation of genomic information into climate change impact assessments and forecasts of climate change responses at the level of populations. Based on model predictions of genomic turnover along environmental gradients, genomic offsets quantify the expected disruption of existing genotype-environment associations (i.e., the degree of maladaptation populations will likely experience under future environmental conditions; Capblancq et al. 2020b). As such, estimates of the genomic offset a population may experience in situ if the local environment changes (local offset, sensu Gougherty et al. 2021) provide a valuable predictive tool for risk assessment at the population level (Rellstab et al. 2021), and have been applied to a variety of taxa (Dauphin et al. 2020; Gougherty et al. 2021; Nielsen et al. 2021).

For red spruce, Lachmuth et al. (2024) used genomic offset analyses in combination with habitat suitability predictions from classical ENMs. Importantly, the offset models incorporated genomic markers that were linked to genetic loci likely involved in climate adaptation (Capblancq et al. 2022), with common garden experiments being used to validate whether genomic offsets based on these loci had a significant negative effect on growth performance (see also Prakash et al. 2022). The forecasted genomic offsets and changes in habitat suitability from ENMs were mostly driven by changes in temperature-related climatic variables and predicted drastic shifts of suitable habitat to areas north of the current range (Fig. 7.9). However, these projected northward shifts in habitat were driven primarily by populations found in the northern portion of red spruce's range and in the isolated margin populations in Pennsylvania. In contrast to many existing forecasts based on ENMs that did not consider local adaptation, genomic offsets suggested that populations in the central and southern Appalachians might be able to persist in situ owing to specific local climate adaptations, at least under moderate climate change (Lachmuth et al. 2024). Such strong local adaptation is common in both climatically marginal populations that experience strong selection pressure as well as at the equatorial edges of species ranges, where local adaptation has had a longer time to develop (Bontrager et al. 2021).

7.3.1.3 Interpreting Model Forecasts

Ecological niche models require a number of simplifying assumptions and are sensitive to the species occurrence and environmental data used in model fitting. For these reasons, ENM-based forecasts should be interpreted with caution. Perhaps most relevant to modeling the future of red spruce in the central and southern Appalachians is that many of the stands occupying the lowest elevations, and therefore the warmest habitats, have been extirpated by logging and land-use change (Adams and Stephenson 1989). For this reason, ENMs may overestimate the extent of future habitat loss since the data available for model fitting are likely skewed somewhat towards occurrences in high-elevation, cooler and more mesic habitats. In addition, ENMs typically use relatively coarse-grained climate data that do not characterize microclimates that could allow persistence in areas that otherwise become unsuitable.

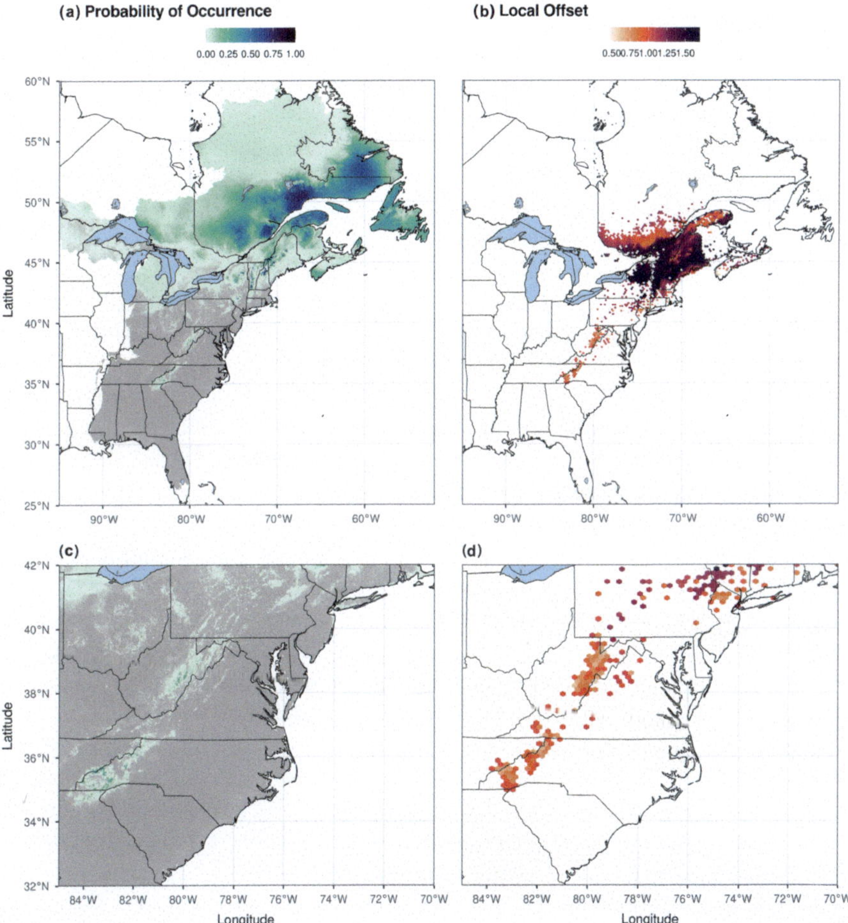

Fig. 7.9 Species versus population-level climate change impact forecasts for red spruce (*Picea rubens*) by the end of the twenty-first century under moderate greenhouse gas emissions (Shared Socio-economic Pathway SSP2-45; Masson-Delmotte et al. 2021). **a** Projection of red spruce occurrence probability (i.e., climate suitability) as predicted by an ensemble of ENMs. Dark gray corresponds to near-zero climate suitability. **b** Genomic offsets of grid cells with contemporary red spruce occurrence (hex map for increased readability). Offsets estimate the disruption of existing genotype-environment associations red spruce populations may experience under climate change. Higher offset values reflect a higher degree of expected mal-adaptation and thus climate change vulnerability. Panels **c** and **d** correspond to those of **a** and **b**, but cropped to the central and southern Appalachians to show greater detail

7.3.2 *Importance of Conserving Genetic Diversity in Situ and in Restoration Strategies for Adaptability to Future Climate Change*

Red spruce genetic variation provides the essential basis for adaptation and resilience to environmental stress and change (Potter et al. 2017), conveying adaptive capacity for future response, the loss of which can be detrimental to the ability of a tree species to survive (Schaberg et al. 2008; Jump et al. 2009). The findings of Prakash et al. (2022) of low to moderate genetic variation and geographic differentiation for phenology and growth traits (Fig. 7.5) suggest that some potential exists to respond to selection and that future climate adaptation will depend on the availability of regional genetic variation in the species. At the same time, the three distinct red spruce ancestry groups, which are associated with different geographic regions of its distribution, exhibit low effective population sizes associated with a sustained population decline across the range of the species (Capblancq et al. 2020a). Lower genetic diversity and higher inbreeding in the fragmented central and southern Appalachian populations of red spruce have been associated with lower germination rate, shorter height, and reduced seedling fitness (Capblancq et al. 2021).

In situ (on site) conservation for tree species such as red spruce is generally considered more efficient for preserving genetic diversity than are ex situ (off site) methods, encompassing the identification of species and areas to protect, the actual protection of an appropriate area, the monitoring of the species or population, and the preparation of management plans (Food and Agriculture Organization of the United Nations 2014). Importantly, protecting or restoring populations in situ is a more evolutionarily dynamic approach because the species or population of interest can maintain its full range of evolutionary and ecological functions and processes while remaining more likely to track the environmental changes to which it must adapt (Rajora and Mosseler 2001). In the medium- to long-term, the restoration of red spruce forests may be necessary in response to past anthropogenic disturbances (e.g., logging, fire, atmospheric pollution), as well as the future potential for widespread mortality resulting from climate change or an insect or disease infestation. Given that red spruce may be vulnerable to as-yet-unknown exotic pests and pathogens, it may be advisable to conserve germplasm that is broadly representative of locally adapted genotypes (Potter et al. 2019). That germplasm could then be used to restore or augment degraded populations in parts of the red spruce distribution. Forest restoration is a complex and relatively young discipline, however (Jacobs et al. 2015).

While forest restoration efforts may successfully conserve the genetic resources of a target tree species, there is also a risk that they could fail. It is important, therefore, that the collection and propagation of germplasm ensure a broad genetic base of restored tree populations, that genetic material is matched to restoration sites where it is likely to be adapted based on current and future site conditions, and that restoration objectives include landscape-level planning (Thomas et al. 2014; Prakash et al. 2024). Specifically, it is critical that planting material represents a minimum level of intraspecific diversity to ensure that the restored plant species will be able to survive and produce viable offspring (Thomas et al. 2014). Such a focus on capturing and even enhancing regional-scale genetic diversity in restoration plantings has become a recent focus in the central Appalachians and should be considered a critical strategy for in situ conservation alongside optimizing local adaptation under current and future climates (Prakash et al. 2024). This maximization of genetic diversity during in situ restoration is especially important for red spruce, which exhibits overall low genetic diversity compared to other similar tree species (e.g., wind-pollinated conifers), likely due to its demographic history of long-term population decline (Capblancq et al. 2020a).

7.3.3 The Potential for Assisted Migration / Gene Flow to Manage Red Spruce Vulnerability to Climate Change

In the past, conservation and reforestation have largely relied on local seed sources following a "local is best" paradigm (O'Neill and Gómez-Pineda 2021). However, it has also become increasingly clear that forest trees, which are characterized by long generation times and limited dispersal, will not be able to realize rapid in situ evolutionary adaptation or track favorable climates by dispersing to new locations given the velocity of the predicted climatic changes (Aitken et al. 2008; Dauphin et al. 2020). This time delay in natural response is expected to result in increased susceptibility to pest infestations, lower productivity, and reduced carbon sequestration (O'Neill and Gómez-Pineda 2021) and will also limit the prospects of in situ management (Dumroese et al. 2015). Considering these threats, assisted migration—an approach that was long mostly considered for rescuing rare and declining species from the brink of extinction (Hoegh-Guldberg et al. 2008)—has received increasing attention in forestry research (Castellanos-Acuña et al. 2018; Benomar et al. 2022) and practice (Park et al. 2018; O'Neill and Gómez-Pineda 2021). The measures summarized as *assisted migration* comprise both assisted gene flow within and assisted colonization beyond the current geographic distribution of a species (Aitken and Whitlock 2013; Aitken and Bemmels 2016). The potential goals of management through assisted migration range widely from re-introductions of species in parts of their historical ranges; reinforcement (building up an existing population), which may or may not include genetic material better adapted to future climatic conditions; assisted range expansion; and assisted long-distance migration (Dumroese et al. 2015).

Given the projected decline in habitat suitability throughout its current range and the projected northward shift of suitable habitats, red spruce could benefit from proactive measures. More than a decade ago, Byers and Norris (2011) suggested in their climate change vulnerability assessment that innovative and unconventional measures such as assisted migration should at least be discussed for highly vulnerable species, including red spruce. Iverson et al. (2019) combined habitat suitability models, migration simulation models, and analyses of biological species attributes and disturbance factors to assess the potential of 125 tree species in the eastern U.S. to migrate or infill naturally into suitable habitats over the next 100 years, with explicit reference to assisted migration planning. Due to its presumed low adaptability, the estimated capability of red spruce to cope with climate change across its U.S. range was estimated as "poor" (Massachusetts, New Hampshire [under RCP8.5], New York [under RCP4.5], Vermont, West Virginia) or "very poor" (Connecticut, North Carolina, New York [under RCP8.5], Pennsylvania, Tennessee, Virginia). Only Maine and New Hampshire (under RCP4.5) reached a ranking of "fair." The only states for which the study saw a potential of filling additional, currently unoccupied habitat were Maine, New York (under RCP4.5) and West Virginia (Fig. 7.10). Georgia, Maryland, and Wisconsin were considered currently unoccupied and of unknown potential. Estimates of current and potential future habitat, coping capability, and migration summaries can be downloaded for a variety of geographies from the USDA Forest Service Climate Change Atlas (https://www.fs.usda.gov/nrs/atlas/), providing a valuable resource for assisted migration planning. However, the Iverson et al. (2019) assessment was based on classical species-level ENMs and did not take local adaptation to climate into account. Just as with in situ risk assessments, we can use genomic offset metrics to leverage the informational value of landscape genomics and select source populations that are likely best adapted to future conditions in a potential recipient area, or conversely, sites where seed from an existing population are likely to thrive in the future (Lachmuth et al. 2024). In addition, landscape genomic analyses allow for identification of the underlying climate adaptations and genetic markers responsible for the matches between donor populations and recipient sites.

For red spruce, landscape genomic analyses by Lachmuth et al. (2024) suggest that both assisted gene flow within the current range and assisted colonization beyond the current northern range limit become necessary to protect substantial parts of the

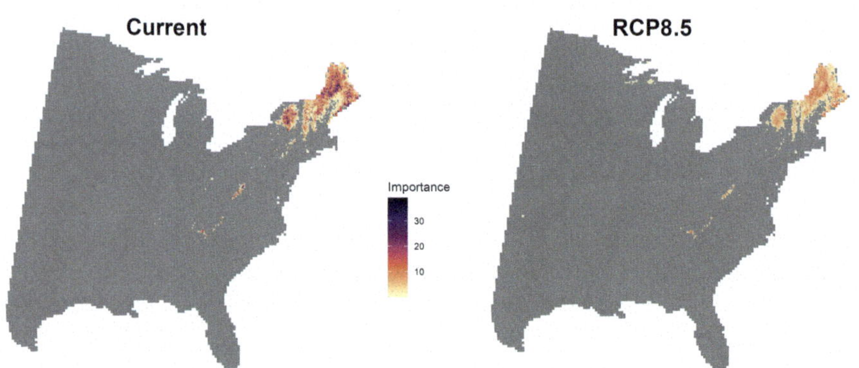

Fig. 7.10 Current predicted (left) and forecasted (right) importance values (a measure of abundance) for red spruce (*Picea rubens*). The forecasted importance values represent a combination among three general circulation models and assuming high emissions (RCP8.5) for the period 2070–2099. Original maps made using data from Peters et al. (2020)

gene pool under climate change. For central and southern Appalachian populations, the analyses confirmed the potential for in situ persistence and potential reforestation or augmentation of existing populations with local seed sources from populations that harbor genotypes adapted to high potential evapotranspiration (PET). The findings further suggest that southern edge populations with adaptations to comparably low PET could serve as donors for assisted migration into the Adirondacks, the Green and White Mountains, and the Canadian Maritime provinces. Moreover, the Maritime provinces may serve as refuges for more southerly populations, including those currently located in Pennsylvania, with adaptations to comparably low chilling degree-days and low precipitation as snow as well as adaptations to low PET and low continentality. The most drastic northward shift of suitable habitat into Ontario and Québec is predicted for red spruce populations with adaptations to high chilling degree-days and high precipitation as snow as well as low PET and high continentality, currently occurring in the northern core of the range. The findings suggest that conserving the genetic resources harbored by these populations would require assisted long-distance migration into a different country and into habitat not currently occupied by the species. This scenario is certainly the most complex case for intervention through assisted migration and requires careful consideration of both risks and benefits, while considering the interests and values of diverse stakeholders (Hagerman and Kozak 2021). Moreover, variability among different climate change emissions scenarios and choice of climatic predictors in modeling add uncertainty to the forecasts, particularly for the southern and central Appalachians, which need to be considered in conservation planning (Lachmuth et al. 2023).

References

Adams HS, Stephenson SL (1989) Old-growth red spruce communities in the mid-Appalachians. Vegetatio 85:45–56

Aitken SN, Bemmels JB (2016) Time to get moving: assisted gene flow of forest trees. Evol Appl 9:271–290

Aitken SN, Whitlock MC (2013) Assisted gene flow to facilitate local adaptation to climate change. Annu Rev Ecol Evol Syst 44:367–388

Aitken SN, Yeaman S, Holliday JA et al (2008) Adaptation, migration or extirpation: climate change outcomes for tree populations. Evol Appl 1:95–111

Alexander JD, Donnelly JR, Shane JB (1995) Photosynthetic and transpirational responses of red spruce understory trees to light and temperature. Tree Physiol 15:393–398

Beane NR, Rentch JS (2015) Using known occurrences to model suitable habitat for a rare forest type in West Virginia under select climate change scenarios. Ecol Restor 33:178–189

Beckage B, Osborne B, Gavin DG et al (2008) A rapid upward shift of a forest ecotone during 40 years of warming in the Green Mountains of Vermont. Proc Natl Acad Sci 105:4197–4202

Benomar L, Elferjani R, Hamilton J et al (2022) Bibliometric analysis of the structure and evolution of research on assisted migration. Current Forestry Reports 8:199–213

Berry ZC, Emery NC, Gotsch SG et al (2019) Foliar water uptake: processes, pathways, and integration into plant water budgets. Plant Cell Environ 42:410–423

Bigras FJ, Colombo SJ (2013) Conifer cold hardiness. Springer, Dordrecht, Netherlands

Blum BM (1990) Red spruce *Picea rubens* Sarg. In: Burns RM, Honkala BH (eds) Silvics of North America. Volume 1. Conifers. USDA Forest Service Handbook 654, Washington, DC, pp 250–259

Bontrager M, Usui T, Lee-Yaw JA et al (2021) Adaptation across geographic ranges is consistent with strong selection in marginal climates and legacies of range expansion. Evolution 75:1316–1333

Butnor JR, Verrico BM, Johnsen KH et al (2019) Phenotypic variation in climate-associated traits of red spruce (*Picea Rubens* Sarg.) along elevation gradients in the southern Appalachian Mountains. Castanea 84:128–143

Byers EA, Vanderhorst JP, Streets BP (2010) Classification and conservation assessment of upland red spruce communities in West Virginia. West Virginia Division of Natural Resources Report, Elkins, West Virginia

Byers E, Norris S (2011) Climate change vulnerability assessment of species of concern in West Virginia. West Virginia Division of Natural Resources Report, Elkins, West Virginia

Capador-Barreto HD, Bernhardsson C, Milesi P et al (2021) Killing two enemies with one stone? Genomics of resistance to two sympatric pathogens in Norway spruce. Mol Ecol 30:4433–4447

Capblancq T, Butnor JR, Deyoung S et al (2020a) Whole-exome sequencing reveals a long-term decline in effective population size of red spruce (*Picea rubens*). Evol Appl 13:2190–2205

Capblancq T, Fitzpatrick MC, Bay RA et al (2020b) Genomic prediction of (mal)adaptation across current and future climatic landscapes. Annu Rev Ecol Evol Syst 51:245–269

Capblancq T, Munson H, Butnor JR et al (2021) Genomic drivers of early-life fitness in *Picea rubens*. Conserv Genet 22:963–976

Capblancq T, Lachmuth S, Fitzpatrick MC et al (2022) From common gardens to candidate genes: exploring local adaptation to climate in red spruce. New Phytol 237:1590–1605

Castellanos-Acuña D, Vance-Borland KW, St. Clair JB et al (2018) Climate-based seed zones for Mexico: guiding reforestation under observed and projected climate change. New Forests 49:297–309

Chang CY, Bräutigam K, Hüner NPA et al (2021) Champions of winter survival: cold acclimation and molecular regulation of cold hardiness in evergreen conifers. New Phytol 229:675–691

Cogbill CV, White PS (1991) The latitude-elevation relationship for spruce-fir forest and treeline along the Appalachian Mountain chain. Vegetatio 94:153–175

Dauphin B, Rellstab C, Schmid M et al (2020) Genomic vulnerability to rapid climate warming in a tree species with a long generation time. Glob Change Biol 27:1181–1195

Davis MB (1981) Quaternary history and the stability of forest communities. In: West DC, Shugart HH, Botkin DB (eds) Forest succession. Springer-Verlag, New York, pp 132–153

Davis RB, Jacobson GL (1985) Late glacial and early Holocene landscapes in northern New England and adjacent areas of Canada. Quatern Res 23:341–368

Day ME (2000) Influence of temperature and leaf-to-air vapor pressure deficit on net photosynthesis and stomatal conductance in red spruce (*Picea Rubens*). Tree Physiol 20:57–63

DeHayes DH, Waite CE, Ingle MA et al (1990) Winter injury susceptibility and cold tolerance of current and year-old needles of red spruce trees from several provenances. Forest Sci 36:982–994

DeHayes DH, Schaberg PG, Hawley GJ et al (1999) Acid rain impacts on calcium nutrition and forest health: alteration of membrane-associated calcium leads to membrane destabilization and foliar injury in red spruce. Bioscience 49:789–800

Delcourt HR, Delcourt PA (1988) Quaternary landscape ecology: relevant scales in space and time. Landsc Ecol 2:23–44

Depardieu C, Gérardi S, Nadeau S et al (2021) Connecting tree-ring phenotypes, genetic associations and transcriptomics to decipher the genomic architecture of drought adaptation in a widespread conifer. Mol Ecol 30:3898–3917

Dourson D, West Virginia Division of Natural Resources (2015) Land Snails of West Virginia. Goatslug Publications, Bakersville, West Virginia

Driscoll CT, Lawrence GB, Bulger AJ et al (2001) Acidic deposition in the northeastern United States: sources and inputs, ecosystem effects, and management strategies. Bioscience 51:180–198

Dumais D, Prévost M (2007) Management for red spruce conservation in Québec: the importance of some physiological and ecological characteristics—A review. Forestry Chron 83:378–392
Dumroese RK, Williams MI, Stanturf JA et al (2015) Considerations for restoring temperate forests of tomorrow: forest restoration, assisted migration, and bioengineering. New Forest 46:947–964
Elith J, Leathwick JR (2009) Species distribution models: ecological explanation and prediction across space and time. Annu Rev Ecol Evol Syst 40:677–697
Fitzpatrick MC, Keller SR (2015) Ecological genomics meets community-level modelling of biodiversity: mapping the genomic landscape of current and future environmental adaptation. Ecol Lett 18:1–16
Foden WB, Young BE, Akçakaya HR et al (2019) Climate change vulnerability assessment of species. Wiley Interdiscip Rev Climate Change 10:e551
Food and Agriculture Organization of the United Nations (2014) The state of the World's forest genetic resources. Commission on Genetic Resources for Food and Agriculture Report, Rome, Italy
Ford WM, Diggins CA, De La Cruz JL et al (2022) Distribution probability of the Virginia northern flying squirrel in the high Allegheny Mountains. J Southeastern Assoc Fish Wildlife Agencies 9:168–175
Foster JR, D'Amato AW (2015) Montane forest ecotones moved downslope in northeastern USA in spite of warming between 1984 and 2011. Glob Change Biol 21:4497–4507
Gougherty AV, Keller SR, Fitzpatrick MC (2021) Maladaptation, migration and extirpation fuel climate change risk in a forest tree species. Nat Clim Chang 11:166–171
Gray AN, Spies TA (1997) Microsite controls on tree seedling establishment in conifer forest canopy gaps. Ecology 78:2458–2473
Guisan A, Thuiller W (2005) Predicting species distribution: offering more than simple habitat models. Ecol Lett 8:993–1009
Gyllenstrand N, Clapham D, Källman T et al (2007) A Norway spruce flowering locus T homolog is implicated in control of growth rhythm in conifers. Plant Physiol 144:248–257
Hagerman S, Kozak R (2021) Disentangling the social complexities of assisted migration through deliberative methods. J Ecol 109:2309–2316
Hamanishi ET, Campbell MM (2011) Genome-wide responses to drought in forest trees. Forestry 84:273–283
Hamburg SP, Cogbill CV (1988) Historical decline of red spruce populations and climatic warming. Nature 331:428–431
Hoegh-Guldberg O, Hughes L, McIntyre S et al (2008) Assisted colonization and rapid climate change. Science 321:345–346
Howe GT, Aitken SN, Neale DB et al (2003) From genotype to phenotype: unraveling the complexities of cold adaptation in forest trees. Can J Bot 81:1247–1266
Iverson L, Prasad A, Matthews S (2008) Modeling potential climate change impacts on the trees of the northeastern United States. Mitig Adapt Strat Glob Change 13:487–516
Iverson LR, Matthews SN, Prasad AM et al (2012) Development of risk matrices for evaluating climatic change responses of forested habitats. Clim Change 114:231–243
Iverson LR, Prasad AM, Peters MP et al (2019) Facilitating adaptive forest management under climate change: a spatially specific synthesis of 125 species for habitat changes and assisted migration over the eastern United States. Forests 10:989
Jackson ST, Overpeck JT (2000) Responses of plant populations and communities to environmental changes of the late quaternary. Paleobiology 26:194–220
Jackson ST, Webb RS, Anderson KH et al (2000) Vegetation and environment in eastern North America during the last glacial maximum. Quatern Sci Rev 19:489–508
Jacobs DF, Oliet JA, Aronson J et al (2015) Restoring forests: what constitutes success in the twenty-first century? New Forest 46:601–614
Johnson AH, Siccama TG (1983) Acid deposition and forest decline. Environ Sci Technol 17:294A-305A

Jump AS, Marchant R, Peñuelas J (2009) Environmental change and the option value of genetic diversity. Trends Plant Sci 14:51–58

Kalberer SR, Wisniewski M, Arora R (2006) Deacclimation and reacclimation of cold-hardy plants: current understanding and emerging concepts. Plant Sci 171:3–16

Kawecki TJ, Ebert D (2004) Conceptual issues in local adaptation. Ecol Lett 7:1225–1241

King PB, Stupka A (1950) The Great Smoky Mountains–their geology and natural history. Sci Monthly 71:31–43

Koo K-A, Patten BC, Creed IF (2011a) *Picea Rubens* growth at high versus low elevations in the Great Smoky Mountains National Park: evaluation by systems modeling. Can J for Res 41:945–962

Koo K-A, Patten BC, Teskey RO (2011b) Assessing environmental factors in red spruce (*Picea Rubens* Sarg.) growth in the Great Smoky Mountains National Park, USA: from conceptual model, envirogram, to simulation model. Ecol Model 222:824–834

Koo KA, Madden M, Patten BC (2014a) Projection of red spruce (*Picea Rubens* Sargent) habitat suitability and distribution in the southern Appalachian Mountains, USA. Ecol Model 293:91–101

Koo KA, Patten BC, Teskey RO et al (2014b) Climate change effects on red spruce decline mitigated by reduction in air pollution within its shrinking habitat range. Ecol Model 293:81–90

Kosiba AM, Schaberg PG, Rayback SA et al (2018) The surprising recovery of red spruce growth shows links to decreased acid deposition and elevated temperature. Sci Total Environ 637–638:1480–1491

Lachmuth S, Capblancq T, Keller SR et al (2023) Assessing uncertainty in genomic offset forecasts from landscape genomic models (and implications for restoration and assisted migration). Front Ecol Evol 11:1155783

Lachmuth S, Capblancq T, Prakash A et al (2024) Novel genomic offset metrics integrate local adaptation into habitat suitability forecasts and inform assisted migration. Ecol Monogr 94:e1593

Lazarus BE, Schaberg PG, DeHayes DH et al (2004) Severe red spruce winter injury in 2003 creates unusual ecological event in the northeastern United States. Can J Res 34:1784–1788

Li W, Kershaw JA Jr, Costanza KKL et al (2020) Evaluating the potential of red spruce (*Picea Rubens* Sarg.) to persist under climate change using historic provenance trials in eastern Canada. Forest Ecol Manag 466:118139

Lindbladh M, O'Connor R, Jacobson GL Jr (2002) Morphometric analysis of pollen grains for paleoecological studies: classification of *Picea* from eastern North America. Am J Bot 89:1459–1467

Lindbladh M, Jacobson GL Jr, Schauffler M (2003) The postglacial history of three *Picea* species in New England, USA. Quat Res 59:61–69

Lindbladh M, Oswald WW, Foster DR et al (2007) A late-glacial transition from *Picea glauca* to *Picea mariana* in southern New England. Quat Res 67:502–508

Major JE, Barsi DC, Mosseler A et al (2003) Light-energy processing and freezing-tolerance traits in red spruce and black spruce: species and seed-source variation. Tree Physiol 23:685–694

Marshall SJ (2009) Glaciations, quaternary. In: Gornitz V (ed) Encyclopedia of paleoclimatology and ancient environments. Springer, New York, pp 389–393

Masson-Delmotte V, Zhai P, Pirani A et al (2021) Climate change 2021: the physical science basis. In: Intergovernmental panel on climate change. Cambridge University Press, Cambridge, England

Mathias JM, Thomas RB (2018) Disentangling the effects of acidic air pollution, atmospheric CO_2, and climate change on recent growth of red spruce trees in the central Appalachian Mountains. Glob Change Biol 24:3938–3953

Menzel JM, Ford WM, Edwards JW et al (2006) Home range and habitat use of the vulnerable Virginia northern flying squirrel *Glaucomys sabrinus fuscus* in the central Appalachian Mountains, USA. Oryx 40:204–210

Milesi P, Berlin M, Chen J et al (2019) Assessing the potential for assisted gene flow using past introduction of Norway spruce in southern Sweden: local adaptation and genetic basis of quantitative traits in trees. Evol Appl 12:1946–1959

Mimura M, Aitken SN (2010) Local adaptation at the range peripheries of Sitka spruce. J Evol Biol 23:249–258

Moran E, Lauder J, Musser C et al (2017) The genetics of drought tolerance in conifers. New Phytol 216:1034–1048

Morgenstern EK, Corriveau AG, Fowler DP (1981) A provenance test of red spruce in nine environments in eastern Canada. Can J for Res 11:124–131

Mosher, D (2020) Past, current, and future potential distributions of red spruce and Fraser fir forests in the southern Appalachians: interpreting possible impacts of climate change. Thesis, East Tennessee State University, Johnson City, Tennessee

National Oceanic and Atmospheric Administration (2023) Greenhouse gases continued to increase rapidly in 2022. https://www.noaa.gov/news-release/greenhouse-gases-continued-to-increase-rapidly-in-2022. Accessed 13 Jun 2023

Nielsen ES, Henriques R, Beger M et al (2021) Distinct interspecific and intraspecific vulnerability of coastal species to global change. Glob Change Biol 27:3415–3431

O'Neill GA, Gómez-Pineda E (2021) Local was best: sourcing tree seed for future climates. Can J Res 51:1432–1439

Opler PA, Lotts T, Naberhaus T (2010) Butterflies and moths of North America. Big Sky Institute, Bozeman, Montana

Park A, Talbot C, Smith R (2018) Trees for tomorrow: an evaluation framework to assess potential candidates for assisted migration to Manitoba's forests. Clim Change 148:591–606

Pauley TK (2022) Forty-years of field notes: the Cheat Mountain salamander (*Plethodon nettingi*). Proc West Virginia Acad Sci 94:1–37

Peters MP, Prasad AM, Matthews SN et al (2020) Climate change tree atlas, ver 4. USDA Forest Service Northern Research Station. https://www.fs.usda.gov/nrs/atlas/tree/. Accessed 1 Aug 2024

Potter KM, Jetton RM, Bower A et al (2017) Banking on the future: progress, challenges and opportunities for the genetic conservation of forest trees. New Forest 48:153–180

Potter KM, Escanferla ME, Jetton RM et al (2019) Prioritizing the conservation needs of United States tree species: evaluating vulnerability to forest insect and disease threats. Glob Ecol Conserv 18:e00622

Prakash A, DeYoung S, Lachmuth S et al (2022) Genotypic variation and plasticity in climate-adaptive traits after range expansion and fragmentation of red spruce (*Picea rubens* Sarg.). Philos Trans R Soc London B: Biol Sci 377:20210008

Prakash A, Capblancq T, Shallows K et al (2024) Bringing genomics to the field: an integrative approach to seed sourcing for forest restoration. Appl Plant Sci 12:e11600

Prasad A, Pedlar J, Peters M et al (2020) Combining US and Canadian forest inventories to assess habitat suitability and migration potential of 25 tree species under climate change. Divers Distrib 26:1142–1159

Rajora OP, Mosseler A (2001) Challenges and opportunities for conservation of forest genetic resources. Euphytica 118:197–212

Reinhardt K, Smith WK (2008) Leaf gas exchange of understory spruce–fir saplings in relict cloud forests, southern Appalachian Mountains, USA. Tree Physiol 28:113–122

Rellstab C, Dauphin B, Exposito-Alonso M (2021) Prospects and limitations of genomic offset in conservation management. Evol Appl 14:1202–1212

Rentch JS, Schuler TM, Ford WM et al (2007) Red spruce stand dynamics, simulations, and restoration opportunities in the central Appalachians. Restor Ecol 15:440–452

Rentch JS, Ford WM, Schuler TS et al (2016) Release of suppressed red spruce using canopy gap creation—Ecological restoration in the central Appalachians. Nat Areas J 36:29–37

Ribbons RR (2014) Disturbance and climatic effects on red spruce community dynamics at its southern continuous range margin. PeerJ 2:e293

Richardson AD, Denny EG, Siccama TG et al (2003) Evidence for a rising cloud ceiling in eastern North America. J Clim 16:2093–2098

Rossi S (2015) Local adaptations and climate change: converging sensitivity of bud break in black spruce provenances. Int J Biometeorol 59:827–835

Savolainen O, Pyhäjärvi T, Knürr T (2007) Gene flow and local adaptation in trees. Annu Rev Ecol Evol Syst 38:595–619

Schaberg PG, Shane JB, Cali PF et al (1998) Photosynthetic capacity of red spruce during winter. Tree Physiol 18:271–276

Schaberg PG (2000) Winter photosynthesis in red spruce (*Picea Rubens* Sarg.): limitations, potential benefits, and risks. Arct Antarct Alp Res 32:375–380

Schaberg PG, Strimbeck GR, Hawley GJ et al (2000) Cold tolerance and photosystem function in a montane red spruce population: physiological relationships with foliar carbohydrates. J Sustain for 10:173–180

Schaberg PG, DeHayes DH, Hawley GJ et al (2008) Anthropogenic alterations of genetic diversity within tree populations: implications for forest ecosystem resilience. For Ecol Manag 256:855–862

Schomburg FM, Patton DA, Meinke DW et al (2001) FPA, a gene involved in floral induction in *Arabidopsis*, encodes a protein containing RNA-recognition motifs. Plant Cell 13:1427–1436

Shortle WC, Smith KT, Minocha R et al (1997) Acidic deposition, cation mobilization, and biochemical indicators of stress in healthy red spruce. J Environ Qual 26:871–876

Stanton JM (2009) Modeled red spruce distribution response to climatic change in Monongahela National Forest. Thesis, Marshall University, Huntington, West Virginia

Strimbeck GR, DeHayes DH, Shane JB et al (1995) Midwinter dehardening of montane red spruce during a natural thaw. Can J for Res 25:2040–2044

Strimbeck GR, Kjellsen TD, Schaberg PG et al (2007) Cold in the common garden: comparative low-temperature tolerance of boreal and temperate conifer foliage. Trees 21:557–567

Strimbeck GR, Kjellsen TD, Schaberg PG et al (2008) Dynamics of low-temperature acclimation in temperate and boreal conifer foliage in a mild winter climate. Tree Physiol 28:1365–1374

Strimbeck GR, Schaberg PG (2009) Going to extremes: low temperature tolerance and acclimation in temperate and boreal conifers. In: Gusta LV, Wisniewski ME, Tanino KK (eds) Plant cold hardiness: from the laboratory to the field. CAB International, Cambridge, England, pp 226–239

Strimbeck GR, Schröder WP, Fossdal CG et al (2015) Extreme low temperature tolerance in woody plants. Front Plant Sci 6:884

Thibault E, Soolanayakanahally R, Keller SR (2020) Latitudinal clines in bud flush phenology reflect genetic variation in chilling requirements in balsam poplar, *Populus Balsamifera*. Am J Bot 107:1597–1605

Thomas E, Jalonen R, Loo J et al (2014) Genetic considerations in ecosystem restoration using native tree species. For Ecol Manag 333:66–75

Vann DR, Strimbeck GR, Johnson AH (1992) Effects of ambient levels of airborne chemicals on freezing resistance of red spruce foliage. For Ecol Manag 51:69–79

Vann DR, Johnson AH, Casper BB (1994) Effect of elevated temperatures on carbon dioxide exchange in *Picea rubens*. Tree Physiol 14:1339–1349

Veloz SD, Williams JW, Blois JL et al (2012) No-analog climates and shifting realized niches during the late quaternary: implications for 21st-century predictions by species distribution models. Glob Change Biol 18:1698–1713

Verrico BM, Weiland J, Perkins TD (2020) Long-term monitoring reveals forest tree community change driven by atmospheric sulphate pollution and contemporary climate change. Divers Distrib 26:270–283

Walter JA, Neblett JC, Atkins JW et al (2017) Regional- and watershed-scale analysis of red spruce habitat in the southeastern United States: implications for future restoration efforts. Plant Ecol 218:305–316

Wason JW, Dovciak D (2017) Tree demography suggests multiple directions and drivers for species range shifts in mountains of northeastern United States. Glob Change Biol 23:3335–3347

Wason JW, Dovciak M, Beier CM et al (2017) Tree growth is more sensitive than species distributions to recent changes in climate and acidic deposition in the northeastern United States. J Appl Ecol 54:1648–1657

Watts WA (1979) Late Quaternary vegetation of central Appalachia and the New Jersey coastal plain. Ecol Monogr 49:427–469

White PS, Cogbill CV (1992) Spruce-fir forests of eastern North America. In: Eagar C, Adams MB (eds) Ecology and decline of red spruce in the eastern United States. Springer, New York, pp 3–39

Wilkinson RC (1990) Effects of winter injury on basal area and height growth of 30-year-old red spruce from 12 provenances growing in northern New Hampshire. Can J for Res 20:1616–1622

Williams JW, Shuman BN, Webb T III et al (2004) Late-Quaternary vegetation dynamics in North America: scaling from taxa to biomes. Ecol Monogr 74:309–334

Yetter E, Chhin S, Brown JP (2021) Dendroclimatic analysis of central Appalachian red spruce in West Virginia. Can J for Res 51:1607–1620

Young B, Byers E, Gravuer K et al (2011) Guidelines for using the NatureServe climate change vulnerability index. NatureServe, Arlington, Virginia

Chapter 8
Ecological Restoration and Adaptive Management

Deborah Landau, Anna M. Branduzzi, Chris D. Barton, David R. Carter, Will Evans, Chad M. Landress, Benjamin M. Rhodes, David Saville, Kathryn M. Shallows, Alexander Silvis, and James A. Thompson

D. Landau (✉)
The Nature Conservancy, Maryland/DC Chapter, Rockville, MD, USA
e-mail: dlandau@tnc.org

A. M. Branduzzi
Green Forests Work, Lexington, KY, USA

C. D. Barton
University of Kentucky, Department of Forestry and Natural Resources, Lexington, KY, USA

D. R. Carter
Michigan State University, Department of Forestry, East Lansing, MI, USA

W. Evans
The Nature Conservancy, West Virginia Chapter, Elkins, WV, USA

C. M. Landress
USDA Forest Service, Monongahela National Forest, Bartow, WV, USA

B. M. Rhodes
Ruffed Grouse Society & American Woodcock Society, Barbourville, KY, USA

D. Saville
Appalachian Forest Restoration LLC, Morgantown, WV, USA

K. M. Shallows
The Nature Conservancy, Appalachians Program, Fairmont, WV, USA

A. Silvis
West Virginia Division of Natural Resources, Elkins, WV, USA

J. A. Thompson
West Virginia University, School of Natural Resources and the Environment, Morgantown, WV, USA

D. J. Brown et al. (eds.), *Ecology and Restoration of Red Spruce Ecosystems in the Central and Southern Appalachians*, https://doi.org/10.1007/978-3-032-19616-3_8

8.1 Ecological Restoration for Resilient Red Spruce and Spruce-fir Ecosystems

In general terms, ecological restoration aims to restore degraded ecosystems to a reference or desired condition that can maintain climate resilience and can encompass multiple elements of the ecosystem, including soils, fungi, plants, and wildlife (Society for Ecological Restoration 2004). Central and southern Appalachian red spruce (*Picea rubens*) forests are considered one of the most endangered ecosystems in the U.S. (Noss et al. 1995), with their decline and degradation largely attributed to industrial logging activities and associated wildfires in the late 1800s and the early 1900s (see Chaps. 1 and 5). For red spruce and spruce-fir forests, ecological restoration endeavors to create a dynamic, diverse, and ecologically functional forested condition. Reaching this condition could take decades, if not centuries, due to the complexities of successional dynamics of this long-lived system. Reestablishing red spruce and ecosystem associates such as balsam fir (*Abies balsamea*) or Fraser fir (*Abies fraseri*) as dominant canopy species is a primary goal across the historical range of red spruce and spruce-fir in the central and southern Appalachians. Additional goals differ based on the variety of reference site conditions and plant communities (see Chap. 4). Restoring the complexity and integrity of red spruce and spruce-fir ecosystems results in more ecologically functional and connected habitats that will expand and enhance habitat for associated wildlife species (see Chap. 6). An intact network of these forested landscapes is expected to be more resilient to changes in temperature and precipitation from climate change (see Chap. 7), increasing the likelihood that red spruce and spruce-fir forest ecosystems will persist into the future, providing climate refugia for the region's critical biodiversity.

In this chapter, we discuss restoration activities implemented by the CASRI and SASRI partnerships to accelerate ecological restoration of red spruce and spruce-fir forests in the central and southern Appalachians. Foundational to ecosystem restoration is to first set target desired ecosystem conditions, which are often a blend of historical references, future potential conditions driven by climate change, and the realities of site limitations for achieving desired conditions. We first outline options for setting desired ecosystem condition goals that will provide guidance for restoration actions and decades-long adaptive management programs. Next, we describe how spatial analyses have guided landscape-scale planning for restoration sites, followed by methods for ecological restoration based on the best available science, covering restoration of plant communities and ecological functions, and finally monitoring and adaptive management approaches to ensure restoration projects reach their ecological goals.

8.2 Defining Desired Ecosystem Conditions as Restoration Targets

Before considering restoration sites and actions, setting restoration goals, at both the landscape- and site-levels, can provide guidelines for restoration methods. Landscape-level goals address the spatial extents of restoration targets and site-level goals address the desired ecosystem conditions. A combination of both landscape-level and site-level assessments can help drive specific goals and actions for restoration planning, resulting in red spruce-dominated stands that are more structurally intact and more connected at a landscape scale.

Landscape-level goals for restoring red spruce and spruce-fir communities are overall to improve the permeability of land and water for the movement of organisms and ecological processes (e.g., hydrological flows; Game et al. 2010). Both vertical (elevation) and latitudinal (north–south) landscape-scale connectivity provide movement corridors and dispersal options for forest species. A large, intact forest landscape is better able to withstand severe weather events that might lead to the destruction of a smaller forest patch. Thus, restoration sites should be consciously integrated into the larger ecological matrix through increasing patch size and connectivity of natural communities and improving ecological conditions. Red spruce restoration projects should aim to create large patch sizes of functional red spruce forests to improve resiliency to more frequent and severe disturbances anticipated under climate change. Restoration actions should focus on locations that increase functional patch size for key wildlife species, increase riparian cover for headwater stream protection, and connect disjunct red spruce-hardwood stands within and among landowners to improve the ecological integrity of the red spruce-northern hardwood ecosystem.

In defining restoration targets, managers and planners should consider the composition and function of a similar undisturbed site, if available, and understand how the ecosystem became degraded and how that degradation has impaired ecosystem function. Here we consider vegetation composition, vegetation structure (vertical, horizontal, tree size, snags, woody debris), and ecosystem processes and functions (soils and hydrology). Management practices required to restore a prior, desired ecological state will differ based on the current ecological state. For example, restoration of logged sites will differ from restoration of sites that have been logged and burned. Implementation of management practices will often be based on the needs and composition of specific forest stands.

Restoration ecology offers many approaches to defining ecosystem restoration targets. For example, the CASRI restoration approach defines restoration targets based on the National Vegetation Classification (NVC) system's reference species assemblages (Byers et al. 2011). That document provides guidance on how to use known plant communities in restoration, with a simple dichotomous key to help guide a restoration practitioner to determine which NVC is appropriate for their project area (Byers et al. 2011). Additionally, the document outlines restoration scenarios for seven potential site conditions, ranging from northern hardwood forests with no red spruce in the understory to surface mines to red spruce monocultures. Another approach for defining restoration targets is to utilize ecological site descriptions (ESD). Both strategies work towards describing reference communities based on extensive evidence for how these communities can form and persist and have been utilized by the red spruce restoration community. We focus here on ESDs as there are well-defined ESDs for several key red spruce and spruce-fir communities. We also provide examples for defining desired conditions and relevant impacts from past disturbances in locations without red spruce-fir ESDs.

8.2.1 Red Spruce Ecological Site Descriptions

Ecological sites (ES) are defined as "a distinctive kind of land based on recurring soil, landform, geological, and climate characteristics that differs from other kinds of land in its ability to produce distinctive kinds and amounts of vegetation and its ability to respond similarly to management actions and natural disturbances" (USDA NRCS 2017). Ecological sites are portrayed in a corresponding state and transition model (Fig. 8.1; Briske et al. 2005; Bestelmeyer et al. 2009). The ESD approach can aid in understanding and integrating historical conditions, disturbances that led to degradation, restoration goals, and the desired future condition. Bestelmeyer and Brown (2010) provide more information on how ESDs can be used as a restoration and conservation tool to help select and implement appropriate management practices to meet goals for a particular landscape.

Currently, there are two fully developed ESDs and six provisional ES (PES; first approximations of ESDs) for red spruce communities in the central and southern Appalachians.[1] Provisional ecological sites are in the developmental process to better assess ecosystem dynamics within the ecological site. Ecological sites and PES share the same functional framework and both can be used for restoration management decision-making.

The two red spruce ESDs that occur in central Appalachia are the Spodic Shale Upland Conifer Forest ES (USDA NRCS 2016a), and the Spodic Intergrade Shale Upland Hardwood and Conifer Forest ES (USDA NRCS 2016b). These two red spruce ESs exist on shale geologies interbedded with sandstone that produce low-pH soils and contain similar vegetation communities, although the two ESs vary in vegetation composition and characteristic soil properties. Both ESs have similar climatic influences, being characterized by frigid soil temperature regimes (mean annual soil temperature 0–8 °C [32–46 °F]) and perudic soil moisture regimes (precipitation exceeds evapotranspiration in all months). Historical management practices are similar across both ESs. Both share similar species associations, with red spruce and eastern hemlock (*Tsuga canadensis*) being major components of the reference community. Black cherry (*Prunus serotina*), red maple (*Acer rubrum*), striped

[1] These ESDs and PESs can be publicly accessed through the Ecosystem Dynamics Interpretive Tool (EDIT) website (https://edit.jornada.nmsu.edu/), managed by the USDA Natural Resources Conservation Service (NRCS).

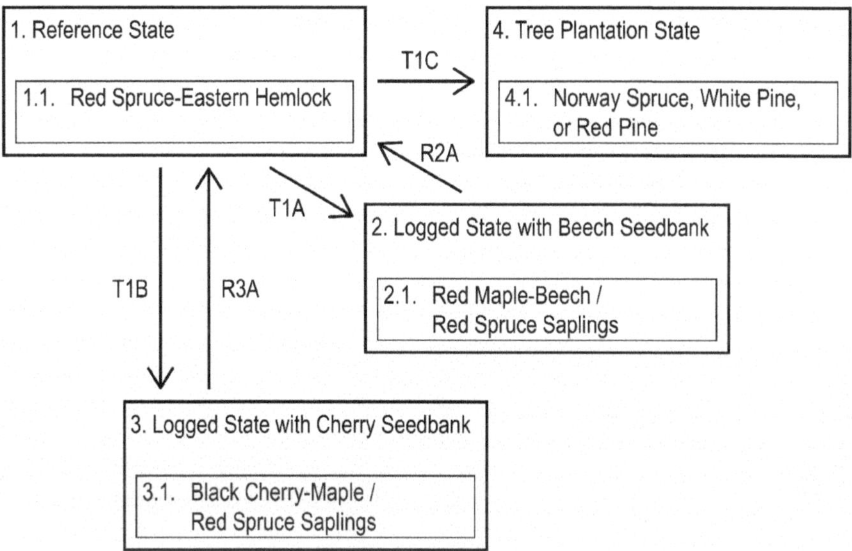

Fig. 8.1 Example of a state and transition model for the Spodic Shale Upland Conifer Forest Ecological Site (ES), developed by Jason Teets (USDA Forest Service, Monongahela NF). Arrows leading to and from different ecological states of the ES represent either a restoration pathway (R) or a transition pathway (T). Restoration pathways represent a suite of management practices that help to restore the specific ecological state back towards the Reference State, or to an ecological state nearer to the Reference State. A transitional pathway relates to any number of natural or anthropogenic disturbances, such as fires, logging, drought, or disease, that may trigger an ecological state to transition to a different ecological state further removed from the Reference State condition. These restoration and disturbance pathways are uniquely related to the ES being examined and are defined in the Ecological Site Description. For more information, see USDA NRCS (2016a)

maple (*Acer pensylvanicum*), mountain maple (*Acer spicatum*), American beech (*Fagus grandifolia*), American basswood (*Tilia americana*), white ash (*Fraxinus americana*), northern red oak (*Quercus rubra*), sweet birch (*Betula lenta*), yellow birch (*Betula alleghaniensis*), Allegheny serviceberry (*Amelanchier laevis*), mountain magnolia (*Magnolia fraseri*), and cucumber magnolia (*Magnolia acuminata*) comprise overstory vegetation within these ESs, but exhibit less overstory dominance than red spruce or eastern hemlock. Understory shrubs for these ESs include *Rhododendron* spp., mountain holly (*Ilex montana*), and mountain laurel (*Kalmia latifolia*), while groundcover often consists of New York fern (*Thelypteris noveboracensis*), intermediate woodfern (*Dryopteris intermedia*), hypnum moss (*Hypnum imponens*), and liverwort (*Bazzania trilobata*). Species may be similar across both ESs, but the degree to which these species exhibit overstory dominance and the forest's community composition varies, alongside characteristic soil properties.

The Rubbly Upland Conifer Forest PES has been developed in West Virginia, while the Frigid Interbedded Sedimentary Residuum, Frigid Colluvium, Frigid Residuum, Shallow Frigid Residuum, and Frigid Mountain Bog PES occur in the southern Appalachians in Tennessee and North Carolina. The Frigid High-Elevation Upland PES occurs within West Virginia at its northernmost extent, and extends southward. These red spruce PESs in the central and southern Appalachians vary in the degree to which they have been conceptualized and developed, with some having very strong ecopedological concepts attributed to them, and others being only initial placeholders for further investigation and verification. The process of ES and PES development is ongoing, with more red spruce PESs and ESs likely being developed into the future as more time is spent collecting data and making ecological-pedological correlations.

All ESDs are geography-dependent based on soil and the ecological communities related to that soil. Consequently, every soil is associated with one and only one ES. Thus, if a soil that is linked to a red spruce ESD is not found in a geographic area, then there are no red spruce ES or PES in that area. Regardless, until more red spruce community ESDs are developed throughout the central and southern Appalachians, many of these communities may be expected to respond similarly to on-the-ground management practices. Generally speaking, these communities were impacted similarly by the intensive logging and subsequent fires of the late 1800s and the early 1900s. In terms of restoration, similar management pathways of restoration are expected to be applicable to the array of red spruce PESs and ESs we have available currently.

8.2.2 Other Methods of Defining Restoration Targets

If ESDs are not available for a restoration site in the central and southern Appalachians, other methods can be used to determine desired ecological conditions as restoration targets, and to incorporate an understanding of past disturbances that have led to current conditions and may impact restoration results. As discussed in Sect. 8.2, the CASRI restoration approach recommends following the NVC system's reference species assemblages to set restoration targets (Byers et al. 2011). The document provides a simple key to help a restoration practitioner determine which NVC is most appropriate for a given restoration project. The LANDFIRE Biophysical Setting Model (https://www.landfire.gov/bps-models.php) can also be used to determine reference conditions. LANDFIRE uses vegetation data that includes natural community occurrence records, estimates of canopy cover and height per plant taxon, and measurements of individual trees. The biophysical settings products represent the vegetation that may have been dominant on the landscape prior to Euro-American settlement. It is based on both the current biophysical environment and an approximation of the historical disturbance regime, and can be utilized to produce a central and southern Appalachian spruce-fir forest model (Medlock 2015).

8.2.3 *Other Considerations for Setting Restoration Targets*

High-elevation systems in the central and southern Appalachians are home to numerous ecosystems in addition to red spruce forests, such as Fraser fir forests, high-elevation rocky outcrops, northern hardwood forests, northern red oak forests, grassy balds, and heath balds. On *sky islands*, for example, multiple unique systems can often be found on one site. As climate change differentially affects these systems and their capacity to self-perpetuate, it can be challenging to balance the needs of multiple ecotypes competing for limited growing space when setting restoration targets. To address this, it is recommended that an inventory of the different ecosystems be made before setting restoration targets. If the spatial extents of the ecosystems prior to major anthropogenic influence are known, this can be used as a reference condition to inform restoration targets. If reference conditions are not available, efforts could be made to balance restoration objectives among ecotypes, particularly to ensure persistence of those that may be range-limited. As climate change continues to shape these regionally rare habitats in potentially unexpected ways, continuing to monitor and enhance drivers of resiliency within and across these systems will be critical to maintaining regional forest diversity.

8.2.3.1 Balds

Throughout the southern Appalachians, there are high-elevation grasslands and shrublands generally known as balds. In some cases, such as at Roan Mountain in Tennessee (Fig. 8.2), these are documented to have been present in the earliest written records and are presumed natural. In other cases, such as the Great Balsam Mountains in North Carolina and Mt. Rogers in Virginia, balds are known to have been created from spruce-fir forests by logging and burning. Often, historical data are lacking and it is not known if openings are naturally occurring balds or historically created, such as through grazing, or logging and subsequent fire (Medlock 2015), or the term is related to the past use of the word bald to mean white-headed, as used for the bald eagle (*Haliaeetus leucocephalus*), because these high-elevation sites tend to hold ice or snow longer (Schafale 2024). Sites that support balds may not be noticeably different from similar sites that support spruce-fir forests, and can appear as appropriate for red spruce restoration. However, grass balds occur on < 1% of the sites suitable for spruce-fir (White and Sutter 1999), and heath balds occur on 4–9% of the sites suitable for spruce-fir (White et al. 2001). Thus, care should be taken when considering rare habitats for red spruce restoration to prevent loss of these natural systems in favor of restoring another.

In the central Appalachians, and specifically the Dolly Sods Plateau in West Virginia (Fig. 8.3), there are shrub and grass-dominated balds that are thought to have a relationship with human and possibly natural disturbance, such as fire. However, evidence is generally lacking on whether these areas were historically forested or a combination of open shrub and grass-dominated balds before the industrial logging

Fig. 8.2 Grassy Bald on Grassy Ridge, Roan Mountain area, North Carolina (photo by Michael Schafale)

period, making it difficult to make management recommendations for this unique landscape (Burkhart 2011). In the absence of fire and grazing, it appears that the shrub balds of the Dolly Sods Plateau are undergoing forest succession such that shrubs are increasing and hardwood species and red spruce are beginning to dot the landscape. It is likely that without continued disturbance to revert natural succession these balds will return to forest, with red spruce forest likely the climax condition (Mike Powell, The Nature Conservancy, written communication, February 2024).

8.3 Spatial Prioritization Tool to Guide Landscape-Scale Restoration Planning

Identifying where to focus restoration activities is a critical component of initiating restoration activities. The representativeness of an area, the resilience potential and current condition of the sites, and the connectivity between sites to allow for movement and adaptation are important to take into consideration (Anderson et al. 2023). Clark et al. (2023) described the importance of aligning evaluation of local restoration opportunities with the goal of sustaining continental-scale movement potential in the central Appalachians. In their prioritization approach, they targeted areas for investments at the regional scale, as informed by the Resilient and Connected Network assessment (Anderson et al. 2023), and compared the potential change in connectivity benefit associated with realistic restoration site opportunities defined by local staff, thus recognizing the local influence of the restored site. Similarly, the broad distribution of historical red spruce communities, along with changes in environmental

Fig. 8.3 Fall scene at Dolly Sods, West Virginia. The Dolly Sods Wilderness is a vast area of plateaus above 1,200 m (4,000 ft) and steep-walled stream valleys in the Allegheny Mountains of eastern West Virginia and part of the Monongahela NF. The area was originally covered with a thick red spruce (*Picea rubens*) forest but was aggressively logged in the early tewntieth century. Today, the area is dominated by broad plains covered with heath and grasses, with many bogs. Hardwoods dominate the lower elevations but the spruce forest is coming back at higher elevations (photo by Hannah Snyder)

conditions over the past century, necessitates careful planning and prioritization to best utilize limited resources for maximum restoration outcomes.

The historical, current, and potential future habitat ranges for red spruce have been modeled for the central Appalachians (LANDFIRE; Byers et al. 2010; Beane et al. 2013). These data layers have provided a foundation for restoration prioritization. In 2013, CASRI partners developed a restoration priority map built from two ecological variables: current red spruce cover data (Byers et al. 2010), and a MaxEnt model of current habitat suitability for red spruce forests (Beane et al. 2013). A simple matrix was produced that intersected areas having current zero cover (but within historic range), low cover, medium cover, and high cover with areas having zero suitability, low suitability, medium suitability, and high suitability. Areas with high current suitability, historical red spruce cover, and zero current cover were designated for red spruce plantings. Areas with high suitability and low or medium current cover were designated for red spruce release, which utilizes naturally occurring advance regeneration to increase red spruce cover. Areas with high current cover were designated as top priority for protection, allowing these red spruce-dominated stands to grow and spread through natural processes. The restoration prioritization has been a critical tool in CASRI's planning processes when combined with local land managers' expertise. In areas not covered by the prioritization, similar spatial datasets can be used to inform land management decisions. New advances to the prioritization tool are being developed to include a broader suite of spatial and ecological variables (Donald Brown, USDA Forest Service, oral communication, 20 Apr 2024).

In the southern Appalachians, historical reference conditions for red spruce forests have been mapped, offering an alternative, but also data-driven, approach to that which was taken by CASRI. Appendix A of the SASRI restoration plan (Medlock 2015) discusses the importance of utilizing information on pre-disturbed ecosystems to guide restoration efforts. The authors begin by determining the reference time (i.e., prior to the beginning of industrial logging in the 1880s) and utilize historical red spruce forest composition, structure, and extent to establish reference conditions. These reference conditions can be used to help identify, set, and prioritize restoration goals, site selection, and evaluation of restoration treatments.

8.4 Methods for Ecological Restoration of Red Spruce and Spruce-Fir

Ecological restoration planning begins with the process of defining current and desired future conditions for species and structure (see Sect. 8.2) within the context of landscape-level restoration priorities (see Sect. 8.3) to describe the suite of desired ecological outcomes. Managers should select the methods most appropriate to achieving the desired outcomes. For example, red spruce release treatments may be effective where red spruce are abundant in the understory, but ineffective in areas with sparse red spruce cover. In general, plantings and silvicultural prescriptions for releasing red spruce advance regeneration to the canopy are the primary methods for restoring Appalachian red spruce and spruce-fir communities. Here we cover red spruce plantings and release treatments that primarily target plant community goals, as well as methods to restore key ecological functions of red spruce ecosystems, with a case study of a mined land reclamation initiative that used several restoration methods to address multiple ecosystem restoration goals. As with any land management and restoration project, managers should consider relevant regulatory, social and environmental factors when selecting the most appropriate methods for their site.

8.4.1 Seed Sourcing, Propagation and Planting

Red spruce seed used to propagate seedlings for restoration purposes has generally been sourced from populations in the same ecoregion as the restoration site so that plantings are adapted to local site conditions and climate. Climate change-driven shifts in current and future habitat suitability require land managers to consider how seed sourcing affects the climate adaptive capacity of a species. Researchers and conservation practitioners are using insights from ecological genomics to test and develop guidance for sourcing red spruce seed to increase the genetic diversity of seed mixes to reduce the climate vulnerability of red spruce plantings in the central Appalachians (see Sect. 9.5.3 of Chap. 9). At the time of publication, this group is exploring how forecast modeling of future red spruce habitat suitability that incorporates genomic offsets can be used to inform strategies for forest-assisted gene flow and migration across ecoregions in the Appalachians, described in Sect. 7.3 of Chap. 7.

Red spruce trees only produce enough viable seeds for collections every 4–8 years, so it is important to capitalize on cone crops when they occur, processing and storing enough seed to last several years until the next crop. Cones are collected in early fall, before they open, and spread on screened racks to finish ripening before further processing (Fig. 8.4). They are then kiln-dried and tumbled, a process that is repeated until the majority of seeds have been released. In nature, cones hanging on the uppermost branches of the tree will open and release a few seeds at a time when it is dry, then close again when it rains. The winged seeds are thus slowly dispersed by the wind over a period of months with each drying and wetting cycle. In mimicking nature to harvest the seed, the cones are kiln-dried to open them, tumbled to knock as many seeds free as possible, soaked in water to close the cones up again, then re-kiln-dried and tumbled, in a process that is repeated until no more seeds are released. Seeds must then be mechanically de-winged. Three steps are used to clean the seed: (1) scalping—removing all matter larger than the seeds), (2) screening—removing all matter smaller than the seed, and (3) winnowing—using an air stream to further remove any material lighter than the seed (e.g., wing fragments). Using an air column machine, winnowing is again used to *purify* the seed by removing hollow and empty seeds. The seed is then tested for moisture content and germination before storage. Once collected and processed, seeds are dried to below 8% moisture and flash frozen to −2.8 °C (27 °F). The seeds must remain at a consistent temperature while frozen; specialized seed bank freezers that do not cycle on and off are required to maintain this constant temperature.

Broad-scale plantings are an effective method for reconnecting and expanding red spruce forests. Red spruce seedlings can be planted with hand trowels, shovels, or dibble bars (as are most commonly used for volunteer planting events), or planting spades and hoedads (which are often used by professional tree-planting crews; Fig. 8.5). A limiting factor in replanting efforts has historically been that few commercial producers of locally-sourced red spruce seedlings exist. Efforts by CASRI to expand seedling availability have resulted in increased production capacity, and are described below (Dave Saville, Appalachian Forest Restoration LLC, oral communication, 13 Jul 2023).

Methods for growing red spruce can include seedbed-produced seedlings, seedbed-produced transplants, plug-grown transplants, or containerized plants. When considering propagation and planting, factors to be considered include cost, survivability, and ease of planting. First- and second-year growth is important as the seedlings need to outcompete surrounding vegetation. Restoration practitioners in

Fig. 8.4 Red spruce cones collected from forest stands in the central Appalachians (left). Dried cones release winged seeds (right; photos by David Saville)

Fig. 8.5 Red spruce (*Picea rubens*) seedling being planted with a dibble bar in an old field in western Maryland (photo by Deborah Landau)

CASRI have found success with 15 cubic-inch containerized plugs. These container-produced plants have minimal root damage during extraction, packing, transport, and transplanting, thus minimizing transplanting shock. Additionally, whereas seedbed-produced plants develop tap roots that consume the plant's energy during development, containers are designed to *air-root-prune* the root systems, avoiding the development of tap roots. The result is a dense, fibrous root system, packed full of stored energy ready to propel the newly planted seedling to greater first-year growth. This leads to greater survivability, not only for the initial planting year, but also greater second- and third-year growth, enabling the plant to outcompete surrounding vegetation. An additional, critical advantage is that these plugs are easier to plant (and require no root pruning), reducing room for error from volunteer plantings (David Saville, Appalachian Forest Restoration LLC, written communication, 2023).

Fall or early spring, as soon as the ground is thawed and snow accumulation has melted to no more than 5 cm (2 in), is an ideal season to plant (Fig. 8.6). Spring plantings have higher success rates for red spruce plug plantings due to reduced risk of frost heave, whereby soil water expands and pushes planted plugs up and out of the ground. Red spruce are shallow-rooted and prefer moist soils. Although they do not tolerate long droughts, red spruce are known to have high survival rates even in highly degraded lands. In fact, red spruce plugs planted on mined lands had higher survival rates relative to those planted in old fields, likely due to availability of soil nutrients (Rhodes and Barton 2024). The highest elevations in the central and southern Appalachians receive the greatest amounts and frequency of precipitation, making them ideal for red spruce growth regardless of site conditions. In lower elevations with less precipitation, site conditions can be selected to maximize survival, such as north-facing slopes, streamsides, riparian areas, and wetlands. One notable exception for lower elevation plantings is within the frost pockets of western Maryland, including the Savage River watershed (approximately 800 m [2,624 ft]). Here, red spruce trees are often found adjacent to bogs or swamps. Planting along these cool, wet sites has been very successful in western Maryland. Eastern hemlock is a good indicator for red spruce persistence. Planting red spruce under eastern hemlocks infested with hemlock woolly adelgid (*Adelges tsugae*) can be a strategy for providing shade to streams after the hemlocks have died. Red spruce seedlings can persist under hemlocks for decades, ready to take advantage of light gaps when the hemlocks ultimately succumb to adelgid infestations.

8.4.2 Release Treatments

Red spruce release treatments emulate natural disturbance by using mechanical or chemical methods to create canopy gaps over suppressed midstory and understory red spruce seedlings and saplings (Fig. 8.7). Release methods are based on an understanding of red spruce's life history which relies on advance regeneration and canopy-gap dynamics of mature forests (see Sect. 5.2.1 Canopy Gap Dynamics). Red spruce seedlings are highly shade-tolerant and adapted to germinate and grow in low-light conditions under a mature forest (Blum 1990). For the first few years of growth, the survival and vigor of red spruce seedlings are inversely correlated with sunlight availability (Westveld 1931). However, once the seedlings reach the sapling stage, this correlation reverses, and optimal growth is then achieved only under nearly full sunlight (Blum 1990). Release from overstory shade spurs accelerated leader growth. In the absence of high-light conditions, red spruce seedlings can remain suppressed in the understory for decades. However, as the tree ages the speed and vigor of their response to release decline (Blum 1990).

Red spruce seedlings that were established and subsequently suppressed by faster-growing hardwood species during the early to the mid-1900s comprise a large cohort of advance red spruce regeneration in many Appalachian landscapes that is now approaching 100 years old (Rentch et al. 2016). Suppressed red spruce that are approaching 100 years in age may take several years to respond to a release event, if they respond at all. Climate change is expected to create increasingly challenging

Fig. 8.6 Red spruce (*Picea rubens*) seedling planted by professional tree-planting crews in early spring in the understory of a hardwood forest in southwestern Virginia (photo by Tal Jacobs)

growing conditions, even for suppressed red spruce in younger age cohorts (Beane and Rentch 2015). Average red spruce age, compounded by climate change, leaves a limited timeframe in which existing advance regeneration might reach the forest canopy. Here we outline ecological goals and strategies that use release treatments, discuss guidelines for gap size and species selection for release treatment approaches developed for ground crews, and present potential additional options from traditional silvicultural methods.

8.4.2.1 Restoration Goals Supported by Release Treatments

Red spruce ecosystem restoration projects often target multiple ecological goals that can be achieved by release treatments. Release treatments can support restoration goals such as restoring reference plant communities, enhancing forest connectivity, enhancing late successional forest conditions, and improving wildlife habitat for both terrestrial and aquatic species. Expanding connectivity of stands with canopy red spruce can support goals for climate refugia and goals for resilience to climate changes and other disturbances and stressors. Strategies to achieve these goals can

Fig. 8.7 Hardwood trees culled with herbicide to accelerate the release of a midstory red spruce (*Picea rubens*) to the canopy (photo by Benjamin Rhodes)

be spatial (i.e., designing treatment units that connect disjunct populations or prioritizing riparian corridors for enhancing stream health), and strategies can address forest stand composition (i.e., enhancing stand heterogeneity and late successional characteristics). Single project areas may benefit from targeting multiple goals and using multiple strategies. Using project goals as a foundation to guide strategies, release treatment design can then be translated to tactics and tailored with on-the-ground decision-making. Tactics may include gap location and size, prioritized tree species for avoiding or culling, and use of either mechanical or chemical means of culling trees.

For a project-site goal to restore reference plant communities the treatment design will aim to accelerate growth of red spruce to the forest canopy at a density that will put the forest on a transitional pathway towards the target forest community composition and structure (see Sects. 8.2.1 and 8.2.2). In most red spruce restoration projects in the central Appalachians, the target species composition emulates second-growth forests described by Byers et al. (2010). The target species composition is designed based on reference conditions, with most aiming for 30–50% red spruce canopy cover and an appropriate mix of red spruce-associated species. For project-site goals to enhance late-successional conditions, release treatments can be designed to increase dead woody material and improve the structural complexity of the stand (i.e., diversity of tree age classes, size classes, and spatial distribution) to be more structurally similar to a late-successional forest, a developmental stage often lacking representation in red spruce communities across the landscape today. These stands can provide improved habitat for wildlife and have increased resistance and resilience to disturbances and stressors (Niedermaier et al. 2022).

8.4.2.2 Gap Size and Species Selection Guidelines for Ground Crews

Red spruce release can be effectively implemented with minimal ground disturbance by crews on foot using mechanical or chemical treatments (Fig. 8.8). Methods with minimal ground disturbance are ideal for sensitive sites, such as those with rare species, steep slopes, and sensitive soils as they can be performed without heavy equipment. Crews may use either mechanical or chemical treatments for release. Mechanical treatments refer to methods that use girdling to cull competing hardwoods with a chainsaw, brush cutter, or other equipment. Felling hardwoods for release is not recommended given the safety concerns, difficulty in avoiding impacts to released red spruce, stump sprouting, and inefficiency in time use. Mechanical treatments are suitable for highly sensitive ecosystems, such as those with rare plants, karst geology, or rare, threatened, or endangered species that may be adversely impacted by herbicide. Chemical treatments refer to methods that use herbicide to cull trees and are often more efficient and effective compared to mechanical treatments. Here we present release guidelines from research and practitioners that should be tailored to match stand-specific considerations (e.g., site index, aspect).

Overall, canopy gaps must be large enough so that adjacent hardwood trees do not close the gap prior to the target red spruce reaching the canopy. For reference, natural gap sizes from single- or double-tree mortality events range in size from approximately 56–186 m^2 or 0.0056–0.0186 ha (600–2,000 ft^2; Rentch et al. 2010). Released seedlings may grow at a rate of 0.3–0.9 m (1–3 ft) per year for several years (Rentch et al. 2016). After an initial burst of post-release growth, growth rates in full sunlight decline to an average 0.23–0.28 m (0.75–0.92 ft) per year (Westveld 1931; Rhodes and Barton 2024). As the released red spruce grow, mature hardwood trees along the gap edge may extend their branches laterally into the newly created gap at a rate of approximately 0.3 m (1 ft) per year. As an example, a 12 m (40 ft) diameter canopy gap (117 m^2 or 0.01 ha [1,260 ft^2]) created by the removal of a single tree will close in approximately 10 years. Assuming average growth rates as described above, a red spruce seedling released by this 12 m (40 ft) gap will grow only 4.5 m (15 ft) before the gap closes. To ensure that released red spruce will reach the canopy, the initial gap must be sufficiently large or reentry for additional release treatments will be needed (Rentch et al. 2010). Given that treatments may take decades to reach the target conditions, priority may be given to locations where a single release entry is sufficient (i.e., prioritizing midstory spruce stands over understory stands). This takes careful planning and tree-by-tree decision-making. Releasing the largest available suppressed red spruce, typically well-established midstory red spruce with long leader lengths, is optimal as these trees will have the greatest chance of reaching the canopy before the gap closes. Releasing multiple red spruce in a single gap

Fig. 8.8 Canopy gap creation by crews on foot using hatchets and herbicide to cull hardwood trees to release a stand of midstory red spruce (*Picea rubens*; photo by Will Evans)

maximizes time efficiency, and it can create localized areas of dense canopy red spruce which will create shade and alter the microclimate more effectively than a single canopy red spruce.

Gap size and the extent of overstory removal are related to the height of the red spruce being released, the horizontal extent and density of stems within the regeneration pool, and the presence of competitors. Generally, practitioners have found that for red spruce 3–9 m (10–30 ft) tall it is recommended to snag all overstory and midstory stems directly atop and up to 15 m (50 ft) away from any spruce targeted for release. For spruce greater than 9 m (30 ft) tall it is recommended to snag all overstory and midstory stems overtopping or competing with the target spruce and stems up to 7.5 m (25 ft) from the targeted spruce. Culling hardwood trees to the south, east, west, and uphill of the gap can increase the amount of light reaching the target spruce. Research in eastern late-successional forests shows that red spruce need, at minimum, canopy gaps of 0.04 ha (0.1 ac) to ascend to the canopy (D'Amato 2021). Practitioners in the central Appalachians, including some authors, aim for gaps averaging 0.05 ha (0.12 ac) and are monitoring to determine if this size is large enough to maintain the gap from a single release treatment and suitable to achieving desired reference conditions at greater scales. Gaps that are too large (> 0.4 ha [1 ac]) may result in the establishment and growth of fast-growing hardwoods, which may overtop red spruce and limit recruitment success if not controlled. However, gap sizes can be increased if reserve trees or overstory are retained.

Reserve trees are chosen based on various local and regional considerations, such as mast-producing species benefitting wildlife, species rarity, relevance to supporting microhabitats, and contribution to structural diversity. Culled species may be prioritized based on high local abundance and relatively low wildlife value. Species with localized rarity should be designated as legacy trees to retain potential genetic diversity. Species on the leading edge of their range may also be prioritized for retention. Overstory red spruce trees should be retained to provide a seed source and shelter regeneration. However, it is best to avoid leaving canopy red spruce within a gap as it exposes them to more windthrow loss; sufficient adjacent hardwood should be left to protect from windthrow. If the regeneration layer beneath the gap contains a high density of red spruce they may impede the growth of competitive vegetation. Regardless of gap size, increased sunlight benefits both red spruce and its competitors; therefore, gaps should be designed to maximize the competitiveness of red spruce by maintaining a red spruce overstory, targeting areas where red spruce is already dominant in the understory and midstory, and reducing potential competitors.

8.4.2.3 Release by Sustainable Silvicultural Methods

In the central and southern Appalachians, the majority of red spruce release has been accomplished with ground crews using herbicide to create gaps. In the northern Appalachians, the core of its range, spruce-fir forests have long been managed with traditional sustainable silvicultural methods. In this section we present silvicultural methods, shelterwoods and variable density thinning, that may have applicability in the central and southern Appalachians for ecological restoration goals. Release prescriptions implemented with heavy equipment, with or without a timber-sale component, will entail soil disturbance, and should be suitably sited, planned, and monitored to ensure ecological goals are met with minimal negative impacts to the harvest areas.

Shelterwood harvests are a regeneration method commonly utilized in spruce-fir forests in New England and Canada (Westveld 1938; Frank and Bjorkbom 1973) which may have utility in the central and southern Appalachians but have not been widely applied. Shelterwood systems are highly modifiable (Raymond et al. 2009). Several common variants of the uniform shelterwood approach include high-density (75–80%; shade-tolerant species promotion) to low-density (25–35%; shade-intolerant species promotion) retention, extended shelterwoods (i.e., extended duration of overstory retention), group shelterwoods (i.e., overstory retained within created gaps), and the broad category of irregular shelterwoods which typically includes a deviation from the uniform shelterwood both spatially (i.e., spatially variable retention) and temporally (i.e., length of time the overstory is retained can vary per stem from short-term to indefinite). There is potential to adapt these methods to the central and southern Appalachians to give restoration practitioners an additional strategy they can employ (Kenefic et al. 2021).

Establishing red spruce regeneration requires retaining a high density of residual trees to serve as a seed source, to regulate understory microclimates, and to mitigate light transmission levels to the forest floor. Dense overstory retention in red spruce shelterwoods prevents the establishment of unwanted shade-intolerant hardwoods while fostering conditions conducive to red spruce establishment. If, for example, the presence of red spruce advance regeneration is patchy and variably sized but abundant in a stand, an irregular shelterwood could be utilized to accelerate the ascension of the red spruce regeneration to the canopy. This is in contrast to the conditions that may warrant the use of a group selection system where midstory or tall understory spruce are present but only in occasional isolated pockets.

An irregular shelterwood could address a variably sized and variably dense regeneration layer. In such a condition, where midstory or tall understory spruce are present in sub-stand groups, a high proportion of the overstory could be removed, akin to a gap. In areas of the stand where little to no understory red spruce are present beneath an overstory of red spruce, removal of midstory or canopy trees of intermediate or suppressed canopy classes could be conducted to initiate understory establishment of red spruce, akin to a high-density shelterwood. In areas where red spruce is not present in the understory or canopy, limited overstory removal in conjunction with planting red spruce may be an option. Monitoring should continue in these systems, and additional reductions in unwanted species in the overstory and the regeneration layer should be completed as needed. Deployment of the irregular shelterwood allows land managers to address the variable condition of the regeneration layer in the stand. The examples of release treatments provided here are intended to demonstrate the flexibility in release treatment options and are not intended to serve as a rigid formula for success. As an advance regeneration-dependent species, the status of the red spruce population should guide decision-making, utilizing an adaptive management framework to guide future actions.

Variable density thinning (VDT) prescriptions address homogeneous, even-aged stands of red spruce with the goal of increasing within-stand structural heterogeneity. These prescriptions are characterized by a diversified intensity of tree removals within a stand (e.g., Willis et al. 2018) and are intended to promote a range of stem diameters and stem densities within the stand to promote structural diversity in mature stands. In a standard thinning (i.e., the opposite of a VDT), trees for removal are selected across the harvest area with an aim toward uniform residual structure and avoiding the creation of larger canopy gaps. In contrast, a VDT prescription is typically accomplished by overlaying *skips* (i.e., portions of the stand which receive no removals) and *gaps* (i.e., portions of the stand that are completely removed) atop a typical uniform thinning. Because thinnings are generally conducted in the stem exclusion stage of stand development, when a homogeneously structured even-aged stand is approaching or currently undergoing density-dependent mortality, VDT is intended to restore structural variability into these stands.

Varied removal intensities result in a spectrum of growing environments for residual stems, ranging from relatively open with ample room to grow to densely stocked conditions that remain highly competitive. This causes the diameter distribution of the stand to widen and flatten, reminiscent of old, uneven-aged stands.

In this case, management accelerates the development of these conditions that are likely to manifest with time. Given the scarcity of late-successional stands containing red spruce, accelerating the development of this forest condition often is desired. A similar approach can be utilized in young, dense, even-aged stands (e.g., thickets of Fraser fir regenerating after infestation of the balsam woolly adelgid [*Adelges piceae*]) to accelerate their development into structurally complex stands by varying the severity of removals within the stand. These structural elements (range of stem diameters and densities) that may otherwise take several additional decades to appear can be achieved in a shortened time frame.

8.4.3 Restoring Red Spruce and Spruce-Fir Ecosystem Functions

The reconstruction of altered ecological systems is an essential component of successful restoration of red spruce and spruce-fir habitats. This section addresses this from a hydrological perspective (i.e., reconnecting streams and floodplains and returning structural complexity to these hydrologic systems) and a soils perspective, through a better understanding of soil properties and functions, and effects of restoration actions.

8.4.3.1 Hydrology: Riparian and Stream Restoration

High-elevation headwater streams and wetland systems in the central Appalachians (and to a lesser extent in the southern Appalachians) are closely associated with red spruce communities. Stream and watershed restoration actions frequently include improvements to the overall red spruce ecosystem, which subsequently improves the health of red spruce forests and associated watersheds. Watershed restoration in the context of red spruce communities generally falls into three broad categories: (1) restoration of stream channels and associated riparian areas, (2) wetland restoration, and (3) hillslope hydrology restoration. Components of watershed restoration that may overlap with red spruce restoration goals include restoring hillslope hydrology, stream and riparian habitat restoration, stream buffering to ameliorate acid mine drainage and acid rain, and floodplain and wetland restoration. See Chap. 1 for a description of how the industrial logging era degraded stream, wetland, and watershed conditions in red spruce and spruce-fir forests.

Physical alterations to stream channels, floodplain connectivity, soil depth, wetland storage capacity, and hillslope hydrology are closely interrelated. Cool, moist microclimates associated with red spruce increase water availability, which is conducive to the redevelopment of organic soil horizons (Nauman et al. 2015a). Restoration targeting these watershed components seeks to restore soil and hydrologic structure and function towards pre-industrial logging conditions. Increased soil moisture retention, water volume storage capacity, and organic matter retention favor red spruce restoration and reestablishment of red spruce communities (Yetter et al. 2021). Restoration activities that incorporate red spruce and watershed concepts concurrently may provide more resiliency in the context of climate change stressors such as drought, floods, non-native invasive species, and disease or pest outbreaks (Wipfli et al. 2007). Reduced fragmentation and increased resiliency associated with watershed functions will directly benefit cold-water-dependent species such as brook trout (*Salvelinus fontinalis*), as well as increase resiliency in cool and warm-water aquatic ecosystems downstream. Where acidic deposition (e.g., acid rain, acid mine drainage) is a concern, high-quality crushed limestone sand can be added to the stream to mitigate deposition effects. Buffering of streams can substantially improve aquatic and riparian habitat conditions within red spruce ecosystems as well as downstream ecosystems.

Restoration of Stream Channels and Associated Riparian Areas

Stream channels, and considerations for restoration, can be broadly categorized by stream size. Here, we discuss the various stream types that are encountered in the central and southern Appalachians, and how to address them. Headwater channels include ephemeral, intermittent, and small perennial streams, primarily 1st and 2nd stream orders (Strahler 1957). While small, these channels typically constitute over 70% of the stream channel length in a watershed and their health generally determines the potential quality of aquatic ecosystems in larger downstream river segments (Lowe and Likens 2005). Headwater streams often are important as corridors for connectivity of red spruce across the landscape (Fig. 8.9). As suggested by their topographic characteristics and microclimate (i.e., cove and sheltered locations), headwater streams are less prone to natural disturbance events such as windthrow and fire (Ruffner and Abrams 2003). Headwater streams in the Appalachians are generally characterized by steeper channels, which limit the lateral extent of the riparian area.

The restoration of headwater channels and larger stream systems follows similar principles and techniques, although the overall objectives may differ. Common objectives are to aggrade the stream channel to reconnect stream and floodplain, ensure adequate riparian canopy coverage to provide shading and reduce stream temperatures, and add structural complexity to the channel and floodplain to increase sediment and organic material retention, reduce flood flow energy, and provide nurse logs for natural tree recruitment. Channels can be aggraded by adding large woody material directly into the channel. Wood can also be added to the floodplain to encourage sediment and organic material retention and provide refugia for riparian-dependent species during flood events (Lassettre and Harris 2001).

In open riparian areas, restoration techniques include planting as well as indirect tree recruitment techniques such as felling nurse logs. Planting red spruce and associated species is the most direct method to expand red spruce and improve future stream conditions. However, the historical context of a riparian opening and the

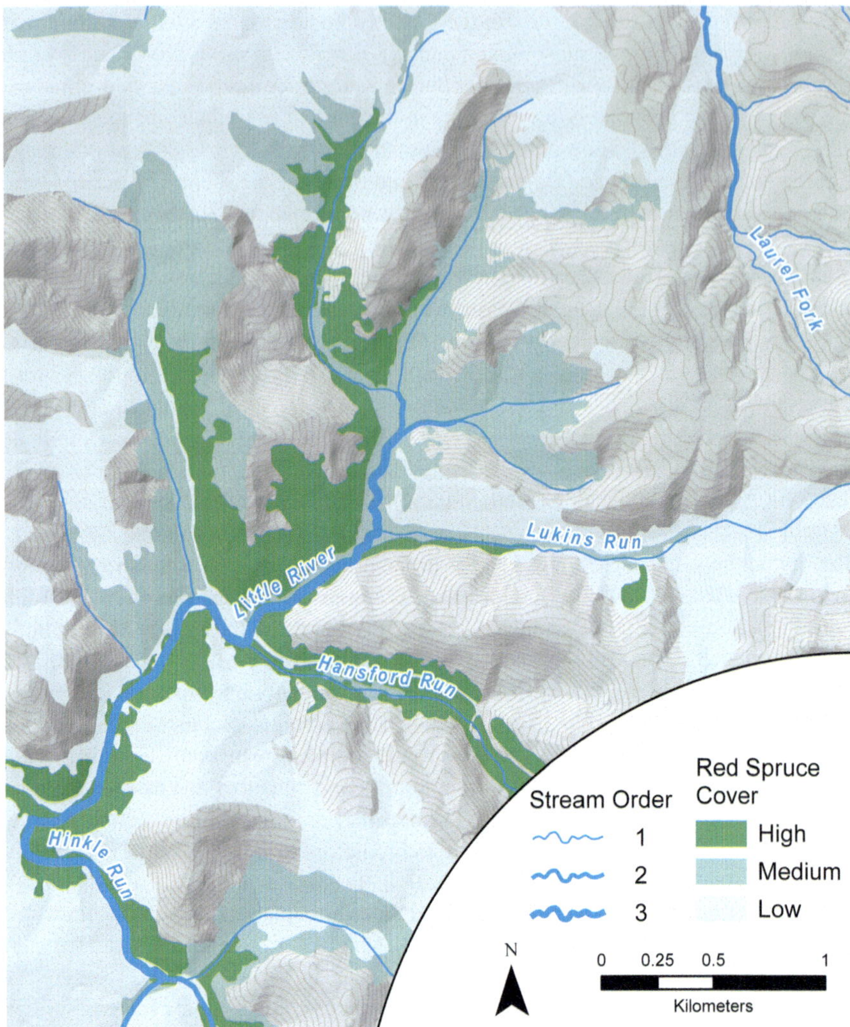

Fig. 8.9 Aerial view of 1st and 2nd order headwater streams highlighting the importance of riparian corridors for red spruce (*Picea rubens*) connectivity

limiting factors that may preclude reestablishment of red spruce are also important considerations to take into account. Openings may be the result of natural conditions, historically human-altered conditions, or more recent impediments to forest succession such as cutting, mowing, and grazing. Natural openings, especially those that contain rare plant communities, can be evaluated for the need to restore red spruce to these areas (e.g., Sect. 8.2.3.1). Historical alterations include human activities that affected the soil, hydrology, and resultant vegetation in riparian and floodplain zones. Examples include sites of mill ponds, splash dams, logging camps, and

Fig. 8.10 Nurse log with naturally recruited red spruce (*Picea rubens*) seedlings (photo by Chad Landress)

historical grazing. These areas often appear to be *naturalizing*, but may be in a state of arrested succession with limited progression to native forest cover and mature seral stages. Natural succession is further complicated by ongoing processes such as deer herbivory, dominance by allelopathic herbaceous species, or increased use by beavers. Evaluating these areas based on ecological and anthropogenic limiting factors, desired conditions, and the efficacy of various methods to overcome such limiting factors will increase the likelihood of achieving the desired conditions.

Appalachian forests are generally not old enough to recruit downed wood at the level of historical baseline conditions (Muller and Liu 1991; Webster and Jenkins 2005). Riparian areas provide optimal growing conditions for many tree species that may live for several centuries and attain relatively large sizes. However, the average age of riparian forests is directly linked to the industrial logging era; these forests are rarely older than a century. Therefore, the recruitment of downed wood into open riparian areas is below historical rates, further reducing successional timelines. Managers can expedite this process in conjunction with red spruce release by selectively felling hardwoods from the edge of the forest into riparian openings. Coarse woody material in both open and closed canopies can function as nurse logs which red spruce are highly adapted to take advantage of as a seed bed (Fig. 8.10). Nurse logs may also be transported from nearby locations into open areas lacking coarse woody material. Nurse logs, in scenarios where there is a locally dispersed seed source, may improve recruitment of red spruce where competition or shading from herbaceous vegetation or periodic flooding from beaver activity inhibits their reestablishment independently.

Red spruce restoration and stream restoration goals can be synergistic. Competing overstory hardwoods can be selectively cut into the stream or floodplain while

Fig. 8.11 Hardwoods cut for stream restoration also released midstory red spruce (*Picea rubens*; photo by Chad Landress)

releasing red spruce at the same time (Fig. 8.11). In the long term, red spruce release in riparian areas should increase stream shading, but project-specific stream temperature monitoring can inform the magnitude and frequency of treatment. In addition to the benefits derived from direct shading and microclimate characteristics of the red spruce ecosystem, reconnecting the floodplain by adding wood to the channel increases hyporheic and shallow groundwater volumes. Long-term monitoring data on the Monongahela NF indicate increased groundwater storage can reduce stream temperatures despite short-term canopy gap creation in stream restoration areas. In similar ecosystems in the Appalachians small canopy gaps generally do not increase stream temperatures (Siderhurst et al. 2010).

Wetland Restoration

Wetland types in the central and southern Appalachians that overlap the native range of red spruce encompass several broad categories, including forested swamps, shrub swamps, herbaceous wetlands, and bryophyte wetlands (Byers et al. 2007; see Chap. 4). Forested and semi-forested wetlands, including those with a red spruce component, are some of the most imperiled wetlands based on state and global conservation ranks (Byers et al. 2007). Wetlands that function within the red spruce forest matrix offer unique closed-canopy, semi-open-canopy, and edge habitats where red spruce is a preferred species for wildlife such as the olive-sided flycatcher (*Contopus cooperi*) and northern saw-whet owl (*Aegolius acadicus*). Given the wide tolerance red spruce exhibits for the range of these characteristics and its ability to withstand mesic and hydric conditions and take advantage of small microhabitats such as nurse logs and hummocks, red spruce is a common component of healthy central Appalachian high-elevation wetland systems. In forested swamps, red spruce is a primary component in all but the most alkaline systems, in which case balsam fir (*Abies balsamea*) becomes the dominant canopy species. More acidic conifer woodland swamps such as Red Spruce—Hemlock/Great Laurel Swamp Forest (see Sect. 4.2.2.2 of Chap. 4) maintain a mostly closed canopy of dominant red spruce. Semi-open-canopy wetlands, such as Red Spruce / Heath Peat Woodland (see Sect. 4.2.2.6 of Chap. 4) are in constant flux, with water table fluctuations controlling recruitment and encroachment of red spruce, as well as intermittent flooding that produces red spruce snags. In open wetland types such as cottongrass fens, shrub peatlands, and wet meadows, red spruce may be reduced in abundance but still important to the ecological function of the system, providing limited cover, shade, nesting areas, and microclimates (Francl et al. 2004; Byers et al. 2007).

Linear forested seeps are a unique forested wetland type of which red spruce is often a dominant component of the overstory (Byers et al. 2007). Forested seeps are characterized by groundwater discharge areas in coves and concavities on mountainsides. They occupy steeper slopes than other Appalachian wetland types and have been almost universally affected by past land uses, which accelerated erosion, caused groundwater to surface, created artificial channel formation, and consequently narrowed the width of these wetland communities. Restoration of forested seeps includes reducing or eliminating ground disturbance in cove positions. Red spruce restoration in cove positions, including selective felling of competing hardwoods and leaving large woody material on the ground, can accelerate forested seep recovery by collecting sediment, slowing water flow, and recreating swales from incised channel conditions (Fig. 8.12).

While most wetland types within the red spruce forest matrix are dynamic, with changing boundaries and canopy openness based on intermittent disturbance and climatic shifts, some wetlands show extreme long-term stability. Cranberry Glades in West Virginia and Cranesville Swamp along the West Virginia–Maryland border have likely persisted in a mostly open wetland state for thousands of years, while other wetlands with similar plant communities but more dynamic hydrology, such as beaver-influenced wetlands, may be more transitory in nature (Moorehead 2003; Francl et al. 2004; Byers et al. 2007; Thompson et al. 2012). Understanding factors controlling vegetation in wetlands, such as soil saturation and pH, provides the context for the extent, method, and potential success of red spruce restoration. Restoration may include actions to raise water levels to increase soil saturation, alter surface topography to interact more directly with groundwater, or increase water-holding capacity. Activities may include adding large woody material or beaver-dam analogs to stream channels and flow paths, excavating vernal pools and depressions, and filling drainage ditches.

Downed trees and pit-and-mound topography from windthrown trees are below baseline historical conditions in and near these wetlands due to historical logging, fire, grazing, and erosional processes (Ruffner and Abrams 2003; Plotkin et al. 2017). Transitional zones from wetlands to upland communities provide opportunities to selectively release red spruce by felling competing hardwoods, which adds nurse

Fig. 8.12 Linear forested seep in abandoned grazing area. Surrounding hardwoods could be cut into the seep area to increase coarse woody material, retain organic material, expand seep, and release red spruce (*Picea rubens*; photo by Chad Landress)

logs to adjacent wetlands. In more heavily impacted floodplain or wetland areas, re-creation of pit-and-mound topography or addition of trees with attached root wads can reinitiate more complex topography conducive to red spruce recruitment (Liechty et al. 1997; Clinton and Baker 2000; Ulanova 2000; Fig. 8.13).

Hillslope Hydrology Restoration

The restoration of hillslope hydrology aims to increase groundwater infiltration and restore the natural flow paths of water through the watershed. Restored hydrology can increase groundwater retention, buffer drought stress, reduce erosion, improve organic soil development, and reduce habitat fragmentation in red spruce ecosystems (Lloyd et al. 2013; Larson and Rew 2022). Roads are probably the most widespread alteration to hillslope hydrology in the Appalachians. This includes open drivable roads, as well as roads closed to traffic, timber skid roads, trails, and any substantial linear feature with alterations to groundwater and surface water flow characteristics. Roads affect hydrologic processes by intercepting rainfall directly on the road surface, bringing groundwater to the surface in the road cutbank, concentrating flow on the surface or in associated ditches, and rerouting water down artificial flow paths.

Fig. 8.13 Recreated pit and mound topography using trees that were removed to release red spruce (*Picea rubens*) in an adjacent area. Red spruce seedlings are planted on the mounds (photo by Todd Miller)

They increase the rate of water leaving the watershed, reduce the function and extent of forested seeps, increase water temperatures, and fragment habitat for terrestrial species dependent on red spruce for the underlying forest floor conditions (Trombulak and Frissell 2000). These effects are most visible on actively used roads, but compacted soils and altered hydrology are persistent on abandoned roads for decades to centuries (Trombulak and Frissell 2000). On heavily compacted roads, red spruce often forms *dog-hair* thickets with impaired growth and shallow root systems that are easily windthrown.

Restoration techniques are dependent on any potential use of the road corridor and may be tailored to fit such uses, although some trade-offs to ecosystem function may be incurred. Varying levels of road treatments range from increased dispersal of water from the road surface to complete obliteration of the road by decompaction and recontouring. Complete road obliteration involves decompacting the road surface with an excavator and moving fill slope (down slope from road) material back into the original roadway to approximate the original hillslope contour. Drainage paths can be restored and naturalized, and groundwater intercepted from the cut slope (up slope from road) can usually be returned as subsurface flow. Complete obliteration provides the most groundwater retention as well as improved soil conditions for red spruce growth (Larson and Rew 2022). Restoration of roads also provides an opportunity to increase red spruce distribution through active planting. In areas where red spruce is absent or below desired conditions, extensive forest road networks treated for hydrologic restoration can be planted to expand the red spruce ecosystem (Fig. 8.14).

8.4.3.2 Soils

Soil conditions are a critical component of red spruce-northern hardwood forests. Although it is very difficult to directly manage soil conditions, it is nevertheless important to understand a site's soil as well as the potential influence that the soil and vegetation have on each other (see Chap. 3). Changes to soil as a result of restoration practices do not happen quickly. Depending on the soil property, these changes can take place on decadal time scales, across numerous centuries, or even across millennia. Yet, there are observable changes to soil properties and soil function that are more immediate, referred to as dynamic soil properties (DSP; Wills et al. 2017). Dynamic soil properties refer to any soil characteristics that change over human timescales in response to land management or disturbance. Properties such as bulk density and permeability are considered DSP, while properties such as rock fragment content are inherent soil properties that do not readily change. Furthermore, many DSP can have a direct or indirect influence on soil functions and resultant ecosystem services (Daily et al. 1997, Robinson et al. 2012, Adhikari and Hartemink 2016a, b), both at the site-level and at larger watershed or regional scales. When we apply restoration practices, we seek not only to improve red spruce extent, but also the DSP and associated ecosystem services derived from soil functions (see Chap. 3).

A key DSP in almost all soils, but especially red spruce soils, is soil organic carbon (SOC). Organic matter is continually incorporated into the soil, both at the surface and deeper below ground. Soil organic carbon is accumulated through this continual addition of organic matter. As restoration adds red spruce into the canopy, the quantity and quality of organic matter additions change and SOC content increases, thus improving soil function and ecosystem services that stem from these DSP changes (Townsend 2010; Ciccarese et al. 2012; Nauman et al. 2015b).

Soil organic carbon influences numerous other soil properties (Jobbágy and Jackson 2000, Drenovsky et al. 2004, Essington 2015; see also Fig. 3.4 in Chap. 3). For example, SOC influences properties like soil structure and bulk density, which further affect soil porosity, infiltration rate, aggregate stability, and water holding capacity. The functioning of these properties combines to provide reduced erosion and sedimentation in rivers and streams, as well as downstream flood mitigation. Soil organic carbon also improves nutrient cycling and cation exchange capacity, which further influence elemental transformations, nutrient availability for soil biota, and the buffering capacity of a soil. The interconnectedness of soil properties with one another, as well as the function of ecosystem services that they provide, is crucial for overall ecosystem health and resilience. Restoration practices that target the improvement of soil properties, or are at minimum are mindful of resulting changes to soil properties, should be considered in restoration and management plans.

Fig. 8.14 A compacted logging road with minimal tree recruitment is being decompacted and outsloped with an excavator (top); newly decompacted and outsloped logging road within a northern hardwood forest provides an excellent opportunity for expanding red spruce coverage (middle); red spruce (*Picea rubens*) planted on the obliterated road a few years after treatment and planting (bottom; photos by Chad Landress)

8.4.4 *Mined Land Restoration to Restore Red Spruce Ecosystems*

Surface coal mining has impacted nearly 600,000 ha (1,482,632 ac) across Appalachia (Zipper et al. 2011), resulting in habitat loss and forest fragmentation. Strip mining at higher elevations has impacted additional spruce forests in the central Appalachians. Mined lands reclaimed after the enactment of the Surface Mining Control and Reclamation Act of 1977 (SMCRA; Congress 1977) are required to be reshaped to approximate original contour, and compacted using heavy machinery to stabilize the soil. Mined sites are often reclaimed to vegetative communities that are not representative of the pre-mining condition (Adams 2017), usually with non-native species that may compete with native flora. Due to the loss of the original topsoil and heavy soil compaction, sites reclaimed to forest post-mining often exhibit poor tree survival and stunted growth (Hall et al. 2009; Burger et al. 2017; Groninger et al. 2017), and exhibit arrested succession, whereby regeneration of native plants and later-successional species is hindered or drastically slowed, impeding long-term recovery (Franklin et al. 2012; Groninger et al. 2017).

To address the ecological recovery limitations from SMCRA restoration methods, a team from the Office of Surface Mining Reclamation and Enforcement, state agencies throughout Appalachia, environmental groups, and reclamation and restoration experts formed the Appalachian Regional Reforestation Initiative (ARRI) developed a method called The Forestry Reclamation Approach, or FRA. The steps of the FRA outline how to create a suitable root medium for trees, reduce compaction, reduce competition from seeded groundcovers, plant early successional trees as well as high-value timber, and plant trees using proper techniques (Adams 2017). For more detailed information about FRA methods see the ARRI program page: www.osmre.gov/programs/arri.

Benefits from this ecological restoration work on mined lands range from watershed improvement to soil formation and nutrient cycling to increased carbon sequestration from the plantings. Additionally, forest fragmentation is reduced, and species-rich native plant communities are established. This also creates early successional wildlife habitat, including pollinator habitat in the short term, and red spruce forests in the long term. Economic benefits of this work include *repurposing* the skills of coal mine industry professionals, contracting regional engineers to create wetlands, and putting money back into local and regional economies through the purchase of seedlings and hiring of tree-planting crews. Volunteer tree-planting events provide local schools with opportunities for educational outreach.

A large-scale restoration project on legacy mined land on Cheat Mountain of the Monongahela NF in West Virginia was initiated in 2009 to restore the historical red spruce ecosystem. The project began on the Mower Tract, a 16,200 ha (40,031 ac) legacy mine in Randolph County near Huttonsville. In the late 1970s, after the enactment of SMCRA, 810 ha (2,000 ac) of the Mower Tract were surface mined for coal. Forty years post-reclamation the impact of mining is still visible on these legacy lands. Restoration at Sharp Knob, located in Pocahontas County near Slaty

Fork, began in 2018. In contrast to the Mower Tract, Sharp Knob is an abandoned mine; it was mined in the 1960s before the enactment of SMCRA. Reclamation pre-SMCRA did not include recontouring or severe compaction, but did include seeding of exotic grasses. Sharp Knob differs most notably from the Mower Tract because it has steep highwalls, flat mine benches with large ponds, and the ground in some areas is so minimally compacted that it allowed for soil formation and unrestricted growth of plantation trees. The Mower Tract and Sharp Knob are part of one of the largest intact remnants of red spruce forest south of Maine, which give them high conservation and preservation value. The mined lands scar the landscape and fragment the surrounding diversity-rich high-elevation red spruce-northern hardwood forests. The Monongahela NF ecological restoration project's goals include decompaction of the mined lands to break the arrested succession and reforestation with red spruce and a diversity of native trees and shrubs, which improve the watershed and wildlife habitat and set the land on a trajectory toward becoming a functioning red spruce forest. The holistic set of site preparation activities described above, including decompacting of the ground, woody debris loading, and construction of wetlands, is implemented prior to reforestation to maximize restoration success. See Sect. 9.6.1.1 in Chap. 9 for more information on ecological restoration outcomes of mined lands on the Monongahela NF.

8.5 Adaptive Management to Achieve Long-Term Restoration Goals

Incorporating long-term monitoring and adaptive management within a restoration program is essential for the successful restoration and maintenance of resilient, long-lived red spruce and spruce-fir ecosystems governed by dynamic and complex interactive processes. Given the difficulty of predicting the trajectory of a restoration project at the outset, monitoring programs must be designed to detect changes resulting from management actions. Examples of adaptive management in red spruce restoration projects could include: (1) enrichment plantings when regeneration methods failed to establish sufficient stocking of red spruce in the regeneration layer; (2) tending the regeneration layer when unwanted species threaten the canopy ascension of released red spruce; (3) expanding created gaps that have closed since the original release treatment; or (4) accepting that adaptive management actions may be too costly and resources could be better applied to an area more likely to respond positively to treatments.

The CASRI partnership developed basic streamlined monitoring approaches for both planting and canopy release of red spruce, and SASRI is working on similar approaches at the time of this publication. The planting protocol consists of measuring survival, height, and health of planted red spruce and associated hardwoods on a representative sample of planting sites. Similarly, the canopy release monitoring protocol has evolved to focus primarily on the most critical efficacy data: health and growth of red spruce before and after release. Monitoring programs developed for adaptive management may benefit from the use of existing standardized methods. Implementing standardized monitoring methodologies across partner organizations and agencies can help ensure that data will be standardized across a restoration landscape, regardless of jurisdictional boundaries.

Examples of established national protocols that can be adapted for red spruce restoration monitoring include the Forest Inventory and Analysis (FIA) National Core Field Guide for the Nationwide Forest Inventory (USDA Forest Service 2023), the Eastern Rivers and Mountains Network (ERMN) Vegetation and Soil Monitoring Program Standard Operating Procedures Version 10.0 (National Park Service 2019), and various documents from the National Ecological Observation Network (NEON). These standardized protocols have been utilized by the scientific community for many years and undergo continual review. In addition, they are used across the continental U.S., and although some adjustments may occur to the protocols to meet specific regional needs, all have the same standard measurements for monitoring and sampling to be comparable across regions.

Developing a core set of measurements is critical to assess the restoration of any ecosystem. Once this core set of measurements is established, other variables may be included to address specific research questions. The simplest monitoring protocols establish small, circular radius plots with only target-tree measurements of height and DBH. Although these plots provide important information about annual growth of target trees, they do not address the responses of associated species to the restoration actions. To better assess ecosystem restoration, additional measurements can be taken based on monitoring protocols like those in the FIA or ERMN documents. Implementation includes mapping and measuring height and DBH of all trees, assessing forest structure from the canopy to nonvascular plant layer (i.e., bryophytes), detailing snags and coarse woody debris, and characterizing the soil (e.g., thickness of the forest floor). Special attention should be placed on the species of specific interest (i.e., red spruce) where tree health is assessed, leader growth of the last few years is determined, and it is determined whether the trees are producing viable cones. Finally, regeneration can be characterized using microplots or transects, and herbaceous responses can be measured to further document community changes post-restoration. By continuously monitoring the effects of restoration, and adjusting management activities based on those observed responses, land managers can maximize the ecological benefits of their restoration actions when time and resources are limited.

8.6 Conclusions

The restoration activities outlined in this chapter cannot, in our lifetimes, bring the land back to its original ecological state before broad-scale anthropogenic disturbance. However, through this work in central and southern Appalachia, we have seen restoration of red spruce-northern hardwood forests, aquatic-focused restoration highlighting red spruce, and reclamation of mined lands result in the beginnings of positive ecological change. These outcomes include the establishment of native plant and tree assemblages initially lost to logging, fire, and mining, habitat development and connectivity coupled with use by countless species, and mitigation of arrested succession. Some improvements in ecological function can be seen immediately, such as water percolation into the ground and native plant volunteers (Branduzzi 2020). Some improvements can be seen within a few years, like amphibian (Lambert et al. 2021; Sherman et al. 2024), bat (Snyder 2022), and bird (Davenport 2023) colonization, tree growth (Rhodes and Barton 2024), and return of soil microbial communities (Swab et al. 2017). Others, such as soil building, will take decades if not centuries. The holistic set of ecological restoration activities implemented across a landscape shaped by red spruce forests will help mitigate the damage caused by centuries of anthropogenic disturbance and allow the land to begin recovering, setting it on a trajectory toward supporting resilient red spruce communities once again.

References

Adams MB (2017) The forestry reclamation approach: guide to successful reforestation of mined lands. USDA forest service northern research station general technical report NRS-169, Newtown Square, Pennsylvania

Anderson MG, Clark M, Olivero AP et al (2023) A resilient and connected network of sites to sustain biodiversity under a changing climate. Proc Natl Acad Sci 120:e2204434119

Beane NR, Rentch JS, Schuler TM (2013) Using maximum entropy modeling to identify and prioritize red spruce forest habitat in West Virginia. USDA forest service northern research station research paper NRS-23, Newtown, Pennsylvania

Beane NR, Rentch JS (2015) Using known occurrences to model suitable habitat for a rare forest type in West Virginia under select climate change scenarios. Ecol Restor 33:178–189

Bestelmeyer BT, Brown JR (2010) An introduction to the special issue on ecological sites. Rangelands 32:3–4

Bestelmeyer BT, Tugel AJ, Peacock GL Jr et al (2009) State-and-transition models for heterogeneous landscapes: a strategy for development and application. Rangel Ecol Manag 62:1–15

Blum BM (1990) Red spruce *Picea rubens* Sarg. In: Burns RM, Honkala BH (eds) Silvics of North America. vol 1. Conifers. USDA forest service handbook 654, Washington, DC, pp 250–259

Branduzzi AM (2020) Enhancing native plant diversity on legacy minelands. University of Kentucky, Lexington, Kentucky, Thesis

Briske DD, Fuhlendorf SD, Smeins FE (2005) State-and-transition models, thresholds, and rangeland health: a synthesis of ecological concepts and perspectives. Rangel Ecol Manag 58:1–10

Burger JA, Zipper CE, Angel PN et al (2017) Establishing native trees on legacy surface mines. In: Adams MB (ed) The forestry reclamation approach: guide to successful reforestation of mined lands. USDA forest service northern research station general technical report NRS-169, Newtown Square, Pennsylvania, p 83–94

Burkhart JQ (2011) A 'devilish queer place': An assessment of forest age structure using dendrochronology at Bear Rocks Preserve. West Virginia University Report, Morgantown, West Virginia, West Virginia

Byers EA, Vanderhorst JP, Streets BP (2007) Classification and conservation assessment of high elevation wetland communities in the Allegheny Mountains of West Virginia. West Virginia division of natural resources report, Elkins, West Virginia

Byers EA, Vanderhorst JP, Streets BP (2010) Classification and conservation assessment of upland red spruce communities in West Virginia. West Virginia division of natural resources report, Elkins, West Virginia

Byers EA, Cimarolli S, Jones K et al (2011) Central Appalachian Spruce Restoration Initiative (CASRI): restoration approach. Central Appalachian Spruce restoration initiative report, Elkins, West Virginia

Ciccarese L, Mattsson A, Pettenella D (2012) Ecosystem services from forest restoration: thinking ahead. New For 43:543–560

Clark M, Hall KR, Martin DM et al (2023) Prioritizing restoration sites that improve connectivity in the Appalachian landscape, USA. Conserv Sci Pract 5:e13046

Clinton BD, Baker CR (2000) Catastrophic windthrow in the southern Appalachians: characteristics of pits and mounds and initial vegetation responses. For Ecol Manag 126:51–60

Congress US (1977) Surface mining control and reclamation act of 1977. Public law 95:87

D'Amato A (2021) Ecological silviculture for northern hardwood ecosystems of northeastern U.S. Ecological silviculture: foundations and applications. In: Brian Palik Anthony W. D'Amato John Wiley & Sons (eds) Long Grove, Illinois, Waveland Press, Inc., p 27

Daily G, Matson PA, Vitousek PM (1997) Ecosystem services supplied by soil. In: Daily GC (ed) Nature services: societal dependence on Natural Ecosystems. Island Press, Washington DC, pp 113–132

Davenport RN (2023) Effects of forest reclamation and landscape features on avian occupancy, species richness, and abundance in Appalachia. University of Kentucky, Lexington, Kentucky, Thesis

Drenovsky RE, Vo D, Graham KJ et al (2004) Soil water content and organic carbon availability are major determinants of soil microbial community composition. Microb Ecol 48:424–430

Essington ME (2015) Soil and water chemistry: an integrative approach. Taylor & Francis, Boca Raton, Florida

Francl KE, Ford WM, Castleberry SB (2004) Characterization of high elevation central Appalachian wetlands. USDA forest service northern research station research paper NE-725, Newtown Square, Pennsylvania

Frank RM, Bjorkbom C (1973) A silvicultural guide for spruce-fir in the Northeast. USDA forest service northeastern forest experiment station general technical report NE-6, Upper Darby, Pennsylvania

Franklin JA, Zipper CE, Burger JA et al (2012) Influence of herbaceous ground cover on forest restoration of eastern U.S. coal surface mines. New For 43:905–924

Game ET, Groves C, Andersen M et al (2010) Incorporating climate change adaptation into regional conservation assessments. The nature conservancy report, Arlington, Virginia

Groninger J, Skousen J, Angel P et al (2017) Mine reclamation practices to enhance forest development through natural succession. In: Adams MB (ed) The forestry reclamation approach: guide to successful reforestation of mined lands. USDA forest service northern research station general technical report NRS-169, Newtown Square, Pennsylvania, pp 65–71

Hall SL, Barton CD, Baskin CC (2009) Seed viability in stockpiled topsoil on a surface mine in Appalachia. Ecol Restor 27:381–383

Jobbágy EG, Jackson RB (2000) The vertical distribution of soil organic carbon and its relation to climate and vegetation. Ecol Appl 10:423–436

Kenefic LS, Kabrick JM, Knapp BO et al (2021) Mixedwood silviculture in North America: the science and art of managing for complex, multi-species temperate forests. Can J for Res 51:921–934

Lambert M, Drayer AN, Leuenberger W et al (2021) Evaluation of created wetlands as amphibian habitat on a reforested surface mine. Ecol Eng 171:106386

Larson CD, Rew LJ (2022) Restoration intensity shapes floristic recovery after forest road decommissioning. J Environ Manag 319:115729

Lassettre NS, Harris RR (2001) The geomorphic and ecological influence of large woody debris in streams and rivers. University of California Cooperative Extension, Paper presented at the Large Woody Debris recruitment monitoring workshop

Liechty HO, Jurgensen MF, Mroz GD et al (1997) Pit and mound topography and its influence on storage of carbon, nitrogen, and organic matter within an old-growth forest. Can J for Res 27:1992–1997

Lloyd RA, Lohse KA, Ferré TPA (2013) Influence of road reclamation techniques on forest ecosystem recovery. Front Ecol Environ 11:75–81

Lowe WH, Likens GE (2005) Moving headwater streams to the head of the class. Bioscience 55:196–197

Medlock K (ed) (2015) Southern Appalachian spruce restoration plan. https://southernspruce.org/wp-content/uploads/2018/07/final_southern-appalachian-spruce-restoration-plan1.pdf. Accessed 26 Jul 2024

Moorehead KK (2003) Effects of drought on the water-table dynamics of a southern Appalachian Mountain floodplain and associated fen. Wetlands 23:792–799

Muller RN, Liu Y (1991) Coarse woody debris in an old-growth deciduous forest on the Cumberland Plateau, southeastern Kentucky. Can J for Res 21:1567–1572

National Park Service (2019) Eastern rivers and mountains network vegetation and soil monitoring program, standard operating procedures, ver 10.0

Nauman TW, Thompson JA, Teets SJ et al (2015a) Ghosts of the forest: mapping pedomemory to guide forest restoration. Geoderma 247–248:51–64

Nauman TW, Thompson JA, Teets JA et al (2015b) Pedoecological modeling to guide forest restoration using ecological site descriptions. Soil Sci Soc Am J 79:1406–1419

Niedermaier KM, Atkins JW, Grigri MS et al (2022) Structural complexity and primary production resistance are coupled in a temperate forest. Front For Glob Chang 5:941851

Noss RF, LaRoe ET, Scott JM (1995) Endangered ecosystems of the United States: a preliminary assessment of loss and degradation. U.S. National biological service biological report 28, Washington, DC

Plotkin AB, Schoonmaker P, Leon B et al (2017) Microtopography and ecology of pit-mound structures in second-growth versus old-growth forests. For Ecol Manag 404:14–23

Raymond P, Bédard S, Roy V et al (2009) The irregular shelterwood system: review, classification, and potential application to forests affected by partial disturbances. J Forest 107:405–413

Rentch JS, Schuler TM, Nowacki GJ et al (2010) Canopy gap dynamics of second-growth red spruce-northern hardwood stands in West Virginia. For Ecol Manag 260:1921–1929

Rentch JS, Ford WM, Schuler TS et al (2016) Release of suppressed red spruce using canopy gap creation—ecological restoration in the central Appalachians. Nat Areas J 36:29–37

Rhodes B, Barton CD (2024) Comparing the response of red spruce plantings on legacy coal mines and old-field restoration sites in the West Virginia highlands. Nat Areas J 44:65–75

Robinson DA, Emmett BA, Reynolds B, Rowe EC, Spurgeon D, Keith AM, Lebron I, Hockley N (2012) Soil natural capital and ecosystem service delivery in a world of global soil change. In: Hester RE, Harrison RM (eds) Soils and food security. Issues in environmental science and technology series, pp 41–68

Ruffner CM, Abrams MD (2003) Disturbance history and stand dynamics along a topographic gradient in old-growth hemlock-northern hardwood forests of the Allegheny Plateau, USA. Nat Areas J 23:98–113

Schafale MP (2024) Classification of the natural communities of North Carolina, fourth approximation. North Carolina department of environment and natural resources natural heritage program, Raleigh, NC

Sherman L, Barton CD, Guzy JC et al (2024) Wetland creation and reforestation of legacy surface mines in the central Appalachian region (USA): a potential climate-adaptation approach for pond-breeding amphibians? Water 16:1202

Siderhurst LA, Griscom HP, Hudy M et al (2010) Changes in light levels and stream temperatures with loss of eastern hemlock (*Tsuga canadensis*) at a southern Appalachian stream: implications for brook trout. For Ecol Manag 260:1677–1688

Snyder BC (2022) Bat activity on West Virginia mined lands restored via the forestry reclamation approach. University of Kentucky, Lexington, Kentucky, Thesis

Society for Ecological Restoration (2004) The SER international primer on ecological restoration. Science and policy working group report, Tucson, Arizona

Strahler AN (1957) Quantitative analysis of watershed geomorphology. Trans Am Geophys Union 38:913–920

Swab RM, Lorenz N, Byrd S et al (2017) Native vegetation in reclamation: improving habitat and ecosystem function through using prairie species in mine land reclamation. Ecol Eng 108:525–536

Thompson Y, D'Angelo EM, Karathanasis AD et al (2012) Plant community composition as a function of geochemistry and hydrology in three Appalachian wetlands. Ecohydrology 5:389–400

Townsend L (2010) Ecological site descriptions: developmental considerations for woodlands and forests. Rangelands 32:37–42

Trombulak SC, Frissell CA (2000) Review of ecological effects of roads on terrestrial and aquatic communities. Conserv Biol 14:18–30

Ulanova NG (2000) The effects of windthrow on forests at different spatial scales: a review. For Ecol Manag 135:155–167

USDA Forest Service (2023). U.S. Forest Inventory and Analysis national core field guide for the nationwide forest inventory, ver 9.3. https://research.fs.usda.gov/understory/nationwide-forest-inventory-field-guide. Accessed 2 Aug 2024

[USDA NRCS] USDA Natural Resources Conservation Service (2017) National ecological site handbook. https://www.nrcs.usda.gov/resources/guides-and-instructions/national-ecological-site-handbook. Accessed 17 Jun 2024

[USDA NRCS] USDA Natural Resources Conservation Service (2016a) Ecologic site F127XY001WV: spodic shale upland conifer forest. Available via the ecosystem dynamics interpretive tool. https://edit.jornada.nmsu.edu. Accessed 17 Jun 2024

[USDA NRCS] USDA Natural Resources Conservation Service (2016b) Ecologic site F127XY002WV: spodic intergrade shale upland hardwood and conifer forest. Available via the ecosystem dynamics interpretive tool. https://edit.jornada.nmsu.edu. Accessed 17 Jun 2024

Webster CR, Jenkins MA (2005) Coarse woody debris dynamics in the southern Appalachians as affected by topographic position and anthropogenic disturbance history. For Ecol Manag 217:319–330

Westveld M (1938) Silvicultural treatment of spruce stands in northeastern United States. J Forest 36:944–956

Westveld M (1931) Reproduction on pulpwood lands in the Northeast. U.S. department of agriculture technical bulletin no. 233, Washington, DC

White PS, Sutter RD (1999) Southern Appalachian grassy balds: lessons for management and regional conservation. In: Peine JD (ed) Ecosystem management: principles and practices illustrated by a regional biosphere cooperative. Lucie Press, Delray Beach, Florida, St, pp 375–396

White PS, Wilds S, Stratton DA (2001) The distribution of heath balds in the Great Smoky mountains, North Carolina and Tennessee. J Veg Sci 12:453–466

Willis JL, Roberts SD, Harrington CA (2018) Variable density thinning promotes variable structural responses 14 years after treatment in the Pacific Northwest. For Ecol Manag 410:114–125

Wills S, Williams C, Seybold C (2017) Assessing dynamic soil properties and soil change. In: Ditzler C, Scheffe K, Monger HC (eds) Soil survey manual. USDA soil science division staff handbook no. 18, Government Printing Office, Washington, DC, pp 481–503

Wipfli MS, Richardson JS, Naiman RJ (2007) Ecological linkages between headwaters and downstream ecosystems: transport of organic matter, invertebrates, and wood down headwater channels. J Am Water Resour Assoc 43:72–85
Yetter E, Chhin S, Brown JP (2021) Sustainable management of central Appalachian red spruce. Sustainability 13:10871
Zipper CE, Burger JA, Skousen JG et al (2011) Restoring forests and associated ecosystem services on Appalachian coal surface mines. Environ Manag 47:751–765

Chapter 9
History and Accomplishments of Red Spruce Restoration Initiatives Across the Central and Southern Appalachians

Alton C. Byers, Melissa A. Thomas-Van Gundy, Katherine Medlock, and Kathryn M. Shallows

9.1 Restoration Community Responds to an Era of Loss

As detailed in Chap. 1, the extent of red spruce (*Picea rubens*) within the central and southern Appalachians was drastically reduced as a result of logging after European colonization (Fig. 9.1). Red spruce forests were logged extensively from the 1880s to the 1930s, with some areas experiencing subsequent catastrophic wildfires fueled by large amounts of logging slash. On many of these sites, hardwood tree species were more competitive and became the dominant species after logging. Red spruce forests are still recovering from this era of unchecked logging followed by extensive wildfire. Additionally, from the 1960s to the late 1980s, these forests experienced widespread declines from acidic deposition (especially sulfur compounds) and mortality of Fraser fir (*Abies fraseri*) from the introduction of the balsam woolly adelgid (*Adelges piceae*; Eagar and Adams 1992). In 1995, a report to the U.S. National Biological Service (now the Biological Resources Division of the U.S. Geological Survey) stated that the red spruce and spruce-fir forests of the central and southern Appalachians were

A. C. Byers (✉)
University of Colorado, Institute of Arctic and Alpine Research, Boulder, CO, USA
e-mail: alton.byers@colorado.edu

M. A. Thomas-Van Gundy
USDA Forest Service, Northern Research Station, Parsons, WV, USA
e-mail: melissa.thomasvangundy@usda.gov

K. Medlock
The Nature Conservancy, Appalachians Program, Knoxville, TN, USA
e-mail: kmedlock@tnc.org

K. M. Shallows
The Nature Conservancy, Appalachians Program, Fairmont, WV, USA
e-mail: katy.shallows@tnc.org

D. J. Brown et al. (eds.), *Ecology and Restoration of Red Spruce Ecosystems in the Central and Southern Appalachians*, https://doi.org/10.1007/978-3-032-19616-3_9

Fig. 9.1 A steam loader on a Shay engine, 1910 (photo by USDA Forest Service)

among the most endangered ecosystems in the U.S. (Noss et al. 1995). With the implementation of air quality standards through the Clean Air Act, the acidic deposition in these high elevation *cloud forests* has declined, reducing stress (Banks 2013; Mathias and Thomas 2018). In recent decades, analysis of inventory data showed increasing red spruce recruitment in the forest understory (Nowacki et al. 2010; Thomas-Van Gundy and Morin 2021), even at lower elevations previously not considered suitable habitat.

The period of broad-scale logging affecting all forests of the eastern U.S. stimulated two general modes of conservation. First, conservationists successfully campaigned for a national park to protect and preserve remaining old-growth forests in perpetuity. Congress authorized the Great Smoky Mountains National Park in 1926 and, after funds were successfully raised and lands acquired, the park was established in 1934. When Great Smoky Mountains National Park was established, about 25% of its area had not yet been logged, including about 17,000 ha (42,000 ac) of forest in which red spruce and spruce-fir were dominant, the largest block of this kind of old-growth forest in the southern Appalachians. Second, among fears that broad-scale logging would impact the future timber supply, the southern Appalachians became the *cradle of scientific forestry* with the founding of the first school of forestry, the Biltmore Forest School, near Asheville, North Carolina in 1896. The USDA Forest Service also began to protect, manage, and restore red spruce forests. Supported in part by the Weeks Act of 1911 the USDA Forest Service became, by far, the largest manager of high elevation red spruce with land acquisitions for the Monongahela NF of West Virginia in 1915, and the acquisitions of the Cherokee (Tennessee), Nantahala and Pisgah (North Carolina), George Washington and Jefferson (Virginia), Chattahoochee (Georgia), and the Sumter (South Carolina) NFs in 1936. These two conservation responses to unsustainable timber harvest were nationally influential in the development of the two dominant conservation value systems (White and Tuttle 2013), for example, National Parks represent the Preservation Ethic, and NFs represent the Resource Conservation Ethic and the mission of forestry schools. This historical context of exploitative timber harvest and legacy of preservation of unlogged red spruce *sky islands* in the Great Smoky Mountains National Park in the southern Appalachians, compared to the more contiguous restoration opportunities from logged and mined red spruce forests across the Monongahela NF in the central Appalachians, contributed to different timelines of SASRI and CASRI partnership formation, as described in the following sections.

Initial red spruce forest restoration work, predominately in the central Appalachians, was facilitated by the Civilian Conservation Corps (CCC) during the 1930s (Fig. 9.2). Beginning around 1928, the USDA Forest Service nursery in Parsons, West Virginia began reducing its production of Norway spruce (*Picea abies*) seedlings and increasing production of red spruce and red pine (*Pinus resinosa*; Fig. 9.3). According to a 1929 planting report for the Monongahela NF, "…red spruce stock from Parsons will be the chief species used" (Robert Whetsell, personal communication). Seed sources for the nursery stock included the Pisgah NF in North Carolina and locations in New Hampshire and Maine. The production of red spruce seedlings soon outpaced that of Norway spruce and was reflected in the plans for seedling production moving into the 1930s. The 1930 spring production plans for the Parsons Nursery anticipated 993,000 red spruce seedlings. The 1935 spring production plan listed the nursery stock on hand as over 7.8 million red spruce, 5.2 million red pine, 1.1 million Norway spruce, and nearly 1 million eastern white pine (*Pinus strobus*) of varying age classes. While the majority of the seedlings produced by the Parsons Nursery were planted on the Monongahela NF, a great number were also shipped to other eastern NFs, such as the Allegheny NF (Pennsylvania) and Pisgah NF (North Carolina), to aid in their reforestation efforts in the late 1920s and much of the 1930s. The CCC were key to the plantings made on the Monongahela NF between 1933 and 1942. Among the earliest CCC plantation work was at Canaan Mountain on the Cheat Ranger District of the Monongahela NF, where they continued reforestation efforts that had begun in 1925. Other locations where the CCC were active in reforestation efforts included Abes Run, Little River, and Laurel Fork on the Greenbrier Ranger District of the Monongahela NF.

Early in its formation, the Monongahela NF did include some high elevation areas that formerly supported red spruce, however, some areas remained in private ownership. For example, a now key landscape for restoration known as the Mower Tract was privately owned and strip mining and logging occurred until it was purchased in 1987 as NF system land. The first Land and Resource Management Plan (Forest Plan) for the Monongahela NF under the National Forest Management Act of 1976 was finalized and implemented in 1986. A species reliant on red spruce-dominated

Fig. 9.2 Civilian conservation corps planting crew in the lower Little River area, Monongahela NF, 1935 (photo by USDA forest service)

forests, the Virginia northern flying squirrel (*Glaucomys sabrinus fuscus*; Fig. 9.4) was listed as federally endangered in 1985 and shaped management of red spruce-dominated forests on the Monongahela NF. Given the lack of information on the species' habitat needs, the general direction was to avoid any areas with overstory red spruce. Much has been learned about the habitat needs of the Virginia northern flying squirrel since then (see Chap. 6), and its federal status has had a large influence on designing active restoration efforts on federal lands, including guides for the amount of canopy red spruce required in an area, as well as for where active restoration may not be appropriate nor necessary.

Targeted efforts to restore former red spruce forest ecosystems in the central and southern Appalachians at scale were significantly advanced with the formation of CASRI in the mid-2000s and SASRI in the 2010s. Partner organizations across both restoration initiatives strive to restore red spruce-dominated communities across administrative boundaries, prioritize restoration efforts, work with partners to obtain funding, and raise public awareness of the unique characteristics of these forests, among other goals.

Fig. 9.3 Women weeding 1-year-old red spruce seedlings at the Parsons, West Virginia nursery in 1940. Women were recruited for this task because of their dexterity (photo by USDA forest service)

9.2 Origins of the Spruce-Fir Restoration Initiatives

Ecoregional restoration initiatives arise in the context of common challenges in defining and implementing the scope and scale of work, but also complex human and organizational dynamics (Adams et al. 2016). Long-term restoration partnerships can wrestle with tensions between restoration goals (e.g., priority places, species, and ecosystem services) and the methods to achieve them (i.e., long-term natural succession versus active management). The continuous need to advocate for staff capacity and funding, and numerous practical issues of logistics to scale on-the-ground impact (e.g., appropriate seed sourcing and weather conditions) can limit success. Maintaining successful individual, organizational, partnership, and stakeholder dynamics within a partnership is complicated by interpersonal to partner-partner and partnership-community trust (Dietsch et al. 2021; Saif et al. 2022), as

Fig. 9.4 A radio-collared Virginia northern flying squirrel (*Glaucomys sabrinus fuscus*) at Mills Run, Pocahontas County, West Virginia (photo by Corinne Diggins)

well as constant leadership changes and shifting priorities. The red spruce restoration initiatives and partnerships that emerged since the early 2000s had elements of many of these factors that have sustained growth and efficacy. Here, we tell the stories of how these restoration initiatives coalesced over several decades around a common vision and succeeded in building momentum despite the challenges. Each was spurred on by the identification of significant conservation concerns and found momentum in collaboration, with the common goal of restoring red spruce ecosystems in the central and southern Appalachians. The partnerships have been maintained by the dedication of individuals across decades from a range of private citizens and restoration businesses, government agencies, non-profit organizations, and universities.

9.2.1 Identification of Spruce-Fir Conservation Issues

Prior to the growing interest in collaborative restoration of red spruce ecosystems, federal and state agencies were beginning to document species losses across the region and the urgency for restoration. Due to the historical loss and limited recovery of red spruce habitat, the Carolina northern flying squirrel (*Glaucomys sabrinus coloratus*), Virginia northern flying squirrel, and Cheat Mountain salamander (*Plethodon nettingi*; Fig. 9.5), all dependent on this ecosystem, were added to the federal list of Threatened and Endangered Species in 1985 (Carolina and Virginia northern flying squirrel) and 1989 (Cheat Mountain salamander). Their listing refocused agency attention on red spruce habitat, and extensive research on their biology and ecological needs was funded by the USDA Forest Service, U.S.

Fig. 9.5 Adult cheat mountain salamander (*Plethodon nettingi*) in a high elevation red spruce (*Picea rubens*) forest in Tucker County, West Virginia (photo by Donald Brown)

Fish and Wildlife Service, and West Virginia Division of Natural Resources. The Virginia northern flying squirrel was delisted in 2008, then re-listed in 2011, and delisted again in 2013. In support of the species recovery plan, researchers and land managers documented individuals and developed habitat suitability models. Many social factors were involved in the change in federal status including the seemingly contradictory idea that a rare species would benefit from active forest management. Recognition of the importance of red spruce forests for this endangered species galvanized action and contributed to partnership formation in both the central and southern Appalachians.

In the early 2000s, the USDA Forest Service Northern Research Station in Parsons, West Virginia added the restoration of red spruce-dominated forests to its 5-year plan of work. Research projects were initiated that addressed silvicultural options and wildlife habitat needs, including the first attempts to predict habitat for the Virginia northern flying squirrel. This was the first time the research lab included restoration of high elevation forests to its formal research portfolio. This paved the way for scientists to be involved in CASRI, continue work on endangered species and active management for restoration, and provide leadership in organizing and publishing the first conference proceedings in 2009 (Rentch and Schuler 2010).

9.2.2 Origins of CASRI—Central Appalachian Spruce Restoration Initiative

The CASRI partnership had its origins in the successful multi-partner Blister Swamp project that restored 20 ha (49 ac) of red spruce wetlands near the Sinks of Gandy in West Virginia from 1999 to 2000 (see Sidebar 9.1). West Virginia University biologists had confirmed that the wetland hosted populations of globally rare and uncommon wildflowers that had been severely impacted by more than 100 years of cattle grazing. The site was privately owned but surrounded by extensive public lands, and therefore represented a critical habitat connectivity barrier and priority for local restoration partners. The Mountain Institute of West Virginia and The Nature Conservancy joined the effort and received a grant from the U.S. Fish and Wildlife Foundation to complete the fence around core areas of the wetlands to save the remaining rare plants. Partners and volunteers completed baseline surveys and planted native trees and shrubs. The Mountain Institute of West Virginia proposed that the federal, state, and non-governmental conservation organizations, who worked collaboratively at Blister Swamp, formalize a partnership to continue protecting and restoring threatened wetlands throughout the state. Then-Monongahela NF Supervisor Clyde Thompson (Fig. 9.6) encouraged the group to envision beyond the protection and restoration of wetlands and consider conservation of all red spruce ecosystems. The U.S. Forest Service committed to expanding the balsam fir (*Abies balsamea*) and Canaan fir (*Abies balsamea* var. *phanerolepis*) seedling planting at Blister Swamp to the public lands downstream. This successful conservation partnership for restoration, founded upon the principles of cooperation and community-based approaches, provided an example to the conservation community. As a result, the High Elevation Conservation Working Group (HECWG) was established around 2006–2007, a precursor of the partnership that would later become CASRI. Conservationists and practitioners focused on building capacity, funding, agency relationships and policies, planning, and other tools to implement restoration at scale. An outreach professional was invited to a working group meeting in 2007 and argued that the unpronounceable acronym HECWG was a barrier to outreach, proposing instead the easier acronym CASRI, for the Central Appalachian Spruce Restoration Initiative, centering the initiative on the ecosystem and not the topography. By 2012, the USDA Forest Service, partners, and volunteers expanded the Blister Swamp restoration efforts to include an adjacent 16 ha (40 ac) on public land, erecting more deer exclosures and planting red spruce and balsam fir seedlings.

Sidebar 9.1 Blister Swamp restoration and the launch of a movement

Blister Swamp, located in Pocahontas County in West Virginia, is one of the last spruce-fir circumneutral wetlands remaining in unglaciated eastern North America, with the calcium-rich water and cool temperatures supporting a rich fen flora characteristic of more northern latitudes (Byers et al. 2010;

Fig. 9.6 Clyde Thompson, retired forest supervisor of the Monongahela NF (Photo by USDA forest service)

Vanderhorst et al. 2012). *Blister* refers to the resin blisters found on the bark of balsam fir trees. West Virginia University (WVU) biologist Roy Clarkson in 1957 wrote "before the turn of the century... Blister Swamp lay serene in the midst of an enormous spruce forest which stretched, at the higher elevations, for many miles in every direction" (Clarkson 1957). By the 1950s conservationists observed that after close to 100 years of cattle grazing in the area the rare plant communities were nearly eliminated (Clarkson 1957; Stephenson and Adams 1986). Beginning in the late 1990s researchers from WVU, West Virginia state agencies, and conservation organizations worked with the private landowners to delineate a 50-acre area for protection and acquired funds to fence and plant locally sourced balsam fir seedlings (Fig. 9.7). Plantings by conservation partners continued into the 2000s, and an additional 40 acres of adjacent public

Fig. 9.7 Blister Swamp restoration site forms the headwaters of the East Fork of the Greenbrier River in Pocahontas County in West Virginia and hosts rare spruce-fir circumneutral wetlands (top). Volunteers construct fences to discourage deer browse of native balsam fir seedlings (*Abies balsamea*) planted within the Blister Swamp exclosure in 2025 (bottom left). Red spruce (*Picea rubens*), planted in 2010, is thriving in the restored wetland margin (bottom right; photos courtesy of Alton and Elizabeth Byers)

lands were fenced and restored. By 2005, the momentum built from collaboration and successes of the Blister Swamp restoration project galvanized the many conservation partners involved to propose formalizing and expanding their restoration efforts, setting in motion the eventual establishment of CASRI as a regional partnership. Red spruce restoration efforts have continued in the area, including an additional 120-acre area fenced and planted with red spruce, balsam fir, and other native species in 2024.

9.2.3 Origins of SASRI—Southern Appalachian Spruce Restoration Initiative

On December 4, 2012, several restoration organizations planned to gather in Asheville, North Carolina to discuss red spruce restoration efforts in the southern Appalachians. What was originally planned to be a meeting of about a dozen people drew over 50 individuals from 15 to 20 organizations. The group discussed everything from red spruce genetics to silvicultural techniques to propagation needs. A member from CASRI also came to the meeting to share their success. At the end of the day, the answer to the question of whether a sister initiative called SASRI should be created was a resounding yes! While the broad participation in the first meeting demonstrated the enormous interest in restoring red spruce, there were three parallel efforts that came together that supported the formation of SASRI: (1) the North Carolina Wildlife Resources Commission, with support from the U.S. Fish and Wildlife Service, had been working to improve and expand habitat for the federally endangered Carolina northern flying squirrel through red spruce planting; (2) the Southern Highlands Reserve—a non-profit organization based in Lake Toxaway, North Carolina—had been working to propagate red spruce for their own property as well as larger restoration efforts; and (3) the Nature Conservancy and the USDA Forest Service were pursuing collaborative efforts to advance restoration of native forests. Specifically, a group of stakeholders on the North Zone of the Cherokee NF had recommended restoration of red spruce forests. The focus of the SASRI partnership is on forests that would be dominated or co-dominated by red spruce, or would have red spruce as a significant component in the overstory, in the absence of anthropogenic disturbance. In areas altered by logging and other disturbances, SASRI endeavors to restore the system to as close to a natural condition as is feasible.

9.3 National Forest Plans Lay the Foundation for Spruce-Fir Restoration Partnerships

In the years leading up to and following the formation of the CASRI and SASRI partnerships, USDA Forest Service-managed lands in the central and southern Appalachians, including the Monongahela NF (West Virginia), George Washington NF (Virginia), Jefferson NF (Virginia), Nantahala-Pisgah NF (North Carolina), and Cherokee NF (Tennessee), had been engaged in planning for restoration of red spruce and spruce-fir communities in their forest plans. Given the large percentage of high elevation forests on USDA Forest Service lands in the regions, this commitment enabled each partnership to meaningfully pursue the goal of restoring red spruce and spruce-fir forests.

Each forest plan addressed the need to maintain and restore red spruce and spruce-fir ecosystems as a critical component of regional biodiversity while generally protecting these ecosystems from conversion to other forest types, thereby setting the trajectory to increasing the extent of future old-growth forests. The forest plans also specified the application of similar management approaches covering natural regeneration, reforestation, and silvicultural prescriptions to enhance current composition and diversity, and to increase the extent of the ecosystem across its historical range. Older forest plans highlighted acid precipitation and balsam woolly adelgid as the current threats, while newer plans added climate change to the list.

The Cherokee NF Land and Resource Management Plan (USDA Forest Service 2004a) is one of the oldest in the region and does not provide vegetation management practices, with the explanation that restoration methods are "not well established." The Plan does address the need to maintain the > 242 ha (600 ac) of historical spruce-fir forests and restore an additional 40 ha (100 ac), excluding natural grassy and heath balds.

The Nantahala and Pisgah NF Land Management Plan (USDA Forest Service 2023) contains geographic area-specific prescriptive language for maintenance and restoration, including the need to develop climate adaptation strategies for the spruce-fir ecosystem currently covering 6,475 ha (16,000 ac) of the forests. The Nantahala and Pisgah NF Land Management Plan supports underplanting and canopy release of red spruce in historical sites, and thinning dense red spruce stands to enhance diversity. The management approaches call out sensitivities to threatened spruce-fir wildlife, such as using hand release of red spruce to create new den trees for the Carolina northern flying squirrel, and maintaining 30 m (100 ft) buffers from rock outcrops with spruce-fir moss spider (*Microhexura monitvaga*; Fig. 9.8) and rock gnome lichen (*Gymnoderma lineare*) habitat.

The Jefferson NF Land and Resource Management Plan (USDA Forest Service 2004b) recognizes 2,353 ha (5,814 ac) of spruce-fir forests primarily on the highest elevation sites of the Crest Zone and Whitetop areas of the Mount Rogers National Recreation Area, with several smaller disjunct populations in other Districts. The Plan identifies at least 469 ha (1,160 ac) across the Forest for red spruce restoration to create habitat corridors and protect rare species, with potentially more restoration opportunities in locations currently in northern hardwood but within the historic red spruce range. The George Washington NF Land and Resource Management Plan (USDA Forest Service 2014) recommends expanding the current 243 ha (600 ac) of

Fig. 9.8 Spruce-fir moss spider (*Microhexura montivaga*) documented on Mt. Mitchell (photo by Gary Peeples)

red spruce to the historical 526 ha (1,300 ac) in the Laurel Fork which is contiguous with spruce-fir forests in the Monongahela NF, and an additional 324 ha (800 ac) currently in northern hardwood. The Plan identifies Norway spruce and red pine plantations as priority locations for restoration.

The Monongahela NF Land and Resource Management Plan (USDA Forest Service 2006) has the most comprehensive guidance and established a new management prescription (i.e., Prescription 4.1: Spruce and Spruce-Hardwood Restoration). National Forest management prescriptions are sub-divisions of the national forest based on a common management emphasis. A management prescription description outlines the management emphasis, current conditions, desired conditions, goals, objectives, standards, and guidelines for resources within the area. Each management prescription also defines spatially where the management prescription is to be applied. Prescription 4.1 aims to conserve red spruce forests at scale, covering 62,160 ha (153,600 ac) for red spruce and northern hardwood forest conservation and restoration. The desired conditions for these areas include a mosaic of red spruce and hardwood forest types largely consisting of multiple age classes at the stand level and an age class distribution across the landscape in this management zone of 60–80% of the red spruce and red spruce-hardwood forests to be late successional stage. The management approaches cover detailed guidelines for reforestation and silvicultural prescriptions.

9.4 CASRI and SASRI Formalize Partnerships

In 2007, CASRI partners, including the U.S. Fish and Wildlife Service, USDA Forest Service, West Virginia Division of Forestry, West Virginia Division of Natural Resources, and The Nature Conservancy, formalized the partnership under a Memorandum of Understanding (MOU) titled "For the Conservation of the Red Spruce-Northern Hardwood Ecosystem." The MOU set the framework from which the partners could carry out their common vision: "...a broad strategic framework and responsibilities for collaboration...for the long-term conservation of the red spruce-northern hardwood ecosystem in the states of West Virginia and Virginia." The MOU outlined a strategy for partner coordination, identification of spatial priorities for protection and restoration including core forest buffers and linkages, identification of red spruce-northern hardwood habitat types, and descriptions and implementation of management and restoration activities. To realize the MOU vision, CARSI partners developed a 2010–2020 action plan, organized to fund the work through organizational budgeting, sought government and private funding, and expanded partnerships and capacity. In 2011, CASRI's Research Committee produced restoration and management guidance detailing how to restore red spruce, including the use of if–then matrices that recommended specific restoration actions based on the range of initial conditions typically encountered in West Virginia (Byers et al. 2011).

By the early 2010s, SASRI partners elevated the need for a southern Appalachian red spruce restoration plan to serve as a guidepost for restoration goals and success.

Development of the plan included mapping existing red spruce extent and partner consensus on criteria for restorable areas. The plan laid out the overarching goal to "…restore red spruce-fir forests in ecologically appropriate locations throughout the Southern Blue Ridge ecoregion" (Medlock 2015). The SASRI partners included four main objectives aimed at building the vision, collaboration, and capacity needed for the scale of restoration. Following the completion of the plan in 2015, the group determined the need to establish a charter reinforcing those objectives, outlining the roles and responsibilities for the steering committee, standing committees, and membership. The SASRI Charter was finalized and adopted in 2016 and amended in 2022 to better address red spruce propagation needs. To address restoration needs SASRI formed *Sky Island* teams covering the Great Smoky Mountains, Plott Balsam Mountains, Great Balsam Mountains, Blacks Mountains, Unaka Mountains, Roan Mountain, Grandfather Mountain, and Grayson Highlands. Partners meet within Sky Island teams to address forest health concerns, develop restoration action and monitoring plans where relevant, and pool resources for management needs.

9.5 Enabling Conditions to Scale Up Science-Informed Restoration

The partnerships' capability to scale up science-informed restoration from inception in the 2000s (CASRI) and 2010s (SASRI) relied on dedication to various types of gatherings, identifying where and what to restore with habitat modeling and restoration prioritization, availability of seedlings to match restoration demand, and a sufficient workforce for plantings and silvicultural treatments.

9.5.1 Galvanizing Gatherings

The CASRI and SASRI partnerships have continued to meet regularly since formalizing the partnerships. The partnerships meet in geographic relevant groupings, such as within Sky Island teams in the southern Appalachians, and in topical sub-committees, such as research, management and monitoring, and education and outreach. Dedication to regular meetings has maintained momentum across decades as funding fluctuates and has enabled the partnerships to maintain focus on the broader vision and goals. Annual partnership conferences, workshops, and field days have fostered shared learning and galvanized new ideas for new initiatives.

The first regional conference was organized in 2009, the "Conference on the Ecology and Management of High-Elevation Forests in the Central and Southern Appalachian Mountains" held at Snowshoe Mountain Resort, West Virginia. The proceedings include 18 peer-reviewed papers and 40 abstracts pertaining to a wide range of topics, such as acid deposition, nutrient cycling, ecological classification, forest dynamics, wildlife, climate change, genetics, and restoration (Rentch and Schuler 2010). The conference included participants and presenters from both the central and southern Appalachians. The emphasis on restoration and practitioners was key. The steering committee sought to create an event where scientists shared their results with managers and managers shared their findings with scientists. While the conference proceedings include the usual scientific papers from designed studies, the conference also included presentations on lessons learned from implementing restoration actions and a panel discussion on building successful partnerships for landscape-scale restoration.

Several additional conferences have been held since 2009, including in 2019 with a focus on new research, virtually in 2020 focused on measuring restoration success, and in 2023 spanning topics covering red spruce soils, mycorrhizal fungi associations, and integrated watershed scale approaches to managing for resilient red spruce forests and cold-water ecosystems. An Appalachians-wide "Red Spruce Climate Change Workshop" was held in Vermont in 2021 to share research perspectives from ecology and natural history, physiology, genetic diversity, local adaptation, seed zones, and applied restoration and silviculture, with the goals of aligning efforts on needed research and practices for climate adaptation for red spruce ecosystems across its geographic distribution. Participants at this workshop were able to set in motion a new collaborative system for sourcing climate-informed red spruce seed for future restoration projects (see Sect. 9.5.3).

9.5.2 Habitat and Restoration Maps Define Need

In the early years of the CASRI partnership, much effort was put to defining the problem, developing messaging, designation of spatial priorities (i.e., core protected areas and buffer areas), and developing annual work plans and annual reports to be used to demonstrate progress. As restoration efforts grew, CASRI partners recognized the need for comprehensive data on potential habitat and restoration opportunities. In 2007, Natural Heritage ecologists at the West Virginia Division of Natural Resources completed a 4-year conservation assessment of the wetlands within the red spruce ecosystem in West Virginia (Byers et al. 2007). The study reinforced the critical importance of high elevation conservation and was followed by a 3-year assessment of upland red spruce ecosystems (Byers et al. 2010), including the first habitat distribution model of the red spruce ecosystem. For the model, split into upland and wetland communities, environmental variables for sites currently supporting red spruce were used to determine potential suitable habitat outside of the sampled areas. The West Virginia Division of Natural Resources used the model to develop a map of current extent, with support from multiple partners including the USDA Forest Service and AmeriCorps, by incorporating thousands of validation points from on-the-ground surveys (Byers et al. 2013). The stand map had a transformative effect on current thinking; although an estimated 90% of red spruce forests had been logged decades ago, 142,583 ha (352,330 ac) of second growth red spruce-dominated forests existed, much more red spruce forest than previously thought. The data was categorized by > 50% "high" cover (21,573 ha [53,308 ac]), 10–50% "medium" cover (74,805 ha [184,848 ac]), trace to 10% "low" cover (46,204 ha [114,174 ac]), and absent within the potential historical range. The suitable habitat model overlain with current cover data on red spruce forests created the first map of restoration opportunities, and this map became a blueprint for CASRI's conservation management activities.

The Canaan Valley National Wildlife Refuge developed spatial models to evaluate locations for planting based on a habitat connectivity-based least cost path analysis, focusing on threatened and endangered species conservation to address potential restoration impacts. Ecologists with the West Virginia Division of Natural Resources Natural Heritage Program developed statewide maps showing areas of high conservation concern based on lists of Species of Greatest Conservation Need from the West Virginia State Wildlife Action Plan (West Virginia Division of Natural Resources 2015).

As a result of the preceding studies and field work, the estimate of the original extent of red spruce in the central Appalachians was now understood to be 404,685 ha (1 million ac), an order of magnitude greater than what remains. With an ambitious vision, CASRI partners could then focus on collaborating for broad-scale impact across ownerships.

In the southern Appalachians, much of the foundational modeling has also been completed to define restoration opportunities for SASRI. Spruce-fir forests have been mapped across much of the southern Appalachians as part of regional ecological zone mapping initiatives (Simon 2011a, b, 2013). SASRI partners, led by the U.S. Fish and Wildlife Service, have produced a map of spruce-fir percent cover in the overstory and understory, and a prioritization scheme based on elevation, red spruce density, and aspect. Elevations > 1,829 m (6,000 ft) on more mesic aspects with < 25% spruce density are priorities for restoration.

9.5.3 Seed Sourcing and Seedling Availability

To achieve reforestation goals, partners needed greater availability of red spruce seedlings. Early efforts in the late 1990s involved transplanting red spruce seedlings from gas pipelines on the Monongahela NF to areas such as The Nature Conservancy's Cranesville Swamp Preserve in Maryland (Fig. 9.9). The Canaan Valley National Wildlife Refuge also partnered with the USDA NRCS Plant Materials Center in West Virginia to store seed and grow red spruce and balsam fir seedlings. To increase the pace and scale of reforestation, partners needed a steady, reliable source of red spruce, balsam fir, and other associated species.

In the late 1990s, a local conservationist and CASRI co-founder, Dave Saville, began collecting balsam fir cones, processing seed, and producing seedlings for habitat restoration projects (Fig. 9.10). Saville collected balsam fir cones from wetlands throughout the West Virginia highlands, avoiding plantations that were

Fig. 9.9 Volunteers collecting red spruce seedlings from gas pipelines on the Monongahela NF in West Virginia to be transplanted to restoration sites such as on The Nature Conservancy's Cranesville Swamp Preserve in Maryland (photo by Deborah Landau)

likely sourced from outside the ecoregion. In the early 2000s, Saville ramped up seedling production by partnering with a Christmas tree grower who collected seed from balsam firs on the Canaan Valley National Wildlife Refuge. Later, Saville began to collect cones and process seeds himself to be shipped to commercial tree growers in the Pacific Northwest and upper Midwest. Growers shipped the seedlings back to Saville who sold them to partners. A continuous source of red spruce seedlings of local origin enabled partners to plan and plant at scale. Saville began in 2000 with approximately 1,000 seedlings, and by 2024 sold 285,000 red spruce and balsam fir to the CASRI partnership for restoration.

In early planting projects, CASRI partners planted red spruce and balsam fir sourced from local populations with the understanding that local environmental adaptations would result in greater survival and growth. This concern evolved within the era of action for climate change impacts. To address climate adaptation concerns, The Nature Conservancy collaborated with Saville and a research group from the University of Maryland and the University of Vermont studying red spruce genetic diversity across the Appalachians. Broadly, the research pointed to low genetic diversity in the central Appalachians (Capblancq et al. 2020), a risk to capacity for climate adaptation. Through a project funded by the Wildlife Conservation Society's Climate

Fig. 9.10 David Saville, local conservationist with CASRI, collecting balsam fir (*Abies balsamea*) cones to produce seedlings for restoration (photo courtesy of David Saville)

Adaptation Fund, the group incorporated insight and values across science, conservation, and practice to source seed from genetically diverse populations to buffer the broader forested landscape of the restoration sites for future climate changes (Prakash et al. 2024). This effort expanded in the following years as more evidence from common garden studies revealed variation in population climate adaptations across the Appalachians and therefore evidence in support of moving seed northward to sites matching predicted future climate (Lachmuth et al. 2024). The CASRI partners now source red spruce seed from populations that will increase genetic diversity, reduce genetic load, and better match future climate adaptation needs of priority restoration landscapes. Practitioners now use mixes of several genotypes to increase the likelihood of restoration success and persistence of the restored forests into the future. In-depth information on climate adaptation considerations is covered in Chap. 7.

In the southern Appalachians, seedlings have been produced since 2008 from local sources by SASRI partners and volunteers supported by the Southern Highlands Reserve. The Southern Highlands Reserve received funding in 2023 to build a larger greenhouse to ramp up production from approximately 1,000 seedling trees annually to at least 50,000 for the coming years.

9.5.4 *Sufficient Restoration Workforce*

Planting and implementing silvicultural treatments on 100s or 1,000s of acres annually requires a consistent, reliable, and knowledgeable workforce. Ecological restoration goals are often complex, requiring an understanding of the ecosystem (i.e., how trees persist in an environment) not just a single component (i.e., the tree); therefore, treatment prescriptions are often nuanced. Rather than plant trees in rows as in a plantation, trees are planted in clusters to mimic natural regeneration, and near rocks or logs for protection from wind. When using silvicultural treatments to release red spruce trees from midstory competition, the practitioner must know which tree species should not be culled for habitat value, which often varies by location and density of the species. Hiring crews with this type of ecological knowledge is essential to achieving red spruce and spruce-fir restoration goals. CASRI partners have slowly built up their own staff capacity and also rely on volunteers and planting contractors. Future workforce needs remain a key consideration for scaling up restoration efforts.

9.6 Scaling Up Restoration

At the time of this publication, partners have planted well over 2.5 million red spruce and associated native trees and shrubs, and improved spruce-fir forest conditions on estimated 3,200 ha (7,907 ac), primarily across the central Appalachians. Planting projects range from small-scale (approximately 0.5–2 ha [1–5 ac]) volunteer events with school students, community groups, and agency professionals, to large (2–40 ha [5–100 ac] or greater) plantings completed by contractors at landscape scales. Red spruce and associated species have been planted for watershed protection in riparian areas and floodplains, on mountainsides and restored surface mined lands. Red spruce have been planted under existing hardwood forests where soil factors revealed evidence of former red spruce influence (see Chap. 3). Here, we present highlights of partner plantings and canopy release projects from the 2000s to early 2020s.

9.6.1 Scaling up Restoration in the Central Appalachians

By the early 2020s, the CASRI partnership had built an active science-based restoration and monitoring program, supported by a wide range of local scientists and managers with expertise in soils, vegetation, and wildlife. Annual activities include tree plantings, native seed collection and propagation, red spruce canopy release, wetland restoration, monitoring and evaluation, research, education, and outreach. Restoration activities spanned focal areas across the historical footprint, including the Canaan Valley National Wildlife Refuge, Monongahela NF, and state and private lands. Here, we cover a few of the larger restoration projects to highlight how partners have dealt with impaired ecosystem functions, mining impacts, invasive species and other competing vegetation, endangered species, and over-browsing. Restoration projects have historically focused on either planting or release, but projects are increasingly integrating both methods as needed, while seeking to restore the integrated functions for forests and freshwater.

9.6.1.1 Planting-Focused Projects

The Mower Tract Restoration Project in Randolph County and the Sharp Knob Restoration Project in Pocahontas County, both on the Monongahela NF, have restored over 700 ha (1,730 ac) of red spruce ecosystems. The Mower Tract sits atop Cheat Mountain, an 80 km (50 mi) north–south forested corridor in central West Virginia. Cheat Mountain hosts a core remnant of red spruce in the Allegheny Mountains of Appalachia and restoring its integrity at scale has been a restoration priority for partners. Mining and poor restoration practices left the area with large retention ponds, compacted soils, monocultures of non-native trees and grasses, and invasive plants. Project goals were to restore hydrologic function as well as desired red spruce plant communities for wildlife habitat. Between 2011 and 2024, the Monongahela NF and Green Forests Work, in collaboration with nearly 40 partners, have planted over 1 million trees and shrubs, 40% of which were red spruce along with 60 other species, and created over 1,800 wetlands (Fig. 9.11). Practitioners used the Forestry Reclamation Approach (Adams 2017) to remove invasive plant cover and decompact soils with heavy machinery. The project's successes have been a galvanizing force for partners to continue to innovate on restoration methods and build partnerships that are driving restoration in other priority red spruce forests.

Another core remnant of red spruce-fir forests in the Alleghany Mountains sits around and throughout Canaan Valley of West Virginia (Fig. 9.12). This landscape is a complex matrix of intersecting federal, state, and private lands, as well as a popular tourist destination. Forests in Canaan Valley were not heavily impacted by mining, but other historical impacts from logging and burning have resulted in depauperate old fields, unique open red spruce wetland communities, and hardwood forests now dramatically impacted by beech bark disease. Since the 2000s, partners have planted close to 2 million red spruce, balsam fir, and other native plants, primarily on old

Fig. 9.11 Mower tract mined land restoration site with downed plantations of Norway spruce (*Picea abies*), red pine (*Pinus resinosa*), and other woody debris prior to planting native seedlings and wetland creation (photo by Matt Barton, University of Kentucky Agricultural Communications)

fields and underplanting in hardwood forests. Given the high deer browse pressure common in Canaan Valley (30 per km^2 [78 per mi^2]), more emphasis has been placed on red spruce which is less tolerant to deer browsing than other native tree species. The federally threatened Cheat Mountain salamander occurs in the area, which has spurred partners here to examine how restoration practices can avoid impacts to threatened and endangered species.

Smaller planting projects have focused on expanding the red spruce extent in Pennsylvania to replace the conifer functions of dying eastern hemlock (*Tsuga canadensis*) stands, restoring low elevation frost pockets in western Maryland, restoring degraded riparian corridors across the Monongahela NF, and expanding and connecting existing stands in the high elevations of southwestern Virginia.

9.6.1.2 Canopy Release-Focused Projects

Silvicultural prescriptions for the conservation and management of red spruce and spruce-fir forests, such as thinnings, shelterwood harvests, and group selections to benefit advance regeneration, have been extensively researched and documented in its northern range (Frank and Bjorkbom 1973; Dumais and Prévost 2008), but have been

Fig. 9.12 Red spruce (*Picea rubens*) and northern hardwoods in Canaan Valley, West Virginia (photo by David Saville)

largely absent in the central and southern Appalachians. In 2005, researchers in West Virginia initiated one of the first canopy gap release experiments of understory red spruce in northern hardwoods in the central Appalachians to assess growth metrics by a range of gap sizes (Rentch et al. 2016). The study focused on small canopy gaps that would replicate windthrow and other natural disturbances characteristic to red spruce and spruce-fir ecosystem dynamics (see Chap. 5). This study spurred more interest from CASRI partners who continued to refine prescriptions, both mechanical and chemical, and have evolved to include larger gap sizes and group selection, described in detail in Chap. 8. The treatments are implemented by crews on foot and have a low disturbance impact. A crew of four can complete around 200 ha (500 ac) per growing season. Given the extent of restoration potential with existing advance regeneration and the increasing threat of climate change impacts, scaling release treatments at a faster pace has become of interest to CASRI partners. By 2020, some partners began testing new prescriptions for larger scale commercial release of red spruce on the Monongahela NF. At the time of this publication, partners continue to refine commercial prescriptions and monitor outcomes to assess the desired goal of accelerating red spruce to achieve a minimum 30–50% canopy cover, or the relevant canopy cover for the desired conditions.

9.6.1.3 Projects with Multiple Ecological Outcomes and Integrated Methods

The main practices of red spruce restoration are typically aimed at plant community outcomes, either planting new seedlings to expand extent or increasing red spruce and balsam fir canopy cover by removing hardwood competition. But increasingly practitioners are designing projects that aim to restore the integrity of the ecosystem, especially the hydrologic and soil functions. This has been critical where red spruce ecosystems are being restored on former mined lands. The practice of planting seedlings on compacted soils alone would not restore the ecosystem functions in the timescale needed to create a functioning habitat.

9.6.2 Scaling up Restoration in the Southern Appalachians

SASRI's restoration efforts have relied heavily on volunteer support, as opposed to contractors. The first broad-scale collaborative restoration project occurred in 2017 at Flat Laurel Creek in the Great Balsams of North Carolina, a project aimed at improving forest conditions for the endangered Carolina northern flying squirrel and rare bird species such as the red crossbill (*Loxia curvirostra*) and northern saw-whet owl (*Aegolius acadicus*). Partners and volunteers transported red spruce seedlings grown at Southern Highlands Reserve to the project site. Students in Haywood Community College's Forestry and Wildlife programs planted the seedlings and helped the North Carolina Wildlife Resource Commission monitor growth and survival. After this first successful project, SASRI formed *Sky Island* teams in 2018 to facilitate the planning and execution of more projects in each of the mountain massifs in the southern Appalachians. These now include projects on Whitetop, Grandfather, Black, Unaka, Plott Balsam, and Roan Mountains aimed at increasing red spruce and Fraser fir in the understory with plantings, accelerating red spruce to the canopy with gap treatments, increasing stand diversity and structure in red spruce plantations or other unnatural site conditions through the creation of canopy gaps, with concurrent ongoing research on gap size and impacts on red spruce regeneration. Project partners on Roan Mountain are testing planting success across sites impacted by beech bark disease, as well as management methods for reducing damage by elk (*Cervus canadensis*), white-tailed deer (*Odocoileus virginianus*), and feral swine (*Sus scrofa*) on Plott Balsam Mountain.

Funding for restoration in the southern Appalachians increased in the early 2020s, in part due to the 2022 selection of the U.S. Capitol Christmas tree, a red spruce from the Pisgah NF, bringing increased visibility to the ecosystem. In 2023, the USDA Forest Service Collaborative Forest Landscape Restoration Program funded the Pisgah Restoration Initiative for $11 million over 10 years, which will include red spruce restoration for the Black, Roan, Unaka, and Great Balsam *sky islands* (see Fig. 5.3 in Chap. 5). The funds will focus on red spruce cone collection, expanding plantings, and silvicultural treatments for canopy release and recruitment. Restoration activities will restore connectivity of disjunct red spruce patches and build resilience of existing stands to increase structural complexity.

9.7 Conclusions

The CASRI and SASRI partnerships formed and evolved under different geographic and historical contexts, but some of the drivers were similar, including the recognition of unique forest communities and a desire to aid natural processes through active restoration. From this slow accretion of problem identification and partnership building, the work of CASRI and SASRI continues to increase in pace and scale. The partnerships, while formalized in current MOUs, remain fluid and flexible for new partners to be included as needed and for individual agencies, organizations, and private individuals to accomplish their own goals and objectives. The partnership framework is essential for sharing knowledge, lessons learned, funding, and staff and volunteers. The partnerships' continued success will rely on maintaining sustainable levels of funding, adapting to climate change impacts, and dedicated individuals. Although the accomplishments to date are significant, the extent of restoration need, likely greater than 121,000 ha (300,000 ac), across the central and southern Appalachians is still extensive. Red spruce is a slow-growing species and implementing restoration initiatives across administrative boundaries can be challenging, but this landscape-scale perspective is critical for restoring ecosystem structure and function in the central and southern Appalachians.

References

Adams WM, Hodge ID, Macgregor NA et al (2016) Creating restoration landscapes: partnerships in large-scale conservation in the UK. Ecol Soc 21:1

Adams MB (ed) (2017) The forestry reclamation approach: guide to successful reforestation of mined lands. USDA Forest Service Northern Research Station General Technical Report NRS-169, Newtown Square, Pennsylvania

Banks SA (2013) Forest response to the U.S. 1990 Clean Air Act: the southern spruce-fir ecosystem. Thesis, North Carolina State University, Raleigh, North Carolina

Byers EA, Vanderhorst JP, Streets BP (2007) Classification and conservation assessment of high elevation wetland communities in the Allegheny mountains of West Virginia. West Virginia division of natural resources report, Elkins, West Virginia

Byers EA, Vanderhorst JP, Streets BP (2010) Classification and conservation assessment of upland red spruce communities in West Virginia. West Virginia division of natural resources report, Elkins, West Virginia

Byers EA, Cimarolli S, Jones K et al (2011) Central Appalachian Spruce Restoration Initiative (CASRI): restoration approach. Central Appalachian Spruce restoration initiative report, Elkins, West Virginia

Byers EA, Love KC, Haider KR et al (2013) Red spruce (*Picea rubens*) cover in West Virginia, ver 1.0. https://wvgis.wvu.edu/data/dataset.php?ID=455. Accessed 31 Jul 2024

Capblancq T, Butnor JR, Deyoung S et al (2020) Whole-exome sequencing reveals a long-term decline in effective population size of red spruce (*Picea rubens*). Evol Appl 13:2190–2205

Clarkson RB (1957) Blister Swamp, West Virginia. Castanea 22:137–138

Dietsch AM, Wald DM, Stern MJ et al (2021) An understanding of trust, identity, and power can enhance equitable and resilient conservation partnerships and processes. Conserv Sci Pract 3:e421

Dumais D, Prévost M (2008) Ecophysiology and growth of advance red spruce and balsam fir regeneration after partial cutting in yellow birch–conifer stands. Tree Physiol 28:1221–1229

Eagar C, Adams MB (eds) (1992) Ecology and decline of red spruce in the eastern United States. Springer-Verlag, New York, New York

Frank RM, Bjorkbom JC (1973) A silvicultural guide for spruce-fir in the Northeast. USDA forest service northern research station general technical report NE-6, Broomall, Pennsylvania

Lachmuth S, Capblancq T, Prakash A et al (2024) Novel genomic offset metrics integrate local adaptation into habitat suitability forecasts and inform assisted migration. Ecol Monogr 94:e1593

Mathias JM, Thomas RB (2018) Disentangling the effects of acidic air pollution, atmospheric CO_2, and climate change on recent growth of red spruce trees in the central Appalachian mountains. Glob Change Biol 24:3938–3953

Medlock K (ed) (2015) Southern Appalachian spruce restoration plan. https://southernspruce.org/wp-content/uploads/2018/07/final_southern-appalachian-spruce-restoration-plan1.pdf. Accessed 26 Jul 2024

Noss RF, LaRoe ET, Scott JM (1995) Endangered ecosystems of the United States: a preliminary assessment of loss and degradation. U.S. National biological service biological report 28, Washington, DC

Nowacki G, Carr R, Van Dyck M (2010) The current status of red spruce in the eastern United States: distribution, population trends, and environmental drivers. In: Rentch JS, Schuler TM (eds) Proceedings from the conference on the ecology and management of high-elevation forests in the central and southern Appalachian mountains. USDA forest service northern research station general technical report NRS-P-64, Newtown Square, Pennsylvania, pp 140–162

Prakash A, Capblancq T, Shallows K et al (2024). Bringing genomics to the field: an integrative approach to seed sourcing for forest restoration. Appl Plant Sci 12:e11600

Rentch JS, Ford WM, Schuler TS et al (2016) Release of suppressed red spruce using canopy gap creation—ecological restoration in the central Appalachians. Nat Areas J 36:29–37

Rentch JS, Schuler TM (eds) (2010) Proceedings from the conference on the ecology and management of high-elevation forests in the central and southern Appalachian Mountains. USDA forest service Northern research station general technical report NRS-P-64, Newtown Square, Pennsylvania

Saif O, Keane A, Staddon S (2022) Making a case for the consideration of trust, justice, and power in conservation relationships. Conserv Biol 36:e13903

Simon SA (2011a) Ecological zones in the southern Blue Ridge: 3rd approximation. Ecological modeling and fire ecology incorporated report, Asheville, North Carolina

Simon SA (2011b) Ecological Zones on the George Washington National Forest: first approximation mapping. The nature conservancy Virginia field office report, Charlottesville, Virginia

Simon SA (2013) Ecological zones on the Jefferson National Forest study area: first approximation. The Nature Conservancy Virginia Field Office Report, Charlottesville, Virginia

Stephenson SL, Adams HS (1986) An ecological study of balsam fir communities in West Virginia. Bull Torrey Bot Club 113:372–381

Thomas-Van Gundy M, Morin R (2021) Change in montane forests of east-central West Virginia over 250 years. For Ecol Manag 479:118604

USDA Forest Service (2004a) Cherokee National Forest revised land and resource management plan. https://www.fs.usda.gov/Internet/FSE_DOCUMENTS/stelprdb5269436.pdf. Accessed 26 Jul 2024

USDA Forest Service (2004b) Jefferson National Forest revised land and resource management plan. https://www.fs.usda.gov/Internet/FSE_DOCUMENTS/stelprd3834582.pdf. Accessed 26 Jul 2024

USDA Forest Service (2006) Monongahela National Forest land and resource management plan. https://www.fs.usda.gov/Internet/FSE_DOCUMENTS/stelprdb5330420.pdf. Accessed 26 Jul 2024

USDA Forest Service (2014) George Washington National Forest revised land and resource management plan. https://www.fs.usda.gov/Internet/FSE_DOCUMENTS/fseprd525098.pdf. Accessed 26 Jul 2024

USDA Forest Service (2023) Nantahala and Pisgah National Forests final land management plan. https://www.fs.usda.gov/Internet/FSE_DOCUMENTS/fseprd1090063.pdf. Accessed 26 Jul 2024

Vanderhorst J, Byers E, Streets B (2012) Natural heritage vegetation database for West Virginia. Biodiversity & Ecology 4:440

West Virginia Division of Natural Resources (2015) 2015 West Virginia state wildlife action plan. https://wvdnr.gov/wp-content/uploads/2021/05/2015-West-Virginia-State-Wildlife-Action-Plan-Submittal-1.pdf. Accessed 26 Jul 2024

White PS, Tuttle AP (2013) Ecological sustainability as the fourth landmark in the development of conservation ethics. Conserv Biol 27:952–957

Glossary

age class (cohort) One of the intervals into which the age range of trees is divided for classification or use. A distinct aggregation of trees originating from a single natural event or regeneration activity, or a grouping of trees, such as a 10-year age class, as used in inventory or management. (USDA Forest Service 2025)

Appalachians (central) Area comprised of the Central Appalachian, western Allegheny plateau, north central Appalachian, and northern Blue Ridge ecoregions, that generally corresponds to central Pennsylvania through southern West Virginia and eastern Kentucky.

Appalachians (northern) Area comprised of the northern Appalachian plateau and uplands and northeastern highlands ecoregions, that generally corresponds to central Pennsylvania through Maine.

Appalachians (southern) Area comprised of the ridge and valley below West Virginia, southern Blue Ridge, and southwestern Appalachian ecoregions, that generally corresponds to southern West Virginia through northern Georgia.

canopy cover The proportion of the forest floor covered by the vertical projection of the tree crowns. (Jennings et al. 1999)

canopy gap A small overstory canopy disturbance caused by the death and felling of an tree or small group of trees in the overstory. (McCarthy 2001)

co-dominant (tree) Trees with crowns forming the upper level of the forest canopy; these trees receive full light from above but comparatively little from the sides, and their medium-sized crowns are usually more or less crowded on the sides. (Stokes et al. 1989)

coarse woody debris Also known as large woody debris, dead woody material from tree trunks, branches, stumps, and root wads. No strict definition of size, but many researchers use at least 10 cm (4 in) in diameter and 1 m (3.3 ft) in length. (Stahl and Dolloff 2002)

constancy (of species) A species common to a particular association or community, but not necessarily confined to that community. A species of high constancy would be present in all, or almost all, of a series of field samples that describe an association or community. (Allaby 2012)

D. J. Brown et al. (eds.), *Ecology and Restoration of Red Spruce Ecosystems in the Central and Southern Appalachians*, https://doi.org/10.1007/978-3-032-19616-3

dendrochronology The science of dating and interpreting past events using growth rings of trees.

distribution (species) Spatial and temporal pattern of a species and its population occurrences relative to habitat conditions. (Soberón 2010)

dominant (tree) Larger-than-average trees with well-developed crowns extending above the general canopy level and receiving full light from above and partial light from the side. (Stokes et al. 1989)

ecosystem A natural unit consisting of all the plants, animals, microorganisms (biotic) factors in a given area, interacting with all the nonliving physical and chemical (abiotic) factors of this environment. (Levin 2009)

ecosystem dynamics Changes in an ecosystem and its processes over time, often related to disturbance regimes, nutrient cycling, climate, and hydrologic processes, and the pattern of these changes. Changes often described in terms of frequency, scale, and magnitude. (Chapin et al. 2011)

ecosystem management Managing areas at various scales in such a way that ecosystem services and biological resources are preserved while appropriate human uses and options for livelihood are sustained. (Brussard et al. 1998)

ecotone A transition area where spatial changes in vegetation structure or ecosystem process rates are more rapid than in the adjoining plant communities. (Levin 2009)

foundation species A species is considered a foundation species if it is widespread and abundant, acts as a structuring element, creates stable conditions, influences composition, and modulates ecosystem processes. (Dudley et al. 2020)

genome An organisms complete set of genetic material. See also genotype. (Mahner and Kary 1997)

genotype Specific allelic composition at one or more genetic loci; also synonymous with genome. (Mahner and Kary 1997)

habitat The resources and conditions present in an area that produce occupancy, including survival and reproduction, by a given organism. (Hall et al. 1997)

hyporheic The region of sediment and rocks beneath and adjacent to a streambed where mixing of the stream water and groundwater occurs. (Woessner 2017)

importance value The importance value is the sum of relative density, relative dominance, and relative frequency for a species in the community. The larger the importance value, the more dominant a species is in the particular community.

land cover The physical state of a land area in terms of its surface features, such as rainforest, cropland, or desert. (Levin 2009)

mixed conifer forest Forests wherein red spruce is co-dominant with eastern hemlock or other conifers besides fir. This forest type is typically found in the central Appalachians.

northern hardwood forest Forests wherein northern hardwood species (e.g., yellow birch, American beech, sugar maple) are dominant.

rarity (species) Species are considered rare if their area of occupancy or their numbers are small when compared to the other species that are taxonomically or ecologically comparable. (Flather and Sieg 2007)

red spruce release Intermediate silvicultural treatment that provides increased growing space to selected trees through the removal of competition from adjacent trees.

refugium/refugia Areas relatively buffered from contemporary climate change over time that enable persistence of valued physical, ecological, and socio-cultural resources. (Morelli et al. 2016)

restoration (ecological) The process of assisting the recovery of a degraded, damaged, or destroyed ecosystem to reflect values regarded as inherent in the ecosystem and to provide goods and services that people value. (Martin 2017)

second growth (forest) Forests that regenerated mainly through natural processes after significant disturbance or removal of the original forest either thought natural or human causes. An extended definition includes the idea that these secondary forests display major differences in forest structure and composition compared to the original forest. (Chokkalingham and de Jong 2001)

selection thinning The removal of trees in the dominant crown class in order to favor the lower crown classes. (Deal 2017)

shade tolerance The ability of a tree to continue to become established, survive, and thrive in shaded conditions, where the availability of direct electromagnetic energy is low. Shade tolerance can also refer to the ability of a species of tree to continuously and successfully compete with other trees for resources and to regenerate under a contiguous canopy. (Grebner et al. 2022)

site index Height of an overstory tree at a reference base age; often displayed as a tree growth at age curve. See also stand productivity. (Brown 1962)

spodic/spodosol Soil with strong accumulation of iron and aluminum, typically occurring in red spruce forests as a result of production of organic acids and podzolization.

spruce forest Those in which species of Picea constitute 50% or more of canopy stems. See Chapter 4 for further community type delineations.

spruce-fir forest Those in which species of Picea and Abies, alone or in combination, constitute 50% or more of canopy stems. See Chapter 4 for further community type delineations. (Eagar and Adams 1992)

spruce-northern hardwood forest Forests wherein red spruce is co-dominant with northern hardwood species (e.g., yellow birch, American beech, sugar maple). See Chapter 4 for further community type delineations.

stand dynamics Change in forest structure over time including before, during, and after disturbances. A stand is a spatially continuous group of trees and other vegetation experiencing the same disturbance and climate. (Oliver and Larson 1996)

stand productivity Rate at which a stand generates biomass in the form of plant matter; often displayed as a tree height at age curve. See also site index.

succession (ecological) The process of vegetation development following a disturbance, often characterized by relatively predictable sequences of species replacement over time. (Levin 2009)

variable retention thinning Thinning used to promote greater heterogeneity in ecological conditions by varying intensity of removal of overstory trees across a stand. This includes retention of dense areas, as well as creation of harvest gaps. (Deal 2017)

witness tree A collective term for trees listed in deeds and surveys. Trees listed may represent actual corners, or be termed bearing, warrant, or pointer trees by surveyors. (White 1983)

References

Allaby M (2012) Dictionary of plant sciences, 3rd edn. Oxford University Press, New York, New York

Brown JH (1962) Success of tree planting on strip-mined areas in West Virginia. West Virginia Agricultural and Forestry Experiment Station Bulletin No. 473. Morgantown, West Virginia

Brussard PF, Reed JM, Tracy CR (1998) Ecosystem management: what is it really? Landscape and Urban Planning 40:9–20

Chapin FS III, Matson PA, Vitousek PM (2011) Principles of terrestrial ecosystem ecology. Springer, New York, New York

Chokkalingham U, de Jong W (2001) Secondary forest: a working definition and typology. International Forestry Review 3:19–26

Deal R (2017) Dictionary of forestry. Society of American Foresters, Washington, DC

Dudley MP, Freeman M, Wenger S, Jackson CR, Pringle CM (2020) Rethinking foundation species in a changing world: the case for Rhododendron maximum as an emerging foundation species in shifting ecosystems of the southern Appalachians. Forest Ecology and Management 472:118240

Eagar C, Adams MB (eds) (1992) Ecology and decline of red spruce in the eastern United States. Springer-Verlag, New York, New York

Flather CH, Sieg CH (2007) Species rarity: definition, causes, and classification. In: Raphael MG, Molina R (eds) Conservation of rare or little-known species: biological, social, and economic considerations. Island Press, Washington, DC, p 40–66

Grebner DL, Bettinger P, Siri J, Boston K (eds) (2022) Introduction to forestry and natural resources, 2nd edn. Academic Press, Cambridge, Massachusetts

Hall LS, Krausman PR, Morrison ML (1997) The habitat concept and a plea for standard terminology. Wildlife Society Bulletin 25:173–182

Jennings SB, Brown ND, Sheil D (1999) Assessing forest canopies and understorey illumination: canopy closure, canopy cover and other measures. Forestry 72:59–74

Levin SA (ed) (2009) The Princeton guide to ecology. Princeton University Press, Princeton, New Jersey

Mahner M, Kary M (1997) What exactly are genomes, genotypes and phenotypes? And what about phenomes? Journal of Theoretical Biology 186:55–63

Martin DM (2017) Ecological restoration should be redefined for the twenty-first century. Restoration Ecology 25:668–673

McCarthy J (2001) Gap dynamics of forest trees: a review with particular attention to boreal forests. Environmental Reviews 9:1–59

Morelli TL, Daly C, Dobrowski SZ et al (2016) Managing climate change refugia for climate adaptation. PLoS ONE 12:e0169725

D. J. Brown et al. (eds.), *Ecology and Restoration of Red Spruce Ecosystems in the Central and Southern Appalachians*, https://doi.org/10.1007/978-3-032-19616-3

Oliver CD, Larson BA (1996) Forest stand dynamics: update edition. EliScholar, Yale University, New Haven, Connecticut

Soberón JM (2010) Niche and area of distribution modeling: a population ecology perspective. Ecography 33:159–167

Stahl G, Dolloff A (2002) Coarse woody debris. In: El-Shaarawi, Piegorsch WW (eds) Encyclopedia of environmetrics, Wiley, New York, p 361–363

Stokes BJ, Ashmore C, Rawlins CL, Sirois DL (1989) Glossary terms used in timber harvesting and forest engineering. USDA Forest Service Southern Research Station General Technical Report SO-73, New Orleans, Louisiana

USDA Forest Service (2025) Reforestation glossary. https://www.fs.usda.gov/restoration/reforestation/glossary.shtml. Accessed 25 Feb 2025

White CA (1983) A history of the rectangular survey system. USDI Bureau of Land Management Government Printing Office, Washington, DC

Woessner WW (2017) Hyporheic zones. In: Hauer FR, Lamberti GA (eds) Methods in stream ecology, Volume 1: ecosystem structure, 3rd edn. Academic Press, Cambridge, Massachusetts, p 129–157

The manufacturer's authorised representative in the EU is Springer Nature Customer Service Centre GmbH, Europaplatz 3, 69115 Heidelberg, Germany. If you have any concerns regarding our products, please contact ProductSafety@springernature.com

Printed and bound by CPI Group (UK) Ltd, Croydon, CR0 4YY
07/07/2026
02160909-0003